Aviation Automation: The Search for A Human-Centered Approach

HUMAN FACTORS IN TRANSPORTATION

DEDICATION

This book is dedicated to the memory of Hugh Patrick Ruffell Smith, who introduced me to the excitement of aviation human factors in 1955 when he was a Royal Air Force Group Captain in charge of the flight test group at the RAF Institute of Aviation Medicine. He was a valued teacher and friend for 24 years.

Dr. Ruffell Smith's research presaged work still in progress today. His studies of visual and auditory display techniques and media, his work to improve navigation displays, and his determined, although not entirely successful, efforts to standardize flight displays and controls stand as monuments to his understanding of the tasks of pilots and the difficulties often placed in their way. His research project when he was a senior postdoctoral associate at the NASA Ames Research Center, "A Simulator Study of the Interaction of Pilot Workload with Errors, Vigilance, and Decisions" (1979), performed with Dr. John K. Lauber, was the primary stimulus for an enormous volume of research on and application of the principles of cockpit and crew resource management which has taken place over the past 15 years. Pat received too little credit during his lifetime for his monumental contributions, but he understood better than most in our profession the importance of what he was doing.

   Lawrence Erlbaum Associates, Inc., Publishers
   10 Industrial Avenue
   Mahwah, NJ 07430

Library of Congress Cataloging-in-Publication Data

Billings, C. E. (Charles E.), 1929-
      Aviation automation : the search for a human-centered approach
   / Charles E. Billings.
         p. cm.
      Includes bibliographical references and index.
      ISBN 0-8058-2126-0 (cloth : alk. paper). — ISBN 0-8058-2127-9
   (pbk. : alk. paper)
      1. Airplanes—Automatic control. 2. Aeronautics—Human
   factors. 3. Air traffic control—Automation. I. Title.
   TL589.4.B52   1997
   629.13—dc20   96-20189
                                                  CIP

Printed in the United States of America

10 9 8 7 6 5 4 3 2 1

Aviation Automation: The Search for A Human-Centered Approach

by

Charles E. Billings
The Ohio State University

LEA LAWRENCE ERLBAUM ASSOCIATES, PUBLISHERS

1997   Mahwah, New Jersey

Contents

Series Foreword

Barry H. Kantowitz
Battelle Human Factors Transportation Center
Seattle, Washington

The domain of transportation is important for both practical and theoretical reasons. All of us are users of transportation systems as operators, passengers, and consumers. From a scientific viewpoint, the transportation domain offers an opportunity to create and test sophisticated models of human behavior and cognition. This series covers both practical and theoretical aspects of human factors in transportation, with an emphasis on their interaction.

The series is intended as a forum for researchers and engineers interested in how people function within transportation systems. All modes of transportation are relevant, and all human factors and ergonomic efforts that have explicit implications for transportation systems fall within the series purview. Analytic efforts are important to link theory and data. The level of analysis can be as small as one person, or international in scope. Empirical data can be from a broad range of methodologies, including laboratory research, simulator studies, test tracks, operational tests, field work, design reviews, or surveys. This broad scope is intended to maximize the utility of the series for readers with diverse backgrounds.

I expect the series to be useful for professionals in the disciplines of human factors, ergonomics, transportation engineering, experimental psychology, cognitive science, sociology, and safety engineering. It is intended to appeal to the transportation specialist in industry, government, or academia, as well as the researcher in need of a testbed for new ideas about the interface between people and complex systems.

This volume is focused on the aviation domain. It offers eloquent and carefully reasoned arguments for a human-centered approach to the human factors of development and implementation of new technology in aviation. Part I is an overview of automation in aviation and explains both the application of automation and the concept of human-centered automation. Part II traces the evolution and course of aviation automation. This covers air traffic control and management as well as aircraft automation. Part III discusses the roles of human operators in the aviation system and human–machine integration in the future system. Part IV looks to the future and expands on novel concepts and requirements for aviation automation and its certification. Forthcoming books in this series will continue this blend of practical and theoretical perspectives on transportation human factors.

PREFACE

ORIGINS OF THIS DOCUMENT

Automation technology, used in the control of aircraft, surface transit vehicles, and industrial processes for many years, has been revolutionized by the development of the digital computer. The invention of the transistor in 1947 and subsequent miniaturization of computer components enabled widespread application of digital technology in aircraft. The period since 1970 has seen an explosion in aviation automation technology. In 1987, the Air Transport Association of America (ATA) Flight Systems Integration Committee established an industry-wide task force to consider aviation human factors issues.

In its "National Plan to Enhance Aviation Safety through Human Factors Improvements," (ATA, 1989) the Human Factors Task Force stated that "During the 1970s and early 1980s ... the concept of automating as much as possible was considered appropriate. The expected benefits were a reduction in pilot workload and increased safety ... Although many of these benefits have been realized, serious questions have arisen and incidents/accidents have occurred which question the underlying assumptions that the maximum available automation is ALWAYS appropriate or that we understand how to design automated systems so that they are fully compatible with the capabilities and limitations of the humans in the system" (p. 4). The ATA report went on:

> The fundamental concern is the lack of a scientifically-based philosophy of automation which describes the circumstances under which tasks are appropriately allocated to the machine and/or to the pilot. Humans will continue to manage and direct the NAS (National Aviation System) through the year 2010. Automation should therefore be designed to assist and augment the capabilities of the human managers ... It is vitally important to develop human-centered automation for the piloted aircraft and controller work station. (p. 5)

During the same year, the National Aeronautics and Space Administration's Office of Aeronautics and Space Technology approved a new research initiative, "Aviation Safety/Automation" (NASA, 1990). Under this initiative, NASA's Ames and Langley Research Centers were to examine human–machine interactions in aviation and future aircraft automation options. As a response to the need for a philosophy of aircraft automation expressed by the ATA, I prepared a NASA Technical Memorandum on this topic (Billings, 1991). Four years later, prompted by increasing operational experience with highly automated aircraft, I revised the document and expanded it to include consideration of air traffic control and management automation (Billings, 1996). The 1996 document is the basis for this book, although I have also attempted to consider lessons learned from other domains and to add explanatory material which may make the document more useful to persons not familiar with aviation.

RATIONALE

One need only look back over the developments of the last 20 years to realize how much has already been done to integrate advanced automation into the aviation system. Advanced, highly automated aircraft are more productive, more reliable and safer than their predecessors when managed properly. The aviation system has been strained beyond its presumed limits, yet remains safe and fairly resilient. The system is carrying more people, in more airplanes, to more places than at any time in its history.

Why, then, is this document needed? If the ATA Human Factors Task Force were beginning its work today, would automation still be at the top of its list of concerns? Is there any substantive evidence that what we have built to date, and what we are planning to build during the remainder of this decade, will not continue to improve upon the progress of the past two decades?

The Task Force's influence has continued. Aviation automation remains as important a topic today as it was 5 or 10 years ago. The stresses on the aviation system have exacted a price; much more of the system's design capacity is being used, and all credible projections indicate more serious capacity problems in the years ahead. The hub-and-spoke system has created much greater traffic concentrations, and thus greater flight crew and controller workloads, at hub terminals at certain times. Hub-and-spoke implementation has also made the system less tolerant of delays and cancellations.

Automation has freed the crews of newer aircraft from dependence on point-to-point systems of navigation aids, but this freedom from defined route constraints has increased air traffic coordination requirements and has complicated conflict prediction. Economic constraints have increased the pressure on every human and machine element of the system. Each of these factors has played a part in increasing the demand for greater system precision and reliability, and each has shaped, and continues to shape, the behavior of the human operators of the system.

Technology improvements have increased aircraft and system complexity and cost. Some, like ground proximity warning systems, have conveyed substantial benefits; for some others, like electronic library systems, the benefits appear thus far to be marginal at best. It can be confidently predicted that other new technology solutions will be proposed in the future if they appear likely to improve safety or utility. It is certain that they will impose additional tasks upon the humans who operate the system. It is only slightly less certain that some or many of these novel technologies will not operate quite as planned, and that humans will be required to adapt to, compensate for, and shape the new artifacts, as they have always had to do when new technology was implemented.

Readers should keep in mind that the "future" aviation system, to a considerable extent, is with us today. It will rely upon aircraft already in line service (and future derivatives of those aircraft), just as the majority of today's modern aircraft are themselves derivatives of machines developed as long as three decades ago. The general outlines of the future system can be seen today at any major airport. Even supersonic transport aircraft, which may represent the most radical future technology departure from today's system, are presaged by the Anglo–French *Concorde*, which has been safely flying trans-Atlantic routes for over 20 years.

This is not to say that the system will necessarily operate as does today's system. Though the vehicles may appear to be similar, today's aircraft represent vast advances over their progenitors. Their automation, in many cases, is two generations advanced, and the requirements upon the humans who operate them are considerably different. The Air Traffic Control system today operates much as it has for the past two decades, but this is about to change. Over the next decade, radical changes in hardware, software and procedures will result in much more highly automated systems for air traffic management. These changes will have profound implications for the pilots, dispatchers and controllers at the "sharp end" of the aviation system. Today's system works well, but significant problems exist, some of which relate directly to the automation that has become an increasingly important element in its operation.

Aviation is not yet one century old, yet the services it provides have grown over this time to become an absolutely essential part of our global economy. Those of us who have been privileged to work within this dynamic, rapidly advancing system know that we can make the system do more, more effectively. The demands that will be placed on aviation during the next two decades make it quite obvious that we *must* do more; we must develop a still safer, more efficient and more productive aviation system, and do it quickly. The increasing requirements of users demand that the system continue to improve. The improvement of the aviation system, and other complex human–machine systems as well, through more effective coordination between humans and automation is the goal of this document.

This book is not only about technology, nor only about the human users of technology. It is about humans *and* technology, working together in highly dynamic and potentially dangerous environments to accomplish social goals, subject to a multitude of social, political and technical constraints.

ACKNOWLEDGMENTS

I am grateful to the National Aeronautics and Space Administration for its support of my studies of aviation automation during and since my tenure at Ames Research Center. Many colleagues at Ames, and at the Langley Research Center, have been most generous with their encouragement and criticism of this work.

Since my return to The Ohio State University in 1992, Professors David D. Woods and Philip J. Smith have fostered my understanding of cognitive engineering and have helped me plan, organize and execute this book. Dr. Nadine Sarter, Research Associate, and Sidney Dekker, my graduate assistant, have shaped my understanding of the cognitive bases of the human–machine interaction process and have been especially helpful in this effort.

Delmar Fadden, of the Boeing Commercial Airplane Group, has been a superb teacher and constructive critic of my attempts to understand why today's aircraft have developed as they have. The Air Transport Association of America's Flight Systems Integration Committee Chair, Capt. Robert Buley, and its Executive Secretary, Will Russell of ATA, provided insights and encouragment over a long period of time. I am also indebted to the others who have reviewed the manuscript, among them Donald Armstrong, Kevin Corker, Ted Demosthenes, Joan Eggspue-hler, John Enders, Victor Lebacqz, Richard Pew, Guy Thiele, and NASA Aviation Safety Reporting System personnel, especially Vincent Mellone. I am particularly grateful for the invaluable guidance provided by John Lauber, who has shared his insights into the realities of aviation operations and aviation safety throughout the 22 years of our collaboration. Finally, I acknowledge with very special gratitude the indispensable support and tolerance of my dear wife, Lillian.

—*Charles E. Billings*

PART I

AVIATION AUTOMATION: PAST, PRESENT, AND FUTURE

This section introduces the domain and theme. Chapter 1 presents the problem that has motivated the book, summarizes its purpose, and provides a few definitions central to its organization. Chapter 2 describes the context of transport aviation and discusses the physical, operational, and organizational environments within which aviation is conducted. Chapter 3 briefly describes problems associated with aviation automation and presents a concept of human-centered automation, as an approach that can assist in the resolution of these problems.

Note for readers: Most of the serious incidents and mishaps cited with a place name and year in the text are summarized in Appendix 1. Mishap investigation reports are also cited in the References. Acronyms, abbreviations, and technical terms are defined in the Glossary.

CHAPTER 1

STATEMENT OF THE PROBLEM

The modern airplane is the product of a program of research, development and refinement in detail that no other structure or mechanism has ever matched. The results have been so remarkable that there is always danger of forgetting that these extraordinary craft still have to be operated by men, and that the most important test they have to meet is still that of being operable without imposing unreasonable demands or unnecessary strains on the flight personnel.

—*Edward P. Warner, quoted by Ross McFarland, 1946*

INTRODUCTION

The motivation for this book can be stated simply. The advent of very compact, very powerful digital computers has made it possible to automate a great many processes that formerly required large, complex machinery (if they could be automated at all). The book describes the development of automation in one domain, civil aviation, its likely evolution in the future, and the effects that these technologies have had and will have on the human operators of the aviation system. It suggests concepts that may be able to enhance human–machine relationships in the future system. The focus is on the interactions of human operators with the constellation of machines they command and control. I have not attempted to consider either the humans or the automation in isolation, because it is the *interactions among* these system elements that result in the system's success or failure.

The aviation system is a technology-intensive, spatially distributed system in which skilled human managers and operators move passengers and cargo from place to place utilizing complex, variably automated machines. In no endeavor has technology been brought to bear more effectively than in the aviation enterprise, and no enterprise has more effectively stimulated the advance of technology. In the space of a century, we have moved from wood and fabric gliders to aircraft carrying hundreds of people and tons of cargo halfway around the earth at near-sonic speeds in comfort and safety.

3

In the course of this development process, we have learned how to automate this remarkable machinery nearly completely. The newest long-range airplanes can operate almost unassisted from shortly after takeoff in New York until they come to rest after a landing in Tokyo. The considerable psychomotor and cognitive skills of their human operators are hardly called on unless some element of the automation fails or unanticipated environmental circumstances arise. Indeed, maintenance of manual flying skills has become a problem in some operations. Instead of the "unreasonable demands" foreseen by Warner, we now worry about whether routine operations impose enough of a "strain" on pilots to keep them alert and involved in long-haul missions.

But when the environment does not behave as expected, or when the very reliable machinery does not function correctly, we expect these human operators to do whatever is required to complete the mission safely. Is it reasonable to expect human operators in a highly automated, dynamic system always to do "the right thing" when they are called on? Are today's aircraft designed to facilitate effective cooperation between the humans and the machines they manage? Aviation automation has conveyed great social and technological benefits, but these benefits have not come without cost. In recent years, we have seen the emergence of new classes of problems, which are due to failures in the human–machine relationship.

In particular, we have seen the appearance of incidents and accidents that indicate failures to understand automation behavior. We have seen errors in choice of operating modes, lack of mode awareness, and inability to determine what automation was doing. Common to these occurrences have been complex, tightly coupled automated systems that have become more autonomous and authoritarian, and that often provide inadequate feedback to their human operators. These are not isolated problems, and they will not be fixed by "local," narrowly focused measures. They are system problems, and they require systematic correction.

It is these accidents and incidents that have motivated this inquiry into aviation automation. I suggest here that a different approach to automation, which I have called *human-centered automation*, offers potential benefits for system performance by enabling a more cooperative human–machine relationship in the control and management of aircraft and air traffic. This approach requires, and encourages, more effective coordination among system elements so that there is less likelihood of human or machine misinterpretation of system state, or misunderstanding of which element is performing each of the many functions necessary for safe mission completion.

BACKGROUND

It has long been an article of faith that from 65% to 80% of air transport accidents are attributable in whole or part to human error. The figure has been relatively stable throughout the jet era. Indeed, one of the motives for increasing automation in transport aircraft has been the desire by manufacturers and operators to decrease the frequency of human errors by automating more of the tasks of the pilot (Wiener, 1989). Similar motives underlie, at least in part, the interest in automating the air traffic control system. Although one can argue with the usefulness or appropriateness of the retrospective attribution of human error in aviation accidents, the aviation community and the public clearly believe that the human is potentially a weak link in the chain of accident causation (Boeing, 1993). What has been our experience with aviation automation to date? Is there good evidence that the considerable amount of automation already in place has affected accident or incident rates? The answer to this question, briefly, is, "yes and no." But the question is also too simplistic, for *automation* is not a single entity.

Examination of aviation incident and accident data from the past two decades reveals two seemingly contrary trends. On the one hand, there have been sharp declines in certain types of accidents that appear almost certainly to be due to the introduction of automatic monitoring and alerting devices such as the ground proximity warning system (GPWS)[1] introduced in 1975. On the other hand, there is clear evidence that despite the application of automation technology to this problem, controlled flight into terrain accidents still represent perhaps our most serious safety problem. Collision avoidance systems (TCAS)[2] and wind shear advisory systems (WSAS) have more recently been installed in transport aircraft. Like GPWS, these devices can detect environmental conditions that may not be obvious to the unaided human senses. All make use of sophisticated sensors and algorithms to detect, evaluate, and provide timely warning of critical threats in order to permit avoidance action by pilots. TCAS has almost certainly prevented collisions, although it, like GPWS, is plagued by nuisance warnings and it has caused serious problems for air traffic controllers. At least 20 fatal wind shear accidents have occurred since the introduction of jet transports, and WSAS has the potential to assist in preventing such catastrophic microburst encounters, although in at least one recent case (Charlotte, North Carolina, 1994) WSAS failed to provide timely warning of a microburst event which caused a disaster.

There is also a contrary trend in mishap data. Several accidents and a larger number of incidents have been associated with, and in some cases may have been caused by, aircraft automation, or more properly by the interaction between automation and the human operators of aircraft. In some cases, automated configuration warning devices have failed or been rendered inoperative and flight crew

---

[1]See Glossary for definition and brief description of this and other abbreviations and acronyms.

[2]Traffic alert and collision avoidance system.

procedures have failed to detect by independent means an unsafe configuration for takeoff. In other cases, automation has operated in accordance with its design specifications, but in a mode incompatible with safe flight under particular circumstances. In still others, automation has not warned, or flight crews have not detected, that the automation was operating at its limits, or was operating unreliably, or was being used beyond its limits. Finally, we have seen incidents and a few accidents in which pilots have simply not understood what automation was doing, or why, or what it was going to do next.

It is clear, as the ATA report stated (see preface), that aviation automation has conveyed important benefits. It is also clear that certain costs have been associated with automation, and that the presence of automated devices has changed human operator behavior (as all tools do), whether for better or, in some cases, for worse. If we observe Wiener's (1993) maxim that automation does not eliminate human error, but rather changes its nature and possibly increases the severity of its consequences, it is necessary to understand how these devices influence the humans who work with them, and how humans use and shape automated devices as tools with which to accomplish their work.

PURPOSE OF THIS BOOK

My purpose in writing this book is to describe and exemplify several classes of problems that are associated with the implementation of advanced automation in the aviation context. These problems interfere with the effective operation of the aviation system and in some cases degrade the safety and reliability of the system. I attempt to show that these problems arise at the conjunction of humans and automated devices in this human–machine system. I propose that a philosophy or construct that I call "human-centered automation"[3] may be of assistance in resolving these problems in the future system by improving the cooperation between humans and machines. Finally, I attempt to draw parallels between aviation and other domains in which automation is an important part of high-risk, real-time systems, and to suggest how lessons learned in aviation may perhaps be helpful in other endeavors.

DEFINITIONS

Automation, as used here, refers to systems or methods in which many of the processes of production are automatically performed or controlled by autonomous

---

[3]This term is defined later in this chapter; its origins are discussed in chapter 3.

machines or electronic devices. I consider automation to be a *tool*, or *resource*—a device, system, or method by which a human operator or manager can accomplish some task that would otherwise be more difficult or impossible, or a device or system that the human can direct to carry out more or less independently a task that would otherwise require increased human attention or effort. As used here, the word *tool* does not foreclose the possibility that the device may have some degree of intelligence—some capacity to learn and then to proceed independently to accomplish a task. Automation is simply one class of resources among many available to the human operator or manager. *Human-centered automation* means automation designed to work cooperatively with human operators in the pursuit of stated objectives.

Piloting is the use by a human operator of a vehicle (an aircraft) to accomplish a *mission* (to deliver passengers or cargo from one point to another). A mission consists of a number of functions, each involving from one to many tasks and subtasks, which are accomplished using a variety of human and machine resources (Fig. 1.1). Most attempts to decompose the piloting function have adopted a functional hierarchical architecture of this sort.

Resources available to pilots include their own perceptual, cognitive, social, and psychomotor skills, the knowledge and skills of other flight and cabin crew members, and the knowledge and information possessed by other persons with whom the pilot may be able to communicate, especially airline flight dispatchers who share with

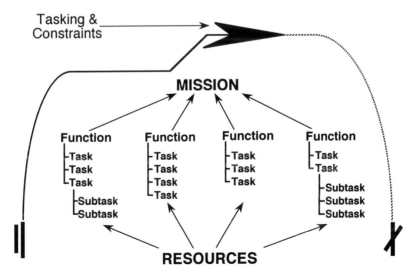

FIG. 1.1.   The piloting task.

the pilots responsibility for the safe planning and conduct of their flights. They are aided by a variety of information sources and control devices, including automated devices, within the aircraft. In aircraft designed for multiple crewmembers, these resources are controlled and managed by a *pilot in command* (PIC), who is ultimately responsible for safe mission accomplishment.

Controlling is the function of directing aircraft, on the ground or in flight, in ways that assist the aircraft to move from point to point in conflict-free trajectories. Control may be *tactical*, involving the direction of specific aircraft through a specified part of the airspace; or *strategic*, involving the provision of general instructions and constraints for the movement of masses of aircraft within a much larger volume of airspace (Fig. 1.2). Tactical control is referred to as air traffic *control* (ATC); strategic control of air traffic is referred to here as air traffic *management* (ATM).

Air traffic controllers are responsible for the safe direction and separation of air traffic. The resources available to them include their perceptual and cognitive skills (psychomotor skills are less important than in piloting), their knowledge of and ability to recall quickly a large body of procedures and regulations governing the control and movement of air traffic, the knowledge, skills, and abilities of other controllers who may be assisting or immediately supervising them, and the support of controllers and team supervisors controlling adjacent airspace. Their material resources include

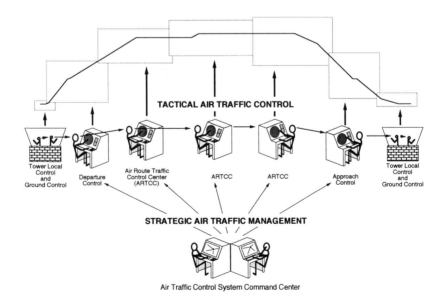

FIG. 1.2.   Air traffic control and management.

the airspace itself, airports, surface or airborne navigation aids, and a variety of surveillance, communications, and data-processing systems and devices.

Strategic *management* and coordination of, and assistance to, tactical air traffic controllers is designed to maximize airspace usage while preventing overload of individual air traffic control facilities. This function (sometimes called "flow control") is provided within the United States by an air traffic management organization called the Air Traffic Control System Command Center (ATCSCC), located near Washington, DC, and connected to all control facilities (and to many air carriers) by various means of voice and data communication.

COMMENT

To summarize: This document is about a human–machine system in which highly skilled human operators use tools of varying complexity to perform cognitively difficult and exacting work in a very demanding, sometimes dangerous, and always highly dynamic physical environment. Their work is tightly constrained by a highly developed, well-integrated operational environment and a complex organizational environment. These environments are also the source of most of the variability within the system, some of it, such as weather, inherently unpredictable and uncontrollable by the humans who control and manage the system. A great deal of automation has been introduced in aviation to assist and support human operators and managers in the performance of their duties (although many of the older aircraft in the system today are not heavily automated and the air traffic control system is still largely unautomated, pending extensive rework now underway). It is highly likely that the airborne and surface components of the aviation system will become much more tightly coupled in the near future. Whether the tighter coupling, as well as the continuing integration, of the various system elements will be accomplished in ways that permit human operators to remain in effective command of the system is not yet clear, and this is one of the major issues raised in this document (see chapter 11).

Most knowledgeable observers agree that future social and political demands on the aviation system cannot be satisfied without more automation. The form of that new automation, how it will interface with the humans who are responsible for system safety, and whether it will materially assist those humans to improve overall system performance are still open questions. I attempt to bound these questions more precisely and to answer those for which principled answers are possible.

CHAPTER 2

THE CONTEXT AND ENVIRONMENTS
OF AVIATION

INTRODUCTION

Human–machine systems do not evolve in a vacuum. To understand such systems, it is necessary to know something of the context in which they operate. In this chapter, I discuss the context of the aviation system: aircraft, people and the tasks they perform, and the environments (physical, operational and organizational) in which aviation is conducted.

THE AIRCRAFT

Although the focus of this book is on transport aircraft, many other aircraft operate in the same environments and interact on a regular basis with transports. Of approximately 192,000 active civil aircraft registered in the United States during 1992, only about 7,300 (less than 4%) were air carrier aircraft. Activity estimates indicate that air carrier flying hours, 16.6 million, constituted 38.5% of the total hours flown during 1992, although 63% of the airplane *miles* flown were by air carriers. Passenger boardings by Part 121[1] carriers operating in scheduled air transport were 473 million; the Aircraft Owners and Pilots Association estimates that an additional 99 million passengers were carried by general aviation aircraft (AOPA, 1994, p. vi).

The Air Carrier Fleet

Except in Alaska and the Caribbean area, nearly all scheduled airline flying today is in turbojet aircraft, ranging from small twin-engine jets carrying perhaps 50 passengers to Boeing 747s carrying 8–10 times that complement. (Commuter airline

---

[1]Part 121 is the section of the Federal Aviation Regulations (FAR) that deals with scheduled air carriers and their operations.

flying involves primarily turbopropeller aircraft.) Airline aircraft range in age from brand-new to a few first-generation transports built during the early 1960s. Many of the older machines are now utilized by overnight cargo carriers.

Older transports have little automation, rely entirely on radio navigation, and carry few flight crew aids aside from simple autopilots. Many older commuter aircraft are not even equipped with autoflight systems, although the newest of them normally have some automation aiding. The newest "third generation" aircraft (see Fig. 5.4) are highly automated machines. All have a flight crew of two persons rather than the two pilots and a flight engineer required in earlier large aircraft. All of these aircraft are flying in the same airspace, and the same air traffic control procedures must be applied to all of them, which has prevented the more automated machines from realizing some of the considerable capabilities of their automation.

Turbojets all fly at high speeds (Mach .76–.85[2]) at high altitudes (up to 45,000 ft), under positive air traffic control on instrument flight rules (IFR) flight plans, between airports certified for transport aircraft. Many spend little or no time outside positive control airspace (see Fig. 7.1). Turbopropeller[3] commuter aircraft, in contrast, spend much more time on very short routes (most less than an hour) in lower altitude strata. Many are still fairly fast (200–350 kt), although slower than jets, and most of their operations are also IFR except at smaller airports that may have part-time control towers. Safety records for commuter aircraft have been less good than for the major carriers, although to some extent this would be expected because they are flying under a broader range of circumstances, under less controlled conditions and in less favorable environments.

General Aviation

This term covers nearly all civil aviation except that conducted by air carriers. General aviation (GA) includes corporate aircraft, some carrying more sophisticated equipment than the airlines, flown by highly experienced professional pilots. It includes a large number of recreational and instructional aircraft. It also includes utility aircraft used for aerial application, water-bombing of forest fires, and other purposes. General aviation is regulated by FAR Part 91.

The aircraft found in general aviation range from antique biplanes built in the 1920s up to modern, high-speed, pressurized single- and multiengine airplanes fully capable of operation under all weather conditions in the airspace system. Virtually all civil flight training is done in small single-engine airplanes, and a great majority

---

[2]*Mach*, after Ernst Mach, is a measure of the speed of sound. It is a function of air temperature, which varies with altitude. Mach .80, a common jet cruising speed, is eight tenths of the speed of sound at the particular altitude at which the airplane is flying.

[3]Aircraft powered by turbojet engines driving conventional propellers.

of the nearly 700,000 active U.S. pilots are involved exclusively in recreational or personal business flying activity in single-engine aircraft. Most U.S. airspace is open to any qualified pilot wishing to utilize it if weather permits, and the national airspace system has been designed to accommodate the broadest possible range of air commerce on a "first-come, first-served" basis. General aviation has always been less safe in terms of accident experience than air carrier flying, but this would be expected based on the fact that many of the missions undertaken by these aircraft involve flight at low altitudes, in the vicinity of obstacles, in uncontrolled airspace.

General aviation aircraft deliveries, formerly a significant part of the aviation economy, have declined disastrously during the past 15 years. From 1982 to 1992, new aircraft shipments declined from 4,266 to 899, a fall of 79%. Active general aviation aircraft have declined by 12% over that period, and hours flown have fallen by 27% (AOPA, 1994).

Despite the considerable costs of participation in aviation activities, many general aviation aircraft are well equipped to participate in today's aviation system. As an example, area navigation equipment (see chapter 5) is found in over 50% of these aircraft, making direct routings possible and alleviating pilot workload. Most of these aircraft are flown by a single pilot whose proficiency may or may not be at a high level; most recreational pilots fly relatively few hours per year (less than 100). The workload involved in single-pilot IFR flight under difficult conditions (whether the difficulties are due to weather, unfamiliar terrain, ATC constraints, or a difficult mission) can be very heavy, and many general aviation pilots are not accustomed to working within a demanding operational environment. Further, there are usually fewer resources available to deal with difficulties, and GA pilots may be less knowledgeable about sources of assistance or less willing to make use of others, including ATC, to help in problem resolution.

Problems Due to the Mix of Aircraft

The wide diversity of aircraft operating in the national airspace system presents pilots and controllers alike with serious problems. They are most pronounced at and around busy airports. Most general aviation aircraft cruise at speeds lower than the final approach speeds for turbojets. Larger jet aircraft produce considerable wake turbulence, especially during approach; this forces controllers to impose greater separation distances between them and smaller following aircraft, which in turn decreases runway capacity if the same runway must be used by both types. General aviation pilots may be less familiar with airspace and airport layouts and may have more difficulty obeying complex instructions; controllers must take this into account in handling these aircraft, both in terminal areas and on the ground at busy airports.

Similarly, the wide mix of older and newer transport aircraft poses problems. Newer transport aircraft have considerably higher power-to-weight ratios than older types; they climb much faster, and their most economical descent profiles differ appreciably from older jets. Controllers in terminal areas find it necessary to have everyone on a given descent path at approximately the same speed and descent rate to preserve spacing in the approach corridor. ATC has generally been unable to accommodate the least-cost descent profiles calculated by today's flight management systems, resulting in higher fuel costs for the new transports.

Small aircraft do not present other pilots with a very prominent visual signature, particularly during less-than-perfect weather or when the two aircraft are on a collision course (during which neither moves in the visual field of the other). GA pilots must take account of the fact that they do not present a prominent target. Although class B airspace (Fig. 7.1) is designed to assist in separation between aircraft by mandated positive control, the areas immediately surrounding this airspace may be relatively crowded with GA aircraft attempting to avoid the delays often associated with such control. Transport pilots are usually climbing or descending when in these areas, and their workload may distract them from required lookout duties.

The principal effects of the traffic mix that characterizes much of our low altitude airspace are thus increased requirements for visual search and vigilance on the part of all pilots flying at those altitudes, and increased planning and execution workload for air traffic controllers. The three-dimensional position information needed by controllers is generally available if, and only if, all aircraft of concern are using altitude-encoding transponders; other targets may or may not be noticed, with sometimes tragic results. The visual information available to pilots may or may not be adequate, depending on environmental variables and to some extent on chance.

Finally, military pilots flying tactical aircraft have, and pose, special problems. Much of their training involves low-altitude flight, often at very high speeds. Military training routes are marked on VFR charts, and en route flight following facilities make every effort to warn pilots when they are in use. VFR aircraft, however, are not required to communicate with ATC facilities nor to take advice when it is offered. Some do neither, presenting the military pilots with the problem of maintaining a constant lookout while controlling a demanding airplane on a complex mission profile close to the ground. Fast-moving camouflaged fighters also are low-visibility targets by design, particularly when viewed from above.

PEOPLE IN THE SYSTEM

A majority of the early air transport entrepreneurs were pilots for whom management duties were often a distraction. The basic motivation for these people was the joy of aviation, whether or not they made money doing it. This attitude still motivates young men and women who remain willing to accept gigantic financial burdens learning to fly, building experience, and even financing their own entry into air carrier flying. A majority of pilots still derive deep pleasure and satisfaction from flying airplanes, and this motivation must be understood by those who support this activity. A similar sense of motivation pervades the air traffic controller community, whether or not they are themselves pilots.

Helmreich, Foushee, and more recently other social psychologists studied effective and less effective air transport pilots in an effort to ascertain personality attributes associated with flying performance (Chidester, 1990; Chidester, Kanki, Foushee, Dickenson, & Bowles, 1990; Helmreich & Wilhelm, 1991; Helmreich & Foushee, 1993). Helmreich, the guiding figure in these investigations, described clusters of attributes that he called "the right stuff," "the wrong stuff," and "no stuff" (Helmreich & Wilhelm, 1989, p. 296; Fig. 2.1). Validation studies of these concepts are underway. Although the jury is still out, it appears likely that air transport flying is a skill in which persons with certain personalities fare much better than others. (For a general discussion of these studies, see Wiener, Kanki, & Helmreich, 1993.)

Although Helmreich's characterizations accurately describe many individual pilots, people in air transport do not ordinarily operate as individuals, but as teams. A full-mission simulation study conducted by Ruffell Smith and Lauber (Ruffell Smith, 1979) yielded results that suggested that management, rather than technical, shortcomings were at the heart of the performance deficiencies observed.

PERSONALITY FACTORS

"Right Stuff"   High levels of positive instrumental traits (task-oriented)
High levels of positive expressive traits (interpersonal)
  ➤ *Consistently higher performance in simulations*

"Wrong Stuff"   High levels of negative instrumental traits (autocratic)
Low levels of positive expressive characteristics
  (uncommunicative)
  ➤ *Lower performance in experimental simulations*

"No Stuff"   Low levels of positive instrumental traits
Low levels of positive expressive traits
  ➤ *Particularly ineffective performance*

FIG. 2.1.   Personality attributes in air transport pilots (Helmreich & Wilhelm, 1989).

Subsequent studies into cockpit or crew resource management (CRM; Foushee, 1984; Chidester, 1990) confirmed the importance of considering and training the air carrier crew as a team, rather than individuals, and have also helped to understand cockpit crew effectiveness under very difficult circumstances (Langer, 1992).

Now there is automation as well, serving as a part of the team—another more-or-less intelligent resource in the cockpit. Very complex automation is often perceived by pilots as having a mind of its own, especially when it does something quite unexpected. Hollnagel (1993), and Sarter and Woods (1992a) called such unexpected actions surprises and discussed how they can erode human operator trust in automation. Sarter and Woods (1994) also discussed the concept of animacy, in which a complex machine, because of its apparently independent actions, appears to act like an animate being. I indicated that automation must be predictable (Billings, 1991; see also chapter 3), if humans are to be able to form a mental model or construct of how it functions. Surprises fundamentally disrupt the integrity of such internal models and are symptomatic of their inadequacy.

Do such factors have implications for the design of aircraft automation? These factors certainly have profound implications for how various organizations will opt to use the technology available to them. Attitudes toward automation, and prescriptions for its use, vary from the flexible approach of Delta Airlines (Byrnes & Black, 1993; see also chapter 6) to more rigid philosophies in some air carriers, especially in Europe.

One of the surest ways to insure that automation will be used is to disempower its users by denying them access to information that would permit them to question it—to reduce its observability. It is to some extent an inherent problem in complex automation because of the narrow "keyhole" (the computer screen) through which a very complex monitored process must be observed (see chapter 13 and Fig. 13.1). Some organizations implementing numerical control in manufacturing insured this by denying machine operators access to the punched cards and codes that instructed the machines (Noble, 1983). In so doing, they also made it difficult for operators to intervene in the machining process when it was obvious to them that cutting speeds needed to be adjusted because of factors beyond the cognizance of the automation (see also chapter 4).

Designers can alleviate this problem to some extent by careful attention to the representations through which human operators are able to visualize the process as it evolves. In the many dynamic processes that can be observed only through that keyhole (and the movement of air traffic is one of these), adequate representations can enable a human operator to access the knowledge required to understand the process and to predict the effects of changes he or she may feel are warranted to ensure the safety of the operation.

On the other hand, some processes are too complex or may evolve too rapidly to be controlled or supervised without substantial machine aiding. The digital computer has made it possible to process a great deal of highly dynamic data quickly and to represent it in ways that efficiently inform the operator of changes within the monitored process. The trick is to provide the operator with relevant information in a form that will support decision making and intervention in time to moderate the process when that is desirable, taking into account both human and system time constants. Operators must have, of course, the knowledge and mental models necessary to "remain ahead" of the processes; the expertise built up over years of learning and practice is not easy to encode within the narrow confines of even a very capable computer.

THE TASKS

Pilots

In chapter 5, I discuss the nature of the pilot's task: to *"aviate"* (control the airplane's path), *navigate* (direct the airplane from its origin to its destination), and *communicate* (provide data and requests and receive instructions and information), and, increasingly in modern aircraft, to *manage* the resources available. These tasks must be carried out in a highly dynamic environment. Although such a complex job is by no means unique, it is one of the more demanding constellations of capabilities required in the work environment. The perceptual requirements are considerable and the cognitive demands even greater, for much of the information required by the pilot must be synthesized from a great amount of raw data, some of it ambiguous under many circumstances.

The psychomotor requirements are considerable, although accident experience suggests that few air carrier accidents involve failures in the "stick-and-rudder" aspects of piloting. Far more mishaps appear to involve failure to manage resources effectively, which in a multiperson cockpit includes, among other things, managing the workload on oneself and on the remainder of the crew.

In any air carrier operation, there is a great deal to be kept track of, for both the airplane and the environment are complex and may change rapidly. The ability to share and allocate attention as required is extremely important. This may not be difficult under ideal circumstances, but flying often demands long hours either of relatively high activity (in short-haul flying) or of very low task demand during long overwater flights. Schedules also impose constraints with respect to both duty hours and availability of time for adequate rest, and studies have indicated that on average, pilots incur sleep deficits that may be considerable during many long schedules (Graeber, Lauber, Connell, & Gander, 1986).

Air Traffic Controllers

Task requirements for air traffic controllers differ in substantial respects from those for pilots. The term *air traffic control* is something of a misnomer. Controllers *manage* air traffic by *directing* pilots (who alone exercise direct control over the movements of individual aircraft) in order to maintain separation between airplanes and an orderly flow of traffic in the airspace assigned to them. The controller is able only to communicate his or her requirements to others, who usually comply, but may or may not do so in a way that meets the controller's needs of the moment. This is a major reason for the tightly circumscribed vocabulary and set of commands controllers are permitted to use, and for the very complex procedures that govern their span of control.

The traffic environment is highly dynamic; the perceptual demands are somewhat less than for pilots because of the radar representations (which are usually unambiguous with current technology), but the cognitive demands may be much higher, for the controller must always maintain a mental image in three dimensions of all traffic under his or her control. Despite the assistance provided by radar and data processing, this task is not simple. If the mental model of traffic is lost (controllers call it "losing the picture"), the controller must rebuild it serially, using flight strips and memory, a time-consuming process (Hopkin, 1994b; see also chapter 7).

Controllers normally work either as individuals or as small teams, almost always under the direction of a team supervisor, although that person will often be involved with another task and may thus be unable to help out when the pace quickens. A fully competent controller must be able, by whatever means are available, to do what needs to be done, with or without assistance, until situations can be resolved.

Memory demands are high and the ability to call to mind appropriate rules, strategies, procedures, and directives is absolutely vital, for each controller is a repository of a very large amount of procedural knowledge, any of which may become important at any time. Controllers operate under very tight human and machine scrutiny; their ability to maintain traffic separation is constantly monitored by ATC automation, and any transgression of separation rules may become a cause for remedial or disciplinary action. Training requirements are stringent; only a minority of those selected for ATC training are able to become full-performance controllers, and that only after several years of training and apprenticeship.

Skill and Training in Aviation

Golaszewski (1983; 1991) and others demonstrated that aviation accident rates in professional pilots decline with increasing recent and total flying experience; recent experience is more important than total experience (Fig. 2.2). Pilots flying more

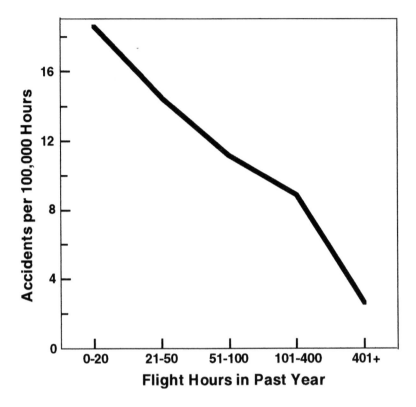

FIG. 2.2.   Accident rates in pilots holding class I and II medical certificates as a function of recent flight experience (data from Golaszewski, 1983).

than 400 hours per year have lower rates than others with equal total experience flying less hours.

The average air carrier pilot flies somewhat more than 500 hours per year, and both military and civil experience indicate that considerable exposure is necessary to remain fully proficient in the demanding tasks of flying complex aircraft. The context of exposure is important, of course: in some extremely long-range aircraft, 600 hours of flying may represent only about 50 to 60 takeoffs and landings per year, which may be shared among three or four crewmembers. Maintenance of adequate motor skills can be a nontrivial problem in such aircraft, and additional "practice" in today's highly capable simulators may be the only answer. A few long-range carriers have mandated such additional training.

The FAA requires annual or semiannual training and proficiency checks as well as periodic observation during line flying. Most pilots, however, spend far more than this amount of time reviewing educational materials, manuals, and regulations, both on the ground and on duty. Air traffic controllers likewise receive periodic training and spend many hours studying when not actually controlling traffic during their duty shifts. This professionalism is a major reason why the aviation system works well and safely.

THE EVOLVING CONTEXT OF AVIATION: CHALLENGES OF THE 1990S AND BEYOND

In the past, most industrial nations supported the growth of air transport fairly heavily. Today, we see instead a growing disinclination to continue public-sector support, however disguised. This tendency, probably initiated by airline deregulation in the United States in 1979, is to let market forces prevail. To a greater extent than ever before, new technology will have to "buy its way onto the airplane." It is against this social and political backdrop that we must examine all other challenges; they are the context of the aviation system during the 1990s.

What challenges do we face during the remainder of this century? At the 9th World Airports Conference in 1991, the Senior Director of the IATA Infrastructure Action Group summarized his view of the situation in this way: "The air transport industry is in the midst of a crisis, the magnitude of which is greater than anything it has faced in the past. . . . The real crisis of the 1990s is the increasing inability of the aviation infrastructure to handle rapidly growing traffic" (Meredith, 1992, p. 109). European airlines are acquiring $5 billion of new aircraft annually; governments are putting only $1.5 billion into infrastructure: air traffic control and airport facilities. Delays in Europe alone cost $5 billion per year; this number may double by 2000, although there is sufficient airspace capacity to accommodate growth to 2010 if a pan-European approach to air traffic control can be worked out. By 2010, 13 of 27 major airports in Europe will be capacity-constrained even with planned and potential improvements to terminals and runways.

In his keynote address to the Human Factors and Ergonomics Society in Seattle on October 12, 1993, Robert Davis, vice president for engineering and product development of the Boeing Commercial Airplane Group, discussed the global marketplace. He began by citing the dismal economic situation of U.S. air carriers, which lost $10 billion over the previous three years, $5 billion more than they had made in their entire history prior to that time. More than $12 billion in aircraft and engine orders were canceled or deferred. Yet the world travel and tourism industry has a cash flow of $3.5 trillion per year; it is the largest industry in the world (Davis, 1993).

Davis projected an average increase in air travel of about 5.4% per year through 2010, requiring an increase from 10,000 to 18,000 airplanes during that period.

(More recent Boeing projections indicate a requirement for 14,000 new aircraft including replacement of retired older machines; Proctor, 1994a). The new aircraft will represent an enormous capital investment; Boeing predicts that the 14,000 new aircraft will represent an outlay of $980 billion.

Finances

Against these projections are the current, and until 1995 generally dismal, financial results for leading U.S. and foreign air carriers. The need to return to sustained profitability is a greater challenge even than the need to serve increasing public demand for air service, for without positive balance sheets world air carriers cannot purchase the new equipment they will need to meet the demand for services during this and the next decade.

Demand for Services

ICAO 1993 air traffic forecasts give a clear picture of world traffic demand during the next decade. Their data indicate that seat demand on international routes will increase by 5.5% per year during this decade, a slightly lower rate than the 5.9% annual rise during the 1980s. *Aviation Week and Space Technology* (Growth projections, 1993) indicated that the 1990s "slower rate of growth is due in part to the 3.5% decline in passenger traffic in 1991, the first decline in commercial aviation's history. Airlines are attempting to recover from that setback, are more than a year behind in doing so, and their efforts are being hampered by the slow-growing economy" (p. 65).

Airports

One of the greatest challenges facing the aviation community as a whole is a lack of capacity, not in aircraft but in airport runways and infrastructure. No matter how efficiently airspace is utilized, airplanes must land at airports. Runway occupancy times are a key factor in airport acceptance rates. Some of the world's major cities are handicapped by single-runway airports, among them Tokyo (Narita), London (Gatwick and Stansted), Osaka (Kansai), and Hong Kong (Kai Tak). The FAA has enforced capacity limits at four high-density airports in the United States since 1969; the agency is presently conducting studies to see whether these acceptance rates can be increased.

Leaving aside the enormous expense and the decades required to plan and build new airports, environmental issues, urban congestion, and local politics make it increasingly difficult to secure consensus that a new airport, or even a new runway, represents a community asset. The NIMBY ("not in my back yard") syndrome is never more evident than in aviation matters, in part because airplanes, even relatively quiet ones, are perceived as significantly noisier than other machines in the environment.

Airspace

It is not only lack of airports that threatens the advance of commercial aviation. In some areas, it is also the lack of airspace. One of the principal aims of the FAA's Air Traffic Automation program is an increase in U.S. airspace capacity. Many air carrier flights are already carried out using direct routing rather than the nation's VHF airways, with savings of fuel, time, and wear and tear on the airplanes. The FAA has facilitated these routes where possible. The task of assessing future conflicts is much more complex when random flight paths are used, however. As more new aircraft enter service, demand for such routings will increase, yet a wholesale conversion to such strategies is not thought to be possible without more advanced ATC automation to detect, and assist in the resolution of potential conflicts.

In Europe, the airspace situation is far more serious because of its fragmented ATC network, which has "51 ATC centres . . . 31 system families, using computers from 18 different manufacturers with 22 operating systems and 33 different programming languages." A program instituted in May 1992 called the European Air Traffic Control Harmonization and Integration Program, EATCHIP, has the ambitious aim of transforming this network into a single integrated air traffic management system by the year 2005 (Philipp, cited in Cooper, 1994a). It is believed that the program can increase airspace capacity by as much as 50%, although Europe's air carriers, which are "paying the price for ATC delays despite traffic decline on many routes" (p. 10), believe the pace of improvement is much too slow.

Aircraft and Equipment Capabilities

The mix of aircraft and avionic equipment will continue to pose major problems for the world's air traffic management systems. An ATC system cannot impose "4-D" (three spatial dimensions plus time) traffic management on aircraft that do not have the capability to make good a specific time of arrival at a metering fix far downstream, yet without 4-D traffic management automation, most efficient use cannot be made of terminal airspace. The International Civil Aviation Organization's Future Air Navigation Systems Committee (FANS) has proposed a functional concept of *required navigation performance* as a way to smooth the technology transition.

One other issue deserves mention. The advent, by the turn of the century, of substantially larger aircraft than the transports in use today (600–1000 passenger machines are under serious consideration) will impose a severe burden on the airports that handle them and the "ground-side" facilities that must serve them. This is another constraint that will involve a great deal of expense in the near future.

In summary, the challenges that will shape the future aviation system are financial (because new aircraft and new technology to cope with the other challenges are expensive), physical (because airspace and airports are constrained physical assets and weather is still an important uncontrollable variable), and organizational (because our present ATC infrastructure is reaching, and in some areas of the world has exceeded, its capacity). Even the most pessimistic observers agree that demand for air services will increase considerably between now and 2010; the only question is by how much. Even the most optimistic observers recognize that without major changes in aircraft, procedures, and infrastructure the world aviation system will be unable to meet the demand. In chapters 6 and 8, I discuss the automation and other technology that have been proposed to assist in meeting these challenges, (except the financial challenge, which is well beyond the scope of this book and well beyond the abilities of its author).

AVIATION ENVIRONMENTS: PHYSICAL, OPERATIONAL, AND ORGANIZATIONAL

As marvelous as are the technological advances that characterize the machines in the aviation system, these machines must function in a disorderly real world. The environments in which aviation is conducted contain many sources of variability over which we have little or no control. Because human–machine systems can only be understood in context, it is vital to understand the environments in which such a system operates. Three environments are discussed here. They are:

- A *physical environment* in which aircraft must fly (and also the physical environments in which the ground elements of the system must function).
- An *operational environment*, by which I mean the procedural environment created by air traffic control and management in the performance of their support functions.
- An *organizational environment* created by the government (whose regulatory and certification elements control aviation), the armed forces, air carriers and other companies that operate aircraft, the labor organizations that negotiate with operators concerning the working conditions of their

members and play important roles in safety surveillance, and other organizations having an interest or a stake in aviation operations.

This section is devoted to brief descriptions of some of the sources of variability that may influence aviation operations, and how they affect the smooth functioning of the complex system under consideration here.

The Physical Environment

Although aircraft have changed dramatically, they are still operated by the same humans in the same physical environment. What has changed is the amount of pressure on airlines to maintain schedule regularity in the face of uncontrollable variability in the physical environment. A number of factors in the physical environment add complexity to the pilot's (and in some cases the controller's) cognitive tasks (Fig. 2.3). They increase the perceptual demands on pilots, add uncertainty to their planning and decision-making processes, and if undetected can require rapid compensatory or evasive action.

Air Turbulence

Turbulence in a mass of air is caused by either vertical or horizontal adjacent air masses moving at different velocities. Close to the ground, turbulence under and in the vicinity of thunderstorms may cause rapid and even extreme changes in lift, a particular danger when an airplane is taking off or approaching a landing at slow airspeed. These vertical currents may move extremely rapidly; in extreme cases, they can exceed the structural limits of even a large airplane. Microbursts have been implicated in more than 20 fatal accidents.

Turbulence in the vicinity of jet streams is due to horizontal air masses moving at sharply different velocities. The differences in air speed may be over 100 kt, sufficient to cause jet aircraft upsets. A third type of turbulence is found over and downwind from high mountain ranges. This phenomenon, known as mountain wave, can likewise produce moderate or severe turbulence, especially for aircraft at relatively low altitudes. Dry air turbulence is commonly encountered over deserts.

Many aircraft have been damaged or lost when turbulence either exceeded their control capabilities or actually caused structural damage in flight. Wake vortices, another type of turbulence, occur behind aircraft in flight. They involve rotary air disturbances of considerable magnitude; NTSB data indicate that a wake vortex encounter has been the probable cause of at least 51 accidents and incidents from 1983 to 1993, with loss or severe damage to 40 aircraft.

HAZARDS TO SAFE FLIGHT

Phenomenon	Probability of Occurrence	Severity of Threat
• Air turbulence		
– Convective turbulence (in thunderstorms)	Frequent	++++
– "Clear air" turbulence	Less frequent	+++
– "Dry air" turbulence	Infrequent	+
– Wake turbulence (from preceding aircraft)	Frequent	++++
• Precipitation		
– Rain	Very frequent	+
– Snow	Frequent	++
– Hail	Infrequent	++++
– Sleet	Frequent	+++
• Ground obstacles		
– Terrain	Invariable	++++
– Towers and other cultural features	Invariable	++++
• Airborne objects		
– Aircraft	Frequent	++++
– Birds	Less frequent	+
– Lighter-than-air craft (balloons, etc.)	Infrequent	++
• Atmospheric debris		
– Volcanic ash	Rare	+++
• Electromagnetic radiation		
– Ground sources	Rare	++
– Sources within airplane	Frequent	++

FIG. 2.3.   Physical environmental hazards to flight safety.

Precipitation

Rain, sleet, snow, and hail are all water or ice. Rain, unless very heavy, is not normally a problem for transport aircraft, although extremely heavy rain can cause jet engines to "flame out" (New Hope, Georgia, 1977).

Moist snow can adhere to aircraft surfaces. Rain at temperatures near freezing can also freeze on very cold surfaces. In properly equipped aircraft, icing is not a serious problem once in flight because of deicing or anti-icing equipment. Prior to takeoff, however, when engines are at or near idle power, little hot bleed air is available from the engines, and during prolonged taxiing or holding on the ground in precipitation ice can form on any part of the aircraft, changing the shape and smoothness of the metal surfaces and severely disrupting airflow during a subsequent takeoff.

Hail, balls of ice generated by violent up- and down-drafts in thunderstorms, can reach considerable size (2 inches or more). Hail showers can very significantly deform the leading edges of wings, destroy projecting antennas, dent or break plastic radomes, and even shatter windshields.

Ground Obstacles

It almost seems unnecessary to mention that an unplanned encounter with the earth, or a man-made obstacle on the earth's surface, constitutes a serious environmental hazard. Such encounters (controlled flight into terrain), however, remain a prominent accident type and threat (Duke, 1996; Russell, 1992), despite the success of GPWS and a high level of awareness of the dangers of this type of mishap. It is a truism that "every airplane comes to rest somewhere at least once during each flight," and insuring that this occurs only on an appropriate runway remains a problem of some magnitude.

Airborne Objects

Air Traffic.   The principal task of the air traffic control system is to maintain separation between controlled aircraft. In class A and B airspace, all aircraft must submit to positive control, and controllers are, or should be, aware of all traffic within their airspace—except aircraft that may blunder into such airspace. Because of these occasional transgressions, pilots are expected to see and avoid other aircraft under visual meteorological conditions regardless of whether they are being controlled by ATC. This is not always easy, especially when haze or smog is present.

Birds.   Birds can be a problem, particularly during their migrations when large flocks may be encountered in flight. Nonmigratory birds are a constant problem around some airports, particularly along coastlines. A Lockheed Electra crashed at Boston, Massachusetts (1960) after its engines ingested a flock of starlings during takeoff, and a Vickers Viscount turboprop crashed near Ellicott City, Maryland (1962), after its horizontal stabilizer was severely damaged by the impact of a whistling swan.

Other Objects.   Much less common than other aircraft, but still a potential threat to air safety, are other objects traversing the atmosphere. ASRS reports describe the variety of such objects, from parachutists jumping from high altitudes on airways, through ultralight aircraft or gliders, to radio-equipped weather balloons sent up from weather stations to take soundings of the lower atmosphere.

Atmospheric Debris

Debris carried aloft in air masses can persist in the atmosphere for considerable periods of time. When such debris is injected into the higher levels of the atmosphere by a volcanic explosion, some fraction of a sometimes huge volume can persist for months. Such contamination is often not visible, especially at night or in cloud, but a heavy concentration can wreak havoc with aircraft. At least two four-engine flame-outs have been ascribed to volcanic ash and dust.

Other Environmental Hazards

One other class of hazards deserves mention here. Certain military devices emit very high power in the microwave portion of the electromagnetic spectrum; civil aircraft avionics are not very effectively insulated against the effects of such radiation. Our growing dependence on miniaturized electronic components, which can be susceptible to such radiation, subjects civil aviation to a real threat of avionics disruption due to electrical transients from such devices.

Small electronic devices carried by passengers can also emit electromagnetic radiation, which under certain circumstances has been suspected of disrupting aircraft navigation units (NASA ASRS, 1994). Air carriers now forbid the use of any such units at altitudes below 10,000 ft, and proscribe the use of cellular telephones and some other devices at any time on board transport aircraft. Having said this, I should mention that in one case, after a total failure of his communications radios, a quick-thinking captain asked whether any passenger had a cellular phone on board, then used the device to reestablish communications with ATC—surely an excellent example of cockpit resource management in action!

The Operational Environment

The air traffic management system manages and controls all movements of air carrier aircraft. Air carrier flight operations systems, operating within constraints imposed by air traffic management, provide a continuing feed of aircraft to the air traffic system (Fig. 2.4). Although tactical air traffic control is still largely a manual system, strategic air traffic management has been automated to a considerable degree. Flow management, designed to insure that the ATC system does not become overloaded, now determines capacity at heavily used airports and redirects the flow of air traffic during contingency operations forced by weather, runway closings, or emergencies in progress.

The work of the FAA System Command Center is largely transparent to individual pilots. Not so the ATC system, which controls literally every movement of every air carrier airplane from gate to gate. Air traffic controllers and pilots together are the operators of the system (Fig. 2.4); they share responsibility for safe mission completion.

Air traffic controllers have been freed to some extent from purely procedural control of air traffic by the advent of radar and altitude-encoding transponders, which provide them with three-dimensional indications of aircraft locations. However, constraints imposed by the increasing volume of air traffic still force them to work largely by inflexible rules. The discrepancy between airborne equipment capabilities and the ability of ATC to permit the use of those capabilities has increased and become more obvious since the introduction of highly automated aircraft whose flight management systems have vertical navigation capability.

The inherently manual nature of air traffic control forces it to operate in a highly orderly manner. The present system is insufficiently tolerant either of disorder or of human error, as was tragically demonstrated in two recent collisions between two aircraft on runways at Detroit (1990) and Los Angeles (1991). Massive automation of the ATC system during the next two decades will permit the limited capacity of U.S. airspace to be utilized to the fullest extent possible, although without new runway capacity the system will continue to be under severe strain into the foreseeable future. The advent of ATC satellite communications capability, together with automatic dependent surveillance (ADS), promises to extend the span of direct control to oceanic routes in the near future (see chapter 8). ATC automation may cause major changes in the processes by which air traffic controllers and pilots work together to accomplish the mission.

Not all of these changes will be bad, by any means; the automated en route system should be able to accommodate pilot and airline route preferences much more often than is now the case. The automated air traffic system by itself will not be able to

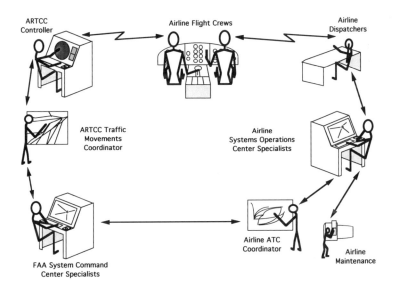

FIG. 2.4.   Management of air traffic is a shared responsibility.

improve terminal area operations appreciably, and research is now underway to assist terminal area traffic management by providing controllers with automated decision aids to improve arrival traffic flows (Erzberger & Nedell, 1989; NASA, 1990). I believe, however, that if a more automated ATC system inhibits the ability of controllers and pilots to work cooperatively to resolve problems, it will severely limit the flexibility of the system, and the loss of that flexibility could undo much of the benefit expected from a more automated system.

The gains in capacity from improved airspace usage will be limited at best without new runways or radical differences in operating methods. The social and political problems posed by new airport construction have thus far seemed nearly insurmountable (FAA/NASA/MIT Symposium, 1988). The fact that this problem has been intractable in many locations is forcing aircraft to operate to tighter and tighter tolerances. Procedural separation standards long considered inviolate have been relaxed; FAA and NASA have begun to examine ways of permitting aircraft to conduct much more closely spaced parallel or converging approaches to landing under instrument meteorological conditions (NASA, 1991; FAA, 1994; Phillips, 1995a). The latter change may be enabled, in part, by new collision avoidance displays, along with better ground radar, but it may also require more automated operations under these conditions, and both changes will certainly require higher levels of vigilance and will probably place higher cognitive demands on pilots and controllers alike.

We are now being forced by increasing traffic congestion to operate to the limits of the rules for air traffic management, and in some carefully considered cases to relax them. This is an exercise fraught with peril and it must be approached with the greatest care, tempered by common sense and careful research and operational testing. Improvements in automation technology can help humans to accomplish new and more difficult tasks, but automation should not be used to increase system throughput beyond the limits of human capability to operate manually in the event of automation failures, if humans are to remain fully responsible for system safety. There is increasing evidence that this could be allowed to happen during the coming decade, particularly in air traffic control.

The Organizational Environment

The Government Role

The third aspect of the aviation environment that requires consideration is the organizational environment. In its early days, aircraft were unfettered by any sort of organizational structure—they were also few in number, not very capable, and not a commercial force. By the late 1920s, the federal government had assigned aviation oversight and promotion to the Department of Commerce, which began to promulgate rules and regulations in the new transportation domain. By the beginning of World War II, the Civil Aeronautics Administration had firm control of all civil aviation. It certificated pilots and aircraft; it maintained an air traffic control and airways infrastructure; and it regulated, tightly, the air carrier industry.

After the war, an International Civil Aviation Organization (ICAO) was established under United Nations auspices, with its seat in Montreal, Quebec, Canada. ICAO served a valuable role in recommending standards and harmonizing the practices of its member states, an important task as aviation became a global enterprise with the onset of a large amount of transoceanic activity enabled in large part by technology developed during the conflict.

Since 1958, civil aviation in the United States has been governed by the provisions of the Federal Aviation Act passed by Congress in that year. Under the act, an administrator, appointed by the President, is responsible for certification and regulation of almost all aircraft, equipment, devices, and persons engaged in aviation. There is no doubt that many of the regulations promulgated by the FAA have caused aviation activities to become much more expensive, as well as more difficult and burdensome to conduct. On the other hand, aviation has flourished in this nation as in few others and our aviation exports have consistently led the world, although other nations are now mounting determined competition to our airframe and avionics industries. The rule structure also imposes a substantial cognitive burden, for all professional activities must be conducted within its constraints, and the rules (and often their interpretation) are not entirely free of ambiguity. Fortunately, in this nation, pilots and controllers are seldom punished for doing their best in a true emergency, especially if their actions save the day.

The FAA, Department of Transportation (DOT) and National Transportation Safety Board (NTSB) are organizations within the executive branch of our government, one of its three coequal branches. Regulations promulgated by the FAA can be appealed to the judicial branch; certificate actions involving individuals or air carriers can be appealed to the NTSB and its decisions can be appealed to the U.S. Court of Appeals, which thus may be placed in the position of arbiter regarding highly technical matters requiring substantial domain knowledge. The U.S. Congress, our legislative branch, is solely responsible for enactment of all Federal laws and for oversight of the executive branch, a responsibility it takes very seriously. GPWS, TCAS, and WSAS were all mandated by Congressional action, and many other safety regulations have been directly influenced by Congressional interest.

Private-Sector Roles

Within this category are virtually all the other actors on the aviation scene except the armed forces (which have also exerted a considerable and often helpful influence on the development of aviation by their support of advanced technology development in this domain). The federal government, through the FAA, may enact only rules that establish minimum certification and performance standards, conditions under which aviation must operate, or, less directly, levels of safety that must be maintained. Beyond these minima, private organizations involved in aviation are free to establish such higher standards as they may feel are necessary or desirable to produce or provide a commercially viable product. They have done so, and the result has been one of the world's safest aviation systems.

Four categories of organizations, broadly speaking, are stakeholders in the aviation process. Organizations in each category have played major roles in bringing aviation in this nation to its current state of development, and each category has a real and profound effect on the operation of the aviation system. The categories are:

- Airframe, avionics, and equipment manufacturers
- Air carriers (major and regional) and other operators
- Labor organizations
- Representative organizations

Each of these categories is discussed briefly as its activities relate to the behavior of the system and its operators, and to the machines with which the operators work.

Airframe, Avionics, and Equipment Manufacturers.   Aircraft and avionics are the tools with which the aviation mission is accomplished. In this country, private-sector corporations are solely responsible for civil (and military) aircraft production, although government organizations (DoD, FAA, and NASA) support their research and development activities. To an increasing degree since 1970, avionics manufacturers have become central to the development and production of new aircraft. Current estimates suggest that up to half the cost of new military aircraft is avionics cost. Although the proportion is probably less for civil aircraft, it is the avionics industry that has provided new airplanes with their advanced automation and their great capabilities. Much of this book is concerned with their products rather than with the aircraft themselves.

During the past decade or so, aircraft manufacturers, and more recently avionics manufacturers also, have become increasingly concerned with customer needs and human factors in the design of their products. Avionics manufacturers have also begun to involve themselves deeply in human factors in the design of their increasingly complex devices and systems.

These developments are encouraging, for they reflect acute awareness of the fact that it is in the line operation of aircraft, not their manufacture, that an airplane's safety record will be accumulated, and that not all pilots of such aircraft are as capable as are the test pilots whose input so influences the design of these aircraft. Unfortunately, the human factors community has not, thus far, provided these manufacturers with enough of the sorts of information that engineers need to improve cockpit design (see also chapter 13). To some extent, this is a communications problem, for we now know a good deal about human behavior in complex systems—but, as this book tries to point out, not enough to answer all of the legitimate questions that engineers must ask, nor to influence the trade-offs that they must make in the design process. Also, avionics technology has evolved much more quickly than our ability to predict or even evaluate its effects on human operator behavior; this is one reason why automation tends to be technology-driven rather than requirements-driven.

Air Carriers and Other Operators.   In a very real sense, it is the operators of aircraft, through their pilots and dispatchers, who are at the "sharp end" of the aviation venture. It is they who must deliver the end product—the transport of people and goods—and they who must set the de facto standards by which that product is delivered. They have done this through the philosophies, policies, and procedures (Degani & Wiener, 1991) under which their personnel fly, maintain, and market the product. In this sense, then, operator rules actually govern the air transport process. Operator training determines whether or not the people have the skills to provide the product safely and efficiently, and operator oversight is the first line of defense against deviation from the standards set for the process.

The corporate operator provides the environment in which its personnel work. That environment may be paternalistic or highly individualistic; it may operate to very tight standards or looser, more flexible rules; it may be authoritarian or it may delegate a great deal of authority to its workers. There have been successful air carriers and other operators that have exemplified each of these extremes. What is more important is that each has been able to motivate its personnel, by whatever means, to do their tasks well and effectively and to work as a team, for it is the performance of those tasks, by motivated workers across a company, that determines whether the operation will be safe and effective.

Hackman (1993) pointed out that

> Resolving [inter-group conflicts within a carrier] requires flight operations managers to use sophisticated political and interpersonal skills in dealing with other departments and in resolving inter-group tensions. . . . Managers who have such skills, and who use them well in negotiating with their peers and bosses, can do much to empower the crews for which they are responsible, enabling those crews to give full attention to their own special responsibilities—without constant distractions and irritations from resource problems that are properly someone else's concern. (pp. 65–66)

Labor Organizations.   These organizations, particularly the pilot representative organizations, deserve mention here because of their leading role in the
maintenance and improvement of safety in air transport. These organizations, while
adhering to their primary responsibility to represent their members, have also been
extremely active in air safety. They have been active participants in industry groups
such as the ATA Human Factors Task Force discussed in the Preface, and they have
been active advocates for improved safety before Congress and the FAA. They have
encouraged their members to become involved in safety research and in the
development of new procedures, devices, and equipment designed to improve
safety. Much of the aviation safety research in this nation depends on their expertise
and participation.

In human factors research, there is no substitute for the knowledge and insights
of the people at the sharp end of any dynamic system. Products designed without
careful attention to their needs are destined to fail or be ignored, as many product
developers have found out to their sorrow. Air carrier pilots and their representative
organizations are a priceless resource for anyone working to improve the design and
operation of aircraft.

Much the same can be said for air traffic controllers, who are the sharp end of
the air traffic management system. The redesign of the air traffic management
system will be a more difficult task than the design and production of any
individual airplane, and operator input to the process is crucial. Line controller
expertise and insights will determine whether the coming automated air traffic
control system will succeed or fall short of its objectives.

Representative Organizations.   Nearly every segment of the aviation community has a representative organization. Many of these organizations have
activities in direct support of aviation safety. A good example is the Aircraft
Owners and Pilots Association (AOPA), whose Air Safety Foundation has been
exceedingly active in general aviation pilot education, the development of
instructional materials for such pilots, and programs to provide flight familiarization for air traffic controllers. The Airline Pilots Association also sponsors a
foundation devoted to aviation safety research and education. As one instance
of their focus on safety matters, virtually all of these organizations have been
directly involved in the oversight and guidance of the NASA Aviation Safety
Reporting System since its inception. As with the labor organizations, the
membership of these organizations represents a good part of the expertise
available to the human factors community and others working to improve
aviation safety. There are also industry-supported organizations such as the
Flight Safety Foundation (FSF), whose sole purpose is the improvement of
aviation safety.

COMMENT

Rasmussen (1988), Reason (1990), Hollnagel (1993), Woods (1993b), and other cognitive scientists and engineers pointed out repeatedly that human cognition can only be studied in the context of real tasks carried out in a real environment. In this respect, they departed from the reductionist model that says that cognition can only be studied effectively in the laboratory. In keeping with their orientation, with which I strongly agree, I have attempted to place aviation professionals in their real world and to discuss, at least briefly, the factors that help or hinder them in the accomplishment of their tasks, using automation as a tool to assist them in coping with the extremely complex context and environment of aviation.

In a document of this sort, we focus primarily on what goes wrong in the system, in order to find ways to improve it. This book cites many mishaps, incidents, and failings within the system, its people, and the machines they use. It must be remembered, however, that we are preoccupied with events that occur at the margins. United States air transports fly 20 million hours per year. This fact alone is strong evidence that the people in the system are doing a very great deal more right than wrong.

Having said this, however, most of these people will admit that few flights or duty shifts are entirely flawless, which motivates them to keep trying to improve their skills. Professionals who support them must recognize that these people at the sharp end are getting the job done effectively and safely despite numerous impediments, some of which we have put in the way.

We owe it to practitioners and to the system to remove such impediments and to improve their tools, both to make their jobs easier and to prevent the few tragic (and usually preventable) catastrophes that continue to plague us. It is only through study of these anomalies that the system can be improved, and aviation, more than any other mode of transportation, has looked critically at its failings throughout its long history. So I make no apologies for looking for trouble at the margins, but we must maintain our perspective by keeping in mind the much larger body of safe and efficient behavior that takes place well inside those margins.

CHAPTER 3

A Concept of Human-Centered Aviation Automation

INTRODUCTION

There are disquieting signs in recent accident investigation reports that in some respects our applications of aircraft automation technology may have gone too far too quickly, without a full understanding of their likely effects on human operators. In this chapter, I summarize some of the shortcomings of today's automation, trying not to lose sight of the benefits conveyed by these remarkable technologies. However, in some cases the automation itself, and procedures governing its use in other cases, appear to have impinged on the authority of human operators. As always happens with new tools, automation has shaped the behavior of those operators, sometimes in ways not foreseen by its designers.

The progress of automation technology will accelerate during this decade, and more rather than less automation will be needed (both in aircraft and in air traffic management) as we confront new capacity demands. Does this new automation (and further development of the automation now in use) have to be as complex, opaque, brittle, and clumsy as the present generation? I hope not, for we have learned a good deal about these problems from observation of today's automation. Can we solve the human–machine interface problems without compromising the utility of these invaluable tools? I believe we can, by carefully examining what we have learned and applying it to the design and operation of new systems. This will not be easy or cheap, but it will be easier and a great deal less expensive than continuing to tolerate aircraft accidents caused by inadequate human–system interfaces.

PROBLEMS ASSOCIATED WITH THE EVOLUTION OF AUTOMATION

Most of this book is concerned with problems associated with aviation automation. In chapter 1, I listed some automation attributes that have been found in aviation

MISHAP	COMMON FACTORS
DC-10 landing in CWS mode	Complexity; mode feedback
B-747 upset over Pacific	Lack of mode awareness
DC-10 overrun at JFK, New York	Trust in autothrust system
B-747 uncommanded roll, Nakina, Ont.	Trust in automation behavior
A320 accident at Mulhouse-Habsheim	System coupling and autonomy
A320 accident at Strasbourg	Inadequate feedback
A300 accident at Nagoya	Complexity and autonomy
A330 accident at Toulouse	Feedback; system complexity
A320 accident at Bangalore	System complexity & autonomy
A320 landing at Hong Kong	System coupling
B-737 wet runway overruns	System coupling
A320 overrun at Warsaw	System coupling
B-757 climbout at Manchester	System coupling
A310 approach at Orly	System autonomy and coupling
B-737 wind shear at Charlotte	System autonomy & complexity
B-757 approach to Cali, Colombia	System complexity & feedback

FIG. 3.1.   Common factors in automation-related aircraft incidents and accidents.

mishaps (Fig. 3.1). To encapsulate them here, they include loss of situation or state awareness, associated with:

- Complexity
- Coupling
- Autonomy
- Inadequate feedback

Complexity makes the details of automation more difficult for the human operator to understand, model, and remember when that understanding is needed to explain automation behavior. This is especially true when a complex automation function is utilized or invoked only rarely. (See Woods, 1996, on the "apparent simplicity, real complexity" of aviation automation.)

Coupling refers to internal relationships or interdependencies between or among automation functions. These interdependencies are rarely obvious; many are not discussed in manuals or other documents accessible to users of the automation. As a result, operators may be surprised by automation behavior, particularly if it is driven by conditional factors and thus does not appear uniformly. (Perrow, 1984, discusses coupling and its potential for surprises.)

Autonomy is a characteristic of advanced automation; the term describes real or apparent self-initiated machine behavior. When autonomous behavior is unexpected by a human monitor, it is often perceived as animate; the automation appears to have a mind of its own. The human must decide, sometimes rather quickly, whether the observed behavior is appropriate or inappropriate. This decision can be difficult, in part because of the coupling just mentioned and in part because the automation may not provide adequate feedback about its activities (Sarter & Woods, 1991, 1992b).

Inadequate feedback describes a situation in which automation does not communicate, or communicates poorly or ambiguously, either what it is doing, or why it is doing it, or, in some cases, why it is about to change or has just changed what it is doing. Without this feedback, the human operator must understand, from memory or a mental model of the automation, the reason for the observed behavior (Norman, 1989). As a pilot remarked, "If you can't *see* what you've gotta know, you gotta *know* what you gotta know" (T. Demosthenes, personal communication, November 1994).

The interposition of more and more automation between the pilot and the vehicle tends to distance pilots from many details of the operation (Fig. 3.2). Although this may have the desirable effect of lessening flight crew workload, it has an undesired effect as well, in that pilots may be, and may feel, less involved in the mission. The newest technologies nearing application—digital data link, automatic

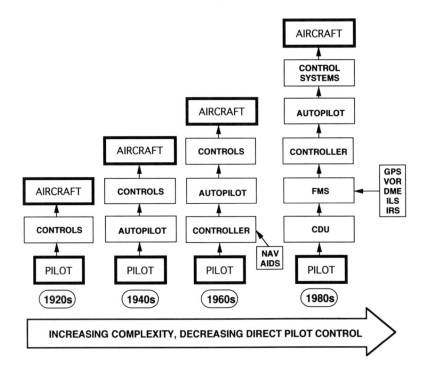

FIG. 3.2.   Effects of increasing complexity on the human–machine (pilot–aircraft) relationship.

dependent surveillance, and direct digital data transfers between flight management system (FMS) and air traffic control (ATC) computers—have the potential to accentuate this tendency toward peripheralization of the flight crew. The effects of less verbal interaction with ATC may also, I believe, tend to distance the flight crew, and the air traffic controller as well, from a sense of immediate involvement in the team venture (as well as depriving pilots of the ability to hear what other pilots are saying, and thus the ability to infer what they are doing).

Recent accidents, among them those listed in Fig. 3.1, have demonstrated that capable pilots can lose track of what is going on in automated aircraft. Although some new aircraft types have had better experience than others, these types differ more in degree than in kind. The mishaps that have occurred serve as a warning of what may lie ahead unless we learn the fundamental conceptual lessons these accidents can teach us. One of the most important lessons is that we must design flight crew, and controller, workstations and tasking so that the human operator is, and cannot perceive himself or herself as other than, at the *locus of control* of the vehicle or system, regardless of the automation or other tools being used to assist in or accomplish that control (Fig. 3.3).

No regulator, aircraft manufacturer, or operator talks aloud about totally replacing the human operator with automation in the aviation domain, and I think that few people in the industry believe it can be done, if only for sociological and political reasons. If pilots and controllers are distanced from their operations by automation,

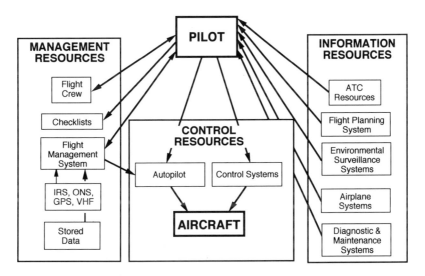

FIG. 3.3.   The pilot must be at the locus of control of the system.

it is an unintended side effect of the way their systems have evolved. I do not believe that a sense of diminished involvement is prevalent—yet—but it may well be if we continue along our present course of automating everything that can be automated, moving the human more and more toward a "backup" or ancillary role. The AERA and free flight concepts of air traffic management now under consideration exemplify this trend (see chapter 8 for discussion of these concepts.).

It is these threats to the loci of control of the system as we know it that lead me to suggest that we need to reevaluate the human–machine interactions in this system at a fairly fundamental level. The concept of "human-centered" automation outlined next is an attempt to do just that. Its thesis is that by beginning with the human and designing tools and artifacts specifically to complement the human's capabilities (Jordan, 1963), we can build more effective and robust systems that will avoid or ameliorate many of the automation problems that now confront us. Most of these problems, of course, are neither automation problems nor human error problems. They are human–machine *system* problems, and they must be attacked as such.

A CONCEPT OF HUMAN-CENTERED AUTOMATION

The remainder of this chapter is devoted to an explanation and defense of some of the principles I believe constitute the essence of *human-centered automation* in aviation. The term is not mine, and I have been unable to find out who first conceived it. Sheridan, Norman, Rouse, Cooley, and many others wrote for many years about human-centered or user-centered technology.

Some people have criticized this term because it appears to emphasize the human rather than the human–machine system; Flach and Dominguez (1995) proposed the term *use-centered technology* to overcome this hurdle. Sheridan (1995) offered penetrating criticism of the concept itself. In the aviation domain, however, in which humans are responsible for the outcome, the human must be the primary focus of our attention. The tools are there to assist human operators in carrying out the mission.

Figure 3.4 is a brief summary of some first principles that I believe are central to this concept. In later chapters, I apply these general principles to specific automation problems and to functions that I think will be implemented in future aircraft and the future air traffic system.

These first principles are stated as absolutes. In reality, of course, they are matters of choice to which system designers may or may not wish to adhere. An aviation system in which pilots and controllers were not at the loci of control is possible, but it would represent a radical departure from today's system. It might convey new

PRINCIPLES OF HUMAN-CENTERED
AVIATION AUTOMATION

PREMISES:

The pilot bears the responsibility for safety of flight.

Controllers bear the responsibility for traffic separation
and safe traffic flow.

AXIOMS:

Pilots must remain in command of their flights.

Controllers must remain in command of air traffic.

COROLLARIES:

The pilot and controller must be actively involved.

Both human operators must be adequately informed.

The operators must be able to monitor the automation
assisting them.

The automated systems must therefore be predictable.

The automated systems must also monitor the
human operators.

Every intelligent system element must know
the intent of other intelligent system elements.

FIG. 3.4.   First principles of human-centered automation in aviation.

benefits, but they would surely be accompanied by new costs and problems. Nonetheless, such radical departures from today's system have been actively considered, and they must be considered here in terms of the role and authority of the human operators. I briefly discuss each of these principles.

Responsibility and Command Authority

In their determinations of the probable causes of several recent aircraft accidents, the members and staff of the National Transportation Safety Board (NTSB) and similar organizations in other nations have made a commendable effort to recognize explicitly that there is much more to aviation system problems than the sharp end: that pilot or controller errors are usually enabled by management, design, and other latent defects in the system (note air traffic control problems in the accident at Cove Neck, New York, 1990; ATC management problems were found in Los Angeles, 1991; a careful analysis of latent factors was done during the investigation of Sydney, 1991; and the investigation of Dryden, Ontario, 1989, found many such factors). There is a growing, although sometimes fragile, consensus that factors throughout the system must be considered before assigning causation for a mishap (Woods, Johannesen, Cook, & Sarter, 1994). By law, however, the human operators—pilots, dispatchers, and controllers—are still responsible for the safety of each flight and for the safety of air traffic movements. The same precept applies in other transportation modes; it is reinforced by an enormous body of statute and case law. The law also provides the responsible operator with very broad discretion in the execution of this heavy responsibility. Although the authority of a pilot or controller operating under normal conditions is circumscribed by a variety of regulatory and procedural constraints, the operator's unfettered authority to use his or her best judgment in an emergency is not usually questioned, even after the fact, if the outcome is successful.

Automation is able to limit the operator's authority, and in some cases it is not obvious to the operator that this has occurred. In chapter 6, I discuss envelope protection or limitation as an example of circumscribing control authority, but homelier examples are found in older aircraft as well. All complex aircraft have "squat" switches on their landing gear struts that sense wheels on ground; the switch, alone or in combination with a wheel spin-up sensor, enables (or disables) a number of important control functions including (in various aircraft) thrust reverser deployment, ground spoiler actuation, and autobraking. It is very important that these functions not occur in flight; the only fatal B-767 accident occurred after a thrust reverser deployed during climb at high altitude (Thailand, 1991). On the other hand, there have been several incidents in which pilots landed gently on a water-contaminated runway and were unable to use these deceleration devices for some time because of delayed wheel spin-up due to hydroplaning (Marthinson & Hagy, 1993a, 1993b; Warsaw, 1993). In the Warsaw case, there was an 8-sec delay during which the airplane traveled almost 2000 ft.

Newer autothrust systems, usually activated early in the takeoff roll, limit engine power to maximum rated thrust or a lower value depending on aircraft weight, runway length, temperature, and other variables. The purpose of this is to minimize engine wear and fuel consumption. The desired takeoff thrust is selected through the thrust management system. Occasionally, an aircraft on takeoff encounters a situation in which all available power reserves are needed to climb over a runway obstacle or to maintain acceleration on a contaminated runway. In older aircraft, pilots simply pushed their throttles fully forward to obtain maximum thrust. Engine overheating usually resulted, but the technique was often successful in avoiding a far more critical threat. (Unfortunately, there are also incidents in which pilots did not utilize all available power when it was needed, perhaps because they feared overheating their engines.) In some of today's aircraft, it is not possible for pilots to obtain more than rated thrust from the engines. Full throttle instructs engine computers to provide rated thrust; no performance reserve is made available. Should a pilot be permitted to "burn up" an engine, or overstress an airplane? It is the pilot, after all, who is responsible for a successful outcome. On the other hand, it is predictable that some pilot will unnecessarily overheat some very expensive engines or "bend" an airplane if given the means with which to do so.

In the air traffic control domain, concepts for advanced en route ATC systems will also be able, either through automation design or procedures, to limit the scope of controller authority appreciably, although the responsibility for a safe operation will remain with the human. If the human operator cannot effectively oversee and retain management authority over his tools, he or she has lost a measure of authority over the entire operation. Will this be a tenable situation?

I believe it comes down to a matter of trust. Will we provide pilots with full authority, train them carefully, and trust them to do "the right thing," whatever it is in particular circumstances? Or will we circumscribe pilot authority by making it impossible to damage the airplane, and in the process perhaps make it impossible to use its ultimate capabilities if the pilot really needs them, or circumscribe controller authority to do whatever is necessary in contingencies? My bias, based on a number of cases in which pilots have been able to recover from extreme emergencies, and other cases in which they did not recover but could have had they used all available resources (e.g., Washington, DC, 1982), is that command authority should be limited only for the most compelling reasons, and only after extensive consultation with both test and line pilots or controllers at "the sharp end" of the system (Woods et al., 1994, pp. 20–21).

Operators Must Be Involved

No one questions the necessity for operator involvement in flight and air traffic operations at some level; the questions relate to the degree of involvement. The tenets of situation awareness, a concept with which the aviation community is much preoccupied, correctly state that it is easy for pilots to become preoccupied with detail at the expense of maintaining the big picture of their operations (Gilson, Garland, & Koonce, 1994). This concept underlies the design philosophy charac-terized as, "If it is technically and economically feasible to automate a function, automate it" (Douglas, 1990, p. 3).

My questions regarding involvement are rather whether pilots of newer aircraft are indeed sufficiently drawn in (the definition of involvement) to their operations by having an active and necessary role apart from simply monitoring the course of the operation. That role may involve active control, decision making, allocation of resources, or evaluation of alternatives, but it should not be passive, as it too often is today. The Flight Safety Foundation ICARUS Committee has also emphasized the need for more disciplined training to ensure that both technical and human factors needs are met (Pinet & Enders, 1994). I believe that pilots must be given meaningful and relevant tasks throughout the conduct of a flight, and that these tasks must be designed into the aircraft automation. This will not be easy, for we have spent the last decade making the automation self-sufficient. The change from passive monitor to active problem solver can be abrupt and difficult, however. *If humans are to remain involved (and without such involvement they will not always remain in command), they must be an essential part of the normal operational flow, not only the resolvers of anomalies.*

One operator has seriously considered asking its long-haul pilots to engage in a continuing optimization of their flight paths to take advantage of changing wind and weather conditions by revising their flight plans while they are being executed. This approach has merit—but technology is now under development to accomplish this automatically on the ground, using automatic dependent surveillance to provide the real-time data! One is reminded of Wiener and Curry's (1980, p. 2) statement: "Any task can be automated. The question is whether it should be..."

This question has not yet come up with respect to controllers, because automation of air traffic control processes is not yet available to them. It will be, however, in the near future. It is hoped that the lessons taught in aircraft by assuming "that the maximum available automation is always appropriate" (ATA, 1989) will not be lost on ATC system architects, but there is little reason thus far for optimism.

Operators Must Be Informed

For many decades, neither pilots nor controllers ever had as much information as they needed to conduct operations optimally under changing and often unpredicted circumstances. During the last two decades, however, there have been quantum increases in the amount of data available in the cockpit and in ATC facilities. Glass cockpit display technology has made it possible to provide much more of this data in aircraft; information management technology has all but erased the problem of insufficient data in the system. *Data, however, is not information. It becomes information only when it is appropriately transformed and presented in a way that is meaningful to a person who needs it in a given context.*

The secret to compressing and transforming data into information lies in a designer's understanding of operators' needs, cognitive models, cultures and operating styles under a wide variety of circumstances. It is absolutely crucial that the designer be able to assume the line pilot's or the controller's role and way of thinking when designing information displays or representations. Further, the designer not only must understand information needs through the minds of the highly experienced test and certification pilots or managers with whom he or she ordinarily interacts, but must also understand the broad range of cultures and capabilities in the population of operators who will fly the airplane in line service, and the full range of environmental circumstances under which it will be flown.

The most capable pilots are able to make do with displays that are far from optimal; it is one measure of their capabilities. But the same displays in service must support fatigued pilots of only average ability operating under difficult conditions: what Charles Schultz's Snoopy calls the "dark and stormy night" in his never-finished novel about World War I flying. Similarly, ATC systems cannot be designed only for the "aces;" they must assist the inexperienced trainee controller as well.

Without adequate information (and what is adequate depends to a great extent on the context), neither pilots nor controllers can make uniformly wise decisions. Without correct and timely information, displayed in a way that minimizes operator cognitive effort, even the best pilots and controllers cannot remain constructively involved in an operation, and thus cannot maintain command of the situation. The designer must ask how he or she is affecting the processes involved in extracting meaning from the data or information provided.

Humans Must Be Able to Monitor the Automation

In automated aircraft, one essential information element is information concerning the status and activities of the automation itself. Just as the pilot must be alert for performance decrements or anomalous behavior in the human crew members (self included), he or she must be equally alert for such decrements in the automated systems that are assisting in the conduct of the operation. The first principles state that "the humans must be able to monitor the automation." This sounds obvious, but it has been observed that advanced automation is often strong and silent (Sarter & Woods, 1994) about its work, leaving humans to wonder about what it is doing, and sometimes why.

In part, this situation reflects the commendable desire of aircraft manufacturers to avoid burdening the pilot with information unless something is wrong. The quiet, dark cockpit concept reflects this philosophy by giving a positive indication only when some system is not operating properly. The equally important issue today is how to inform the pilot (or controller) that the automation is correctly performing each of the functions it has been commanded to perform.

When automation performed only tactical chores in response to direct commands from pilots, it was reasoned that the pilots could monitor the automation by simply observing the correctness of the airplane's responses to autopilot inputs. Today's automation, however, is far more capable and ubiquitous; it accomplishes more functions over a longer period of time, often with only strategic guidance from pilot inputs to the FMS. The number and seriousness of mode errors (Boston, 1973; Luxembourg, 1979; Strasbourg, 1992; Paris, 1994) that occur despite information on the flight mode annunciator panel on the primary flight display suggest strongly that pilots of modern aircraft must be given more salient *affirmative* evidence that their automation is indeed doing what they told it to do, perhaps many hours earlier (Sarter & Woods, 1992a, 1996).

Automation Must Therefore Be Predictable

In many redundant aircraft systems, the human operator is informed only if there is a discrepancy between or among the units sufficient to disrupt or disable the performance of their functions. In those cases, the operator is usually instructed to take over control of that function. *To be able to assume control without delay, the pilot must be aware on a continuing basis both of the function (or dysfunction) of each critical automated system and of the results of its labors to that point, as well as what it was going to do next and when.* This, of course, requires that the pilot have an accurate mental model of how the automation is supposed to behave.

As used here, a *mental model* is a construct, or representation in memory, of the way in which an object or system functions or responds to various inputs. It may be analogous to an analytical model of the system, or it may be a grossly simplified or metaphorical representation, expecially if the actual system is very complex. The formation of such internal models occurs, or should occur, during training, when the pilot learns what "the book" says about particular automation functions and then uses those functions in a simulator or part-task training device. The models are reinforced when the pilot successfully invokes the functions in line operations. They may be modified if the functions are found to be "buggy" or to work in ways not expected, and such behavior, which is fortunately rare, can cause severe disruption to the pilot's mental image of the system. An example in the 757/767 was an occasional turn to an outbound instead of the desired inbound heading when converging on a localizer course.

The pilot's mental model may be an accurate representation of a function, or it may be a drastically simplified construct of a very complex function. If accurate and reasonably complete, the model will help the pilot to detect and diagnose aberrant automation behavior if it occurs. If the model is a grossly simplified or metaphorical representation, the pilot is more likely to be surprised by anomalous behavior of the real system, because its detailed behavior is not a part of his or her mental model.

Because of the logical complexity of modern digital systems, they may fail in ways that are quite different from "physical" systems. This increases the probability that the pilot's mental model will not fully account for its actual performance. Only if the automation's normal behavior is understood, given a certain input or circumstance, will pilots be able to detect subtle signs of failure. It is for this reason that automation must be predictable, so that pilots will be able to observe and respond to unpredicted behavior of these systems. This fact also emphasizes the importance of helping pilots to build adequate mental models of automated systems during training, and the importance of simplicity in functional design. It is difficult for pilots to remember the "normal behavior" of functions that are used only infrequently.

Automation Must Monitor the Human

Just as machines are prone to failure, so are the human components of the human–machine system. Human error is thought to contribute to roughly 80% of aviation accidents (Lauber, 1989). Although we now recognize that a great many of these human failures are enabled by other system factors, there is clearly a need to monitor human behavior at the "sharp end" of the system. Indeed, much of our elaborate safety surveillance apparatus is designed specifically for this purpose.

One of the major reasons air transport is so safe is the ongoing monitoring of the flying pilot by a nonflying pilot in the cockpit. This duty is spelled out in the operating procedures of every air carrier and nearly all other organizations that conduct multiple-pilot operations. Pilots monitor the actions of air traffic controllers, and those controllers monitor the behavior of the aircraft they control. ATC automation monitors both pilot and controller actions. Error detection, diagnosis, and correction are integral parts of the aviation system, and a great deal of effort has gone into making all parts of the system redundant.

Despite everyone's best efforts, however, human errors continue to occur, are missed, and occasionally propagate into a catastrophic system failure. There are many reasons for this; one is that humans are not very good monitors of infrequent events (Mackworth, 1950; Broadbent, 1971) and may fail to detect them when they occur. This is an area in which automation technology can be extremely useful, for computers do not become fatigued or relax their vigilance when a very long period elapses between events of interest, and they fail less frequently than do human operators.

Automated devices already perform a variety of monitoring tasks in aircraft, as indicated throughout this book. Incident reports confirm their effectiveness in preventing mishaps (and also confirm, unfortunately, the failure of pilots to detect problems when the automated monitors fail, as in takeoffs without flaps in the face of an undetected configuration warning system failure; Detroit, 1987; Dallas-Fort Worth, 1988). Designing warning systems to detect failures of warning systems can be an endless task, but it is necessary that we recognize the human tendency to rely on reliable assistance and consider how much redundancy is therefore required in flight-critical warning systems. The trade-off, of course, is increased automation complexity and decreased reliability.

Data now resident in flight management systems and other aircraft systems can be used to provide more comprehensive and effective monitoring of both pilots and controllers, if specific attention is given to potential failure points that have been well documented in aviation operations. Automation, in the air and on the ground, can and should be thought of as a primary monitor of human behavior in exactly the same way that humans are the primary monitors of machine behavior. In the more tightly integrated system of the future, such cross-monitoring will be the key to improved system safety. The use of aircraft automation, especially the FMS, for flight crew monitoring has not been given the attention it deserves.

Communication of Intent

Cross-monitoring (of machines by humans, of humans by machines, and ultimately of human–machine systems by other such systems) can only be effective if the monitoring agent, whether a person or a machine, knows what the monitored agent is trying to accomplish, and in some cases, why. The intentions of the automated systems and the human operators must be explicit, and they must be communicated to the other intelligent agents in the system.

A great deal of this goes on already. Pilots (or airline systems operations centers, AOCs) communicate their intent to air traffic control (ATC) by filing a flight plan. Pilots communicate their intent to the flight management system (FMS) by inserting the flight plan into the computer or calling it up from the navigation data base. Controllers, in turn, communicate their intent to pilots by granting a clearance to proceed; in the near future, data link will transmit this information to the flight management computer (FMC) as well. During flight, clearance changes are communicated to pilots by ATC; pilots acknowledge their understanding of ATC's intentions by reading back the revised clearance as they heard it (though on busy voice communication channels, this procedure is far from faultless; Monan, 1986).

It is when circumstances become abnormal, due either to problems in the physical or operating environment or to in-flight anomalies, that communication of intent among the various human and machine agents becomes less certain. An Avianca B707 accident (Cove Neck, New York, 1990) was a classic example of failure to communicate need and intent between pilots and air traffic controllers, but there have been many others, some as serious. Further, the communication of intent makes it possible for all system participants to work cooperatively to solve the problem. Many traffic control problems occur simply because pilots do not understand what the controller is trying to accomplish, and the converse is also true, as in the Avianca case. Finally, automation cannot monitor pilot performance effectively unless it "understands" the pilot's intent, and this is most important when the operation departs from normality. A shared frame of reference is absolutely necessary for cooperative problem-solving. This problem has the potential to become more pressing as new ATC automation is introduced, for there will be linked human and machine systems both in flight and on the ground, all of which will have to work harmoniously to resolve tactical problems as they arise.

COMMENT

Although humans are far from perfect sensors, decision makers, and controllers, they possess three invaluable attributes. They are excellent detectors of signals in the midst of noise, they can reason effectively in the face of uncertainty, and they are capable of abstraction and conceptual organization. Humans thus provide to the aviation system a degree of flexibility that cannot be attained by computers. Human experts can cope with failures not envisioned by aircraft and aviation system designers. They are intelligent: They possess the ability to learn from experience and thus the ability to respond adaptively to new situations. Computers cannot do this except in narrowly defined, completely understood domains and situations.

The ability of humans to recognize and bound the expected, to cope with the unexpected, to innovate, and to reason by analogy when previous experience does not cover a new problem are what has made the aviation system robust, for there are still many circumstances, especially in the weather domain, that are neither controllable nor fully predictable. Each of these uniquely human attributes is a compelling reason to retain human operators in a central position in aircraft and in the aviation system. Those humans can function effectively, however, only if the system is designed and structured to assist them to accomplish the required tasks. As technology continues to advance, it will become increasingly urgent that its applications on the flight deck be designed specifically around the human who must command them; in short, future aviation automation must be human-centered if it is to be a maximally effective tool.

At the same time, many machines today are capable of tasks that unaided humans simply cannot accomplish. This is true in both the perceptual and cognitive realms. An example today is the calculation of optimal orbital trajectories for systems such as the Space Shuttle; another is the determination of a great circle navigation route. For these tasks, computers and automated systems are an absolute requirement. Competitive pressures in aviation being what they are, it is likely that still more complex automation will be offered in the marketplace, and there will be a tendency to accept it.

If this tendency toward greater complexity is to be countered, it must be by the customers: Airlines and other operators must decide whether the potential gains are worth the certain costs. Some air carriers, among them Southwest Airlines, have decided that simpler is better with regard to cockpit automation. Others, like United Airlines, as recommended by the ICARUS Committee (Pinet & Enders, 1994), are "minimiz[ing] crew confusion by selecting the automation options and methods best suited to their own operations, and training for those options/methods as 'preferred' methods" (p. 11), rather than requiring that the full capabilities of their flight management systems be used in line operations. Given the problems associated with automation complexity, this seems a prudent approach.

Part II

The Evolution and Course
of Aviation Automation

In chapter 4, Sidney W. A. Dekker and I discuss some facets of the history and evolution of industrial automation, to set the stage for a presentation of aircraft (chapter 5) and aviation system (chapter 7) automation. These discussions include the effects that the introduction of specific automation elements have had on the humans in the system, and on the system itself.

Likely future trends in aircraft and air traffic management automation are described in chapter 6 and 8. Chapter 8 also contains a discussion of rapidly evolving trends in aviation system design, and the implications of these new concepts for future system development. These chapters have been written at a time of intense planning activity in the domain of air traffic management (ATM), and ATM concepts will have evolved further by the time this book is published, but some of the changes that are being proposed are so fundamental as to demand consideration here.

CHAPTER 4

HUMANS AND THE EVOLUTION OF INDUSTRIAL AUTOMATION

Sidney W. A. Dekker and Charles E. Billings
The Ohio State University

INTRODUCTION

It is a truism that "those who forget the past are doomed to repeat it." This chapter has been written to remind us where automation came from, its effects on human operators, and some of the mistakes that have been made in conceiving, developing, and implementing industrial automation, in the hope that we may avoid them as we go on to consider the future of aviation automation.

Rather than simply recounting a necessarily incomplete history of industrial automation, we explore some beliefs (more properly, myths) about automation that remain prevalent today. We illustrate them with historical examples. These cases can teach us many lessons, from which we have extracted only a few. On the other hand, this has been the history of technology as well; lessons that could have been learned centuries ago are not always embodied in our present knowledge base.

PREVALENT MYTHS ABOUT TECHNOLOGY AND TECHNOLOGY CHANGE

Myth 1: Technology Change Is Preordained, Predictable, and Imperative

To what extent can technology change be given direction beforehand? Is techno-logical change predictable? Is it ultimately unavoidable? Looking at some of the breakthroughs that marked technological advances during the industrial revolution, we see different patterns. First, technology change often has a long incubation period. Weaving looms were not just a product of the industrial age: Techniques for making cloth had reached a high level of development by the dawn of the industrial revolution. Egyptians made extensive use of weaving looms around 1500 B.C., and other traces of looms go back as far as 4400 B.C. Efforts to automate the loom itself were found in sketches by Leonardo da Vinci. The first real attempt to automate the process was the later (early 1600s) Dutch bar loom.

Second, insights are often more serendipitous than preplanned. The Montgolfier brothers, for instance, burned hay in their hot-air balloon in the belief that as the smoke rose it would lift the balloon. Most technology of that period was developed empirically, or came from a serendipitous insight; nearly all such insights (like many today) were based on porous knowledge, despite an improving scientific base. Do we really understand why cancer chemotherapy works?

And third, technology change is not a self-propelled imperative; rather, it is a function of the interplay between societal constraints and opportunities on the one hand and those technological evolutions and serendipitous insights on the other. Mokyr (1990) likened the evolution of technology to biological evolution. He postulated that "technology is epistemological in nature" (p. 276). Techniques (the knowledge of how to produce a good or a service) are like species: They arise, or speciate, are reproduced, adapt, or fail to adapt and become extinct. New ideas are like mutations; they represent deviations from the norm. Many are stillborn or do not survive infancy (pp. 277–278); some survive and adapt, later to become a new norm.

This implies that not only is a new technique's inception unpredictable; so is its effect within a field of practice. Because an invention represents an attack upon a "constraint that everybody else took as a given" (Mokyr, 1990, p. 9), the new technology will in turn change practices, often in unpredictable ways. Changed technology is likely to have unanticipated, distant, and sometimes uncontrollable effects, including effects in other domains. Improved weaving looms (the water frame and self-acting mule) improved productivity, but required more power, first delivered by water, but by steam engines as early as 1785 (Reeve, 1971). By the 1800s, British industry had attained superiority in delivering power to many kinds of factories.

The motto of the 1933 Chicago World's Fair was, "Science finds, Industry applies, Man conforms" (Norman, 1993). Humans were assumed to adapt to technology, not technology to humans and the various systems that represented them: organizations, trades, societies, and the body politic. The teleology behind the motto was simplistic and oblivious to the evolutionary complexity of technology change. It reinforced the view that technology change is causal and that the direction of technology change is preordained. In fact, technology change fails to follow predictable paths (except in hindsight, which makes it all too easy to attribute causality to its meanderings). The notion of a technological imperative is tempting, but it oversimplifies the complexity of technological change and the haphazard opportunism that often appears to give it direction. The motto of the 1933 World's Fair was misguided, or more properly, inverted, as Norman suggested. Rather, it should be "People propose, Science studies, Technology conforms" (p. 253). But this still implies that technology conforms to human needs without forcing its own constraints on humans, which is manifestly incorrect. Perhaps the motto should have been, "People adapt; Science adapts; Technology adapts."

Myth 2: Technology Can Help Us to Supplant the Unreliable Human

Charles Babbage, the 19th century inventor of huge "engines" that could calculate and compute, often dwelled on the impact his machines could have on a society that still relied on manual labor. Human intelligence, to a great extent, could be replaced by machines: "the wondrous pulp and fibre of the brain ... substituted by brass and iron" (Swade, 1993, p. 88). Could machines indeed replace unreliable human beings and could untrained workers operate them without mistakes? Babbage's machines were never built during his lifetime, so we cannot know the answer. Automating human error out of systems, however, is still "assumed by many to be the prescription of choice to cure an organization's 'human error problem'" (Woods et al., 1994, p. 23). Let us review some historical evidence to see whether automation indeed (a) did eradicate human errors and (b) did reduce claims on human ingenuity.

The changes in work practices brought about by industrial automation conveyed, in hindsight, many benefits, but human error was far from eradicated. Instead, its nature changed (see Wiener, 1993). In 1803, the Frenchman Joseph Marie Jacquard was commissioned to improve a loom built by Jacques de Vaucanson. Without any guidance (documentation was unavailable), Jacquard developed an attachment to the loom in 1805. His device represented the first instance of an important aspect of automation—that of machine programmability.

The Jaquard attachment made the loom automatic, in that it was capable of producing complex weaving patterns in textiles by controlling the motions of many shuttles of different colored threads. The program for the pattern was contained on steel cards in which holes were punched. The attachment was an instant success. By 1812, 11,000 Jacquard looms were working in France alone.

The programmability enabled new creativity for the pattern developers or programmers, who were freed from the constraints that manual weaving had imposed on their creativity, but also enabled new errors (recall Ernst Mach's 1905 adage that error and expertise stem from the same source). Errors changed in nature: for example, from manually shoving the shuttle in the wrong direction, to making programming errors during card punching. Note that the unwanted conse- quences of such inadvertent actions can grow hand in hand with increased automation. A manual shoving error produced just one thread out of place; an error in programming the card could easily ruin the whole woven product. Human errors were neither eradicated nor contained by automation. In this case, they were amplified.

Not only does automation not eradicate human error, the myth that an auto- mated world will make smaller claims on human ingenuity is also palpably false. Both ingenuity and fallibility are fundamentally part of human performance, however automated the system being controlled. Mach's adage is true: Expertise and errors are cut from the same cloth. Changes in technology require the exercise of new human ingenuity. Yet automation—although sometimes requiring human ingenuity to work—often limits the exercise of human creativity. What we see is the necessity for "tailoring" of technology on the one hand, and the limited opportunity to do so through peripheralization of the human role on the other hand. Let us examine this contradiction historically.

First, consider peripheralization. Automation reduces the need to follow all details of a process (Hollnagel, 1993) and often removes the human from hands-on contact with whatever is produced. Not only that, the speed of the process is no longer constrained by what the unaided human can handle. As long as the individual weaver operated the hand loom, the individual produced the whole cotton product. Control and coordination of the various subtasks of the weaving process were at the weaver's discretion. With the introduction of the power loom, pacing and various other aspects of control of the task became external to the operator:

> The most novel feature of factory work was its continuity and regularity; the machines had to be kept running. The pace of work was set by the water-wheel or steam-engine, not by man's physical endurance or dexterity. While the tasks themselves were monotonous, and often simple, the factory worker had to be alert and reliable. It was for this, rather than skill, that he was paid. To punish lateness, absenteeism, casualness and inattention, the factory owners applied a series of scaled fines and wage reductions. (Reeve, 1971, p. 72)

Automation took control away from the human, but made new demands at the same time. This is the other side of the contradiction: the need for tailoring. Practitioners adapt the technology provided for them in a locally pragmatic way, developing a variety of strategies to cope with the automation's brittleness or complexity. Workers always find it necessary to tailor, or adapt, automated devices and their work processes to accommodate the changed technology and to insulate the larger system from automation's idiosyncrasies and deficiencies (Woods et al., 1994). Especially during development and early application, fine control over the process may be needed to make it work at all.

Historical reports of such tailoring are sparse. Given the ubiquity of technological change in the cotton industry in the period between 1750 and 1850, acts of tailoring were probably not sufficiently noteworthy to document. Tailoring was not looked at as a separate phenomenon, simply because of its essential role in technological change. Emery (1977) elegantly described the importance (and at the same time anonymity) of the tailoring process for cotton weaving:

> Cotton was fortunate: its tougher, more predictable fibres withstood the jerkiness of the early machines remarkably well. Nevertheless much patience and skill was required to make the best of the early machines and in the progress of the industry as much credit should go to the anonymous, piecemeal improvers as to the initial innovators. Many of the early wooden machines [thus] lasted well on into the nineteenth century and no doubt prolonged the age of improvisation. (p. 25)

In other words, the changed technology simply would not have worked without various new tailoring activities, enabled by the ingenuity of people.

Logic would dictate that the more mature a technology (the less "jerky" the looms), the less tailoring will be required, and thus the less human ingenuity matters. But does this argument hold? The *law of requisite variety* (Ashby, 1956) states that the variety of a controller should match the variety of the system to be controlled. Effective control will not be possible if the controller has less variety than the system. However, that—exacerbated by the de-skilling that occurs—is exactly what happens here.

A child of what sometimes is called the second Industrial Revolution, *numerical control* (NC) illustrates both sides of the peripheralization–tailoring contradiction and the violation of the law of requisite variety eloquently. Numerical control was another instance of machine programmability. The first numerical control machine tool was demonstrated in 1952 at the Massachusetts Institute of Technology (Reintjes, 1991). Originated for the aerospace industry, numerical control developed into a way to drive machines of all sorts, removing considerable control over production from individual tradesmen and transferring it to management and programmers. The repercussions on individuals were significant. A Boeing machinist, transferred to a numerical control machine, said:

> I felt so stifled, my brain wasn't needed anymore. You just sit there like a dummy and stare at the damn thing. I'm used to being in control, doing my own planning.

Now I feel like someone else has made all the decisions for me. I feel downgraded, depressed. I couldn't eat. When I went back to the conventional milling machine I worked like crazy to get it out of my system. I like to feel like I'm responsible for the whole thing—beginning to end. I don't like anybody doing my thinking for me. With numerical control I feel like my head's asleep. (Noble, 1983, p. 242)

The "central contradiction of numerical control" (Noble, 1983, p. 269) was that it frustrated, yet depended on, the skills and motivation of the people that worked with it. Not only did numerical control de-skill workers, although still requiring those old skills when the automation failed; it also required the acquisition of new skills at the same time, thus defeating much of its claimed efficiency. Paper-tape numerical control machines proved extremely brittle and inflexible. For instance, operators had to be able to read the tape (a skill in which they had not been trained) when it jammed or some other difficulty arose (Noble, 1983; Reintjes, 1991). Automation proved fallible and expensive. Human ingenuity, once again, had to be relied on to keep production going.

Myth 3: Throughout the Ages, Technology Change Has Been Motivated by High Wage Costs

This belief is known as the Habakkuk thesis (after Habakkuk, 1962). Whether wage costs are an incentive to automate remains an issue of considerable debate in many domains. During the industrial revolution, however, we find ample evidence against the Habakkuk thesis.

Wages in Western countries, although generally increasing after 1850, have varied greatly during the past two centuries and do not seem to correlate statistically with technological change (Mokyr, 1990). The Habakkuk thesis would further reinforce the idea that technology change is causal (see myth 1), that it can be given direction based on desires and interests. Again, that ignores the complexity of technological change. The Habakkuk thesis also assumes that self-acting machines represent a commensurate alternative to human labor. This may be hard to maintain in the face of inventions that led to uncertain, brittle, and perhaps expensive technologies requiring continuous human innovation in order to work efficiently, as we have seen in the previous section. Indeed, many weaving mills continued to depend heavily upon the labor of children and women: In the woolen industry hand-loom weaving survived well into the 1870s. One manufacturer from Calverly, near Leeds, recalled in 1902: "I went out of business in 1876, and our firm never had a single piece woven by power." (Emery, 1977, p. 45).

Whether wage costs were a sufficient reason to automate or not, the fear of technological unemployment has doubtless been felt by numerous professions throughout the ages. Although benefits were putatively spread over the whole human condition, the costs of automation in terms of loss of jobs were always borne by smaller, clearly defined groups such as silk weavers, or in aviation, radio operators, navigators, and, more recently, flight engineers.

Much of the resistance against technology during the industrial revolution, however, did not come from threatened professionals, but rather from the seats of power in societies that depended on labor or agriculture—rather than industry—for status and wealth. (It has been suggested, for instance, that the Greeks or Romans did not invent the wheelbarrow because two slaves carrying the load meant more status for their owner.) In 1397, tailors in Cologne were forbidden to use machines that pressed pinheads. In 1551, the English Parliament prohibited the use of mechanical gig mills, used in the process of making woolen cloth. The ribbon loom was invented in Danzig in 1579, but its inventor was reportedly secretly drowned by orders of the city council (Mokyr, 1990). By 1769, however, the British Parliament adopted a law by which willful destruction of machinery was made a felony punishable by death or deportation (Mokyr, 1990). This law signified that a shift had occurred in the interests of the power structure in Europe, which could now accommodate new technology without resistance and could actually benefit from its existence.

In other cases, technology provided an easy target for venting worker dissatisfaction over many issues (Mokyr, 1990). Organized gangs of dissatisfied citizens, called Luddites, regularly attacked machinery in Great Britain during the early 1800s. Luddism, however, was a reaction more to low wages and high prices than to new technology per se. General prosperity in a society tended to suppress Luddism, as in England after the 1815 peace was signed.

Myth 4: Humans and Machines Are Equivalent

Divisions of labor have been prevalent in Western societies since long before mechanization or automation; the specialist guilds of the Middle Ages are examples. It can be argued, however, that before mechanization such division was along vertical more than horizontal lines. Although the various trades were vertically separated from one another, control over pacing and other aspects of the work was still in the hands of the tradesman himself. As mechanization substituted power from mechanical sources for animate power, skilled work became subject to pacing by high capital equipment. The perceived necessity for "chunking" work in more horizontal layers, as well as narrowing tasks down vertically, increased. In the centuries that followed, the dominant themes became fractionation of work, de-skilling, and the emergence of a new class of work: supervisory control.

One of the initial reasons to partition work horizontally as well as vertically was unfamiliarity with factory work and the unwillingness of people to undertake it. As Robert Owen described in *A New View of Society* (1830/1927), the recruitment problems faced by a spinning mill owner were considerable. In the face of the natural environment that farmers and outworkers enjoyed while working at home, the apprehension of factory work was such that "only persons destitute of friends, employment and character, were found willing to try the experiment" (Owen, 1830/1927, p. 27). Factories may indeed have required unskilled workers. But that, or worse, in the eyes of contemporaries of Owen, was also all they could get: "They are a striking instance of mere brute force with brute appetites. This class of labour is as unskilled as the hurricane. Mere muscle is all that is needed; hence every human locomotive is capable of working" (Mayhew's description of the "versatile" London "street people" in the late 1840s, in Emery, 1977, p. 26).

The comparison of people with locomotives makes the point. People are commensurate with and comparable to a machine in their replaceability and in the way they work. Tasks had to become devoid of responsibility, complexity, and skills in order to enable unskilled workers to do them. One can ponder a laborer's acceptance of his or her unenviable fate during those times. But the fear of dismissal was undoubtedly fueled by the ease of replacement, the scanty provisions for the unemployed, and the limited alternatives available to unskilled urban workers.

This philosophy has occupied a central place in the history of mechanization and automation. It is known under a variety of labels, the most expressive of which is probably that of *redundancy of parts*. People and machines are seen as interchangeable components in a production system. Labor is fractionated to its maximum extent, rendering the worker's task predictable, quantifiable, and without a requirement for skills. Not only can people be easily replaced by other people, but, once technology allows, by machines. Fractionated tasks are extremely vulnerable to automation once that becomes technologically possible.

For this paradigm to work, however, certain other changes were necessary. Who, for instance, should coordinate the fractionated microtasks, whether performed by a human or a machine? Supervisory control emerged as a whole new class of work. Supervisors set the initial conditions for, intermittently adjusted, and received information in some form about the details of, activity on the shop floor (Sheridan, 1987). One of the biggest challenges facing supervisors was balancing the line. As fragmented microtasks became located on assembly lines, first in Chicago slaughterhouses and in Henry Ford's factories, to reduce transportation times and costs between the work stations, it became clear that workers were unable to help clear bottlenecks or make up for shortfalls on other parts of the line.

Taylor (1906, 1911) and Gilbreth (1985) reduced the challenge of balancing the line to a quantitative science: *scientific management*. The intelligence of the worker, assumed to be miniscule in the redundancy of parts paradigm (see, e.g., Taylor, 1911), determined the involvement of the supervisor in the process. In scientific management, that involvement was total: "Almost every act of the workman should be preceded by one or more preparatory acts of the management which enable him to do his work better and quicker than he otherwise could. [A worker should not] be left to his own unaided devices." (Taylor, 1911, p. 26). Also, "The workman is told minutely just what he is to do and how he is to do it and any improvement he makes upon the instructions given to him is fatal to success." (Taylor, 1906, p. 9).

Careful observations and measurements, the "scientific" methods in scientific management, not only provided a way to determine the labor requirements of larger parts of the line, but could also reveal the required skill levels and expected work loads at each individual work station. Not only did management control all parameters of an individual's job, it could easily replace the human going through those mundane motions by a machine, once technology for that was available.

Automation of the monotonous microtasks previously performed by human operators has become more feasible with time, especially in so-called "process-control" industries. This wedding of assembly line and automation has done little, however, to make a more satisfying work environment for the remaining workers, whose tasks are no longer repetitive toil but monotonous monitoring. Supervisory control is now exercised over machines and computers, rather than over humans (Sheridan, 1987). Labor, however, consonant with the redundancy of parts paradigm, is still perceived as an interchangeable part in an industrial process. Trade-offs regarding which tasks should be allocated to a machine and which should stay in the hands of a human remain inspired by putative cost savings, although politics can play an equal role (as Noble argued concerning the "lust for power, associated with a primitive enchantment with automation," 1983, p. 58, where "automation not only provoked strikes but undermined them as well," p. 65).

Myth 5: Task Allocation Between Human and Machine Is Necessary and Appropriate

From Taylor onward, the "garden path" of the redundancy of parts paradigm has led to a fundamentally misguided question: Which microtasks should be given to the machine and which should be left to the human? As more microtasks could be automated, this question became more pressing. In 1951, responding to questions that arose about the allocation of functions between humans and machines, Paul Fitts developed what is now referred to as a *Fitts list*. To give guidance as to how and what to automate and what not, Fitts presented general functions that can either be better performed by a machine, or best be left to a human (Fig. 4.1).

Why has the Fitts list concept had so little impact on engineering design? Was it because it contained criteria that are overly general, nonquantitative, and incompatible with engineering concepts (see Price, 1985)? No, not quite. The argument that the Fitts list did not work because it was nonquantitative, is trivial. In fact, the Fitts list affords quantification of the functions that should be allocated to humans and machines fairly well. From the Fitts list it is only a small step to quantification. But that is the point. Following the paradigm of redundancy of parts, Fitts relied on the mathematical representation of a human. In the footsteps of Craik (1947a, 1947b), the Fitts list concept assumed a human to be comparable to a machine.

Jordan (1963) averred that, "Speaking mathematically, a man is best when doing least" (p. 162). Jordan's point is clear: If we are talking about microtasks, we may as well give everything to the machine, because it is always going to do better than the human. Jordan suggested that allocation of functions between humans and machines would become effective only if humans and machines were looked at as *complementary*, rather than comparable. Numerical analysis of microtasks can lead only to designing the human out of the system, relegating people at most to a monitoring task. This creates a system that is in principle brittle because it forecloses

Humans appear to surpass present-day machines with respect to the following:

1. Ability to detect small amounts of visual or acoustic energy;
2. Ability to perceive patterns of light or sound;
3. Ability to improvise and use flexible procedures;
4. Ability to store very large amounts of information for long periods and to recall relevant facts at the appropriate time;
5. Ability to reason inductively;
6. Ability to exercise judgment.

Present-day machines appear to surpass humans with respect to the following:

1. Ability to respond quickly to control signals, and to apply great force smoothly and precisely;
2. Ability to perform repetitive, routine tasks;
3. Ability to store information briefly and then to erase it completely;
4. Ability to reason deductively, including computational ability;
5. Ability to handle highly complex operations, that is, to do many different things at once.

FIG. 4.1. The original "Fitts list" (Fitts, 1951).

human flexibility and creativity. The Fitts list cannot sort out the dilemma of meaningless toil and soporific monitoring.

But the redundancy of parts paradigm was too pervasive to account for human ingenuity. Numerous other attempts to construct Fitts lists (or MABA–MABA lists: "Men are better at/Machines are better at") have been published since the original one (e.g., Chapanis 1965; Mertes & Jenney, 1974; Edwards & Lees, 1972; and Swain & Guttman, 1980). All of these lists were putatively valuable heuristics for engineering designers during the past four decades of the history of industrial automation. Let us see how they, and the redundancy of parts paradigm, have fared.

Plagued by poor quality and low quantity production and "virtual shop warfare" (Noble, 1983, p. 269), General Electric felt compelled to introduce its Pilot Program in order to achieve effective use of its costly numerically controlled equipment. The pilot program was essentially a job enrichment scheme that granted workers greater responsibilities, increased their control over production, and gave them more room for initiative and creativity (Noble, 1983; Reintjes, 1991). The pilot program was an efficiency-motivated effort to reduce the vertical and horizontal chunking of tasks. Eventually, NC itself had to become tailored to the actual demands and contingencies of the fields in which it was implemented. Technology-in-principle (as conceived in the MIT laboratory) did not work as technology-in-practice. NC simply did not obey the law of requisite variety; more control for the operators, in all respects, was actually necessary to make the machines work.

And so, cracks started to appear. Despite the attractiveness of scientific management principles, the balancing problem was never really solved. Kilbridge and Wester (1963) reported that the U.S. automobile industry wasted about 25% of assembly workers' time through uneven work assignment. Ten years later, in 1973, the European Economic Community decreed that the assembly line would have to be abolished from the European auto industry (Emery, 1977). Maximum fractionation of work ultimately proved unworkable. The paradigm of redundancy of parts led to an insoluble dilemma. In the end, there was no more reserve, or "stretch," available to beef up production or cut costs:

> They realized it was not at all like the engineering problem of pursuing maximum aircraft speed by reducing friction and drag. They realized it was not a problem to be solved by the grease of yet higher relative pay, by featherbedding or by any of those things that Walter Reuther of the United Automobile Workers' Union bitterly referred to as "gold-plating the sweat-shop." (Emery, 1977, p. 109)

The violation of the law of requisite variety by the redundancy of parts paradigm suggests that the pursuit of maximum fragmentation of work is likely to be self-defeating.

Myth 6: Task Allocation in a Complex System Is the Task of the Designer

Even if not all tasks are quantifiable and we grant humans all the creativity they desire, it is still the designer of automated devices who must decide which task should be automated and which should not. Or is it? A new method of retrieving coal was introduced in England after the Second World War: the "longwall method of coalgetting." Teams of 40 to 50 workers were sent across the full length of a coal front in functional succession: first a drill and explosives group, then a team that broke and transported the coal, and finally a group that braced the hole that had been made, thus preparing the front for the next cycle. This new strategy replaced the old, where small teams of 6 to 9 men covered the entire collection of tasks associated with delving coal over a modest section of the coal front. There had been no specialization within these teams; the groups were tightly knit, retained their own control over work assignments, and selected newcomers.

The new longwall method severely undermined the unity within the groups, forced specialization on workers, and marked an end to the smooth adaptation and mutual coordination that had been the hallmark of the smaller production groups. The Tavistock Institute of Human Relations was called in to sort out the resulting chaos. Based on common sense and practical insight, they recommended reinstallation of the old self-reliant groups, although they equipped them with some technological improvements (de Sitter, 1989).

The implications of the longwall episode for thinking about organizing manufacturing were not lost on the staff of the Tavistock Institute. Attempts to improve human relations across organizational strata without regard for the way work was partitioned in the end proved as bankrupt as the fractionation of work itself. The longwall lesson showed that social and technical issues in a manufacturing organization are intrinsically related. New theories that followed were grouped under what has become known as the sociotechnical approach (van Beinum, 1989). Instead of a paradigm of redundancy of parts, the sociotechnical approach assumes redundancy (or overcapacity) of functions. The human is viewed as having multiple capacities and a willingness to learn.

Sociotechnicians strive for minimum fractionation of work, leaving the units of the production process—local groups—with considerable autonomy. The sociotechnical approach underlines that it is fallacious to assume that increased automation will eradicate the need to take social variables in an organization seriously. Automation in part means that responsibility for the execution of a task comes to lie with fewer and fewer people. This places an even greater burden on the designer to bring humans and machines into harmony with each other, focusing even more on the characteristics of the people and their distribution over their new work.

The benefits of direct participation and locally autonomous work organization seem to be slowly percolating down through the cracks in the redundancy of parts paradigm. Aided by the pressure of the global economic recession, the new paradigm of redundancy of functions is gaining ground, enabling converted organizations to be more effective and adaptive with the application of new technology. Implementations of small, self-regulating production groups, or *Fertigungsinseln* (literally, "islands that finish a product"), are now found in most Western manufacturing nations. *Business Week* reported improvements in American productivity of 30% to 50% in 1986, and specifically, a considerably better and more efficient use of numerical control machines by these semiautonomous groups (Kuipers, 1989).

The longwall episode illustrates clearly that inflexible allocation of functions during the development stages of new technology is not likely to be effective. Allocation of functions occurs in the dynamic richness of the field of practice in which actual practitioners (as much as they can) decide what they will let their machines do, given that they themselves are ultimately responsible for safety of the system and completion of the task. Allocation of functions is done by practitioners who tailor new technology to the particular demands of their domain and of the situations that may arise during practice. Allocation of functions is utopian for technology-in-principle; it is up to the practitioners, effective only for technology-in-practice.

Jordan's (1963) complementarity notion gets back to looking at artifacts and tools as *instrumentalities providing assistance to human beings,* rather than supplanting them. Artifacts, automated or not, become tools only in practice. It is ultimately responsible practitioners who decide what allocation of functions is really appropriate in the rich context of their problem-solving world. This concept of artifacts as tools is central to the development of automation in aviation, and specifically to the concepts of human-centered automation set forth here (see also chapters 10 and 15).

COMMENT

Have we learned the right lessons from history? Is automation no longer seen as the cure for human error? Have we stopped trying to design systems that marginalize human intelligence and tacit knowledge? Are we no longer allocating all possible tasks to a machine and leaving whatever we cannot yet automate to the human? We doubt it, and the statements of manufacturers and operators too often bear us out.

With the advent of computerization, it has become possible to automate cognitive, as well as manual, tasks. Creative talents, knowledge, and experience provide no more of a sanctuary against the proletarization of cognitive work than the manual skills of tool-making protected machinists against numerical control lathes. The challenge, as we have seen, is to look at computerized tools as aids to the creative and responsible human practitioner, and not to try to fragment his or her work so we can supplant human performance of microtasks (whether cognitive or skill-based) by automation.

Here is the distinction, in context. The cognitive work involved in synthesizing aircraft position and track information from star tracking and astronomical charts would certainly be seen by today's pilots as laborious, repetitive, and potentially distracting. Few pilots who have flown with a flight management system and an integrated flat-panel navigation display would trade its assistance for the old electromechanical instruments or a sextant and charting board (Wiener, 1989). Navigation automation functions as a very effective aiding tool for the pilot. On the other hand, every self-respecting pilot would agree that a cognitively difficult, complex flight followed by a skillful manually flown approach and a safe, gentle landing provides an intellectual satisfaction second to few others. What a waste to hand that over to the machine.

CHAPTER 5

THE EVOLUTION
OF AIRCRAFT AUTOMATION

INTRODUCTION

This chapter discusses aircraft automation from its origins to the present time. It is not possible to discuss the interaction of humans with the machines they control without some understanding of the machines themselves, which is why this discussion is oriented around the technology. But the overriding issue, as noted in chapter 1, is not just the machines, nor the people; it is the processes by which they interact to get the job done.

The earliest flying machines were extremely unstable and often barely controllable. Aircraft automation was invented to complement and assist human operators in carrying out tasks that were difficult or even impossible without machine assistance. Later, it became obvious that automation could appreciably offload pilots, who had increasing numbers of tasks to perform as aircraft utility and aviation system complexity increased. Aircraft automation was conceived as a complement to human operators (Jordan, 1963) from an early point in its development. In this respect, it differed considerably from industrial automation, although it is true that it later came to supplant certain skills, among them those of the navigator and flight engineer.

Until the late 1960s, automation was largely devoted to maintaining aircraft control, leaving navigation, communications, and management functions to the flight crew. The 1970s saw the onset of a technological revolution as the expanding utility of digital computers stimulated the development of miniaturized microprocessors with new solid-state circuitry based on the transistor. In aviation, the changes enabled by the new technology were as revolutionary as had been those during the previous 15 years when faster, larger, higher-flying jets began to replace propeller-driven transport aircraft. Microprocessors have had profound effects on the ways aircraft are flown, on the ways the aviation system is managed, and on the human pilots and air traffic controllers who operate the system.

AIRCRAFT FUNCTIONS

The range of functions that an airplane can perform is really quite limited. Properly controlled, an airplane can move about on a prepared surface. It can take off from that surface and once above the earth's surface, it is free to move in all spatial axes. It can be directed from one location to another, where it can land and again move about on a prepared surface, coming to rest at a predetermined spot.

Several categories of tasks must be performed by pilots in pursuit of their objectives. They must *control* the airplane in three translational and three angular axes. The autopilots just discussed were designed to assist with this task, which requires nearly continuous adjustment of the airplane's control surfaces unless the air is perfectly smooth and air speed remains constant. The pilots must remain cognizant of their airplane's position relative to their objectives, and must direct, or *navigate*, the airplane from one location to another. These functions may be performed by reference to external objects on the ground or to celestial bodies, by dead reckoning, by use of data from radio-frequency navigation aids, or by making use of geographic reference information from onboard aircraft sensors or satellites. In today's operational environment, pilots must also *communicate* with air traffic control, airline operations control, and other facilities to receive and acknowledge instructions, consult regarding changes, and give and receive advice concerning malfunctions or changes in the external environment.

These three invariant requirements are often referred to colloquially in flight safety literature as *aviate, navigate, communicate*. Their successful accomplishment under all circumstances is the hallmark of the capable pilot. To these three functions must be added another, that of aircraft flight and systems *management*. This became a major task as reciprocating-engine transport aircraft became larger and more complex and their engines became more powerful and temperamental during the 1930s, requiring the full-time attention of flight engineers who became an essential part of the cockpit crew. For overwater flights, navigators and radio operators were also carried, although newer technology developments have made all flight crew except pilots superfluous.

THE BEGINNINGS OF AIRCRAFT AUTOMATION

In 1908, Sir Hiram Maxim published a book discussing his experiments in aeronautics. He described a gyroscopic stability augmentation device connected to the fore and aft elevators of a large, highly unstable airplane built and tested while tethered during the 1890s (Fig. 5.1). This device, believed to be the first example of aircraft automation, was patented in England in 1891.

FIG. 5.1.   Gyroscopic stability augmentation system. From *Artificial and Natural Flight* (p. 94), by H. S. Maxim, 1908, New York: Macmillan. Copyright 1908.

In their flight experiments, Orville and Wilbur Wright also recognized the extreme instability of their aircraft. They independently developed stability aug-mentation devices for their machines, for which they received the Collier trophy in 1913. Lawrence Sperry developed a more advanced gyroscopic stability augmenta-tion system, which was demonstrated in flight (while a "mechanician" walked back and forth on the lower wing of a seaplane and the pilot stood with both hands over his head; Fig. 5.2) at the *Concours l'Union pour la Securité en Aeroplane* in France in the summer of 1914. The "automatic pilot" was awarded first prize at the event.

The Sperry name was associated with aircraft automation for the next 60 years. Sperry automatic pilots (called *autopilots*) became available during the 1920s. In 1918, H. J. Taplin patented a nongyroscopic two-axis stabilization device that relied on differential aerodynamic pressures. His device was successfully flown in the United States in 1926 (Taplin, 1969). With this exception, as far as is known, all successful autopilots during this period are believed to have utilized the gyroscopic principle.

A capable three-axis autopilot actuated solely by hydraulic and pneumatic power was an essential part of the equipment installed in Wiley Post's Lockheed Vega, *Winnie Mae*, for his solo round-the-world flight in 1933 (Fig. 5.3; Mohler & Johnson, 1971). The flight's successful conclusion was marked by the *New York Times* with the observation that "By winning a victory with the use of gyrostats, a variable-pitch propeller and a radio compass, Post definitely ushers in a new stage of long-distance aviation.... Commercial flying in the future will be automatic" (July 24, 1933 as cited in Mohler & Johnson, 1971, p. 66).

Long-range civil aircraft during the late 1930s, and military transport and bomber aircraft throughout World War II, were similarly equipped. By that time,

FIG. 5.2.   Flight demonstration of Sperry Automatic Pilot in France, 1914 (Sperry Company, undated).

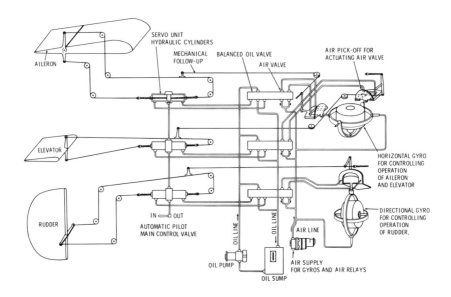

FIG. 5.3.   Diagram of autopilot used in *Winnie Mae*. From Sperry data; in *Wiley Post, His Winnie Mae, and the World's First Pressure Suit* (p. 46), by S. R. Mohler and B. H. Johnson, 1971, Washington DC: Smithsonian Institution Press. Reproduced by courtesy of Dr. S. R. Mohler.

aerodynamic understanding and engineering practice had improved; most of these aircraft were relatively stable platforms under normal conditions. The automatic devices were installed to relieve pilots of the manual labor of hand flying on long flights. They provided inner-loop control (discussed later) of the aircraft in response to direct instructions but left the pilot to perform all navigation and other essential piloting tasks. Nearly all reciprocating-engine transport aircraft introduced after World War II were equipped with autopilots of this sort.

THE JET ERA

The introduction of jet aircraft into civil aviation marked the beginning of a technological revolution (Fig. 5.4). The DeHavilland *Comet*, introduced in 1954, provided air passengers with transportation at much higher altitudes and greater speeds than had been available previously. It was followed in 1958 by the Boeing 707, an outgrowth of the military C-135 transport and tanker. Douglas was not far behind with its DC-8, introduced in 1960. During the early 1960s, both American manufacturers introduced smaller jets, the DC-9, B-720, and the second-generation B-727.

In 1967, the second generation Boeing 737 entered line service. Its systems were generally similar to those of the larger 727 introduced 3 years earlier, but to keep cockpit workload within reasonable limits for a crew of two rather than three persons, Boeing automated the operation of a number of airplane systems and simplified other systems. During the 1970s, the reliability of microprocessors im-

EVOLUTION OF JET TRANSPORT AIRCRAFT

First generation
  • Simple systems
  • Many manual tasks
  • Manual navigation

DeHavilland Comet
Boeing 707
Douglas DC-8
Douglas DC-9

Second generation
  • Systems redundancy
  • Pilot navigation

Boeing 727
Boeing 737-100, 200
Boeing 747-100, 200, 300
Douglas DC-10
Lockheed L-1011
Airbus A-300

Third generation
  • Digital systems
  • Two-person cockpit crews
  • Graphic displays
  • Flight management systems
  • Integrated alerting

Boeing 767/757, 747-400
McDonnell-Douglas MD-80
Airbus A-310, 300-600
Fokker F-28-100
McDonnell-Douglas MD-11
  (transitional to 4th generation)

Fourth generation
  • Fly by wire
  • Integrated systems operation

Airbus A-319/320/321
Airbus A-330, A-340
Boeing 777

FIG. 5.4.   Evolution of civil jet transports (Fadden, personal communication, November, 1995).

proved to the point that Douglas, Boeing, and the new Airbus Industrie consortium all felt themselves ready to take advantage of digital technology in the design of new airplanes. Douglas enlarged on its DC-9 series with the 135-passenger DC-9-80, introduced in 1978. Although the airplane made use of conventional electrome-chanical cockpit instruments, the manufacturer introduced considerably more automation of aircraft systems than in previous models.

Boeing introduced its 767 wide-body twin in 1982 (Fig. 5.5). The findings of the President's Task Force on Crew Complement (1981) allowed the airplane to be certified for two-person operation, and this crewing was adopted as the standard for all new types. Boeing also put into production its 757 series, a narrow-body airplane whose cockpit and systems were virtually identical to the 767; a common type rating covered both. The latter type caught on more slowly but is now in wide use throughout the world, as are various models of the larger 767. These aircraft and the Airbus A310, introduced slightly earlier, were the first "glass cockpit" aircraft in civil service. Their primary flight and aircraft systems displays were on cathode-ray tubes (CRTs), which motivated the *glass cockpit* description, although all three types used some electromechanical instruments as well. They made extensive use of digital microprocessors (the 767 and 757 had over 100).

During the 1980s, considerable operational experience was gained with these third-generation aircraft. As manufacturers gained confidence in the new auto-mation technology, it was incorporated and its uses were extended in new designs. This decade saw the development and introduction of the Airbus A320 (1989) (the first of the all-glass cockpit airplanes), the Boeing 747-400 (Fig. 5.6), a greatly advanced two-person crew version of the venerable 747 in service since 1970, the McDonnell-Douglas MD-11, a two-person crew DC-10 variant that entered serv-ice in 1991, and the Fokker F-100, an enlarged and highly automated outgrowth of the earlier F-28 regional jet.

Fadden (1990) described two categories of aircraft automation; he called them *control automation* (automation whose functions are the control and direction of an airplane) and *information automation* (automation devoted to the management and presentation of relevant information to flight crew members; this category includes communications automation). To these categories, I have added a third, *manage-ment automation* (automation designed to permit strategic, rather than tactical, control of an operation). When management automation is available, the pilot has the option of acting as a supervisory controller (Sheridan, 1987). In aircraft, automation is directed by the pilot to accomplish the tactical control functions necessary to accomplish the objective. This useful taxonomy is used throughout this book. Under each category, I describe the technologies, discuss benefits and problems associated with them, and try to characterize their effects on human operators.

FIG. 5.5.   Boeing 767-300ER (Reproduced by permission of The Boeing Company).

FIG. 5.6.   Boeing 747-400 cockpit (Reproduced by permission of The Boeing Company).

CONTROL AUTOMATION

Throughout most of the history of aviation, automation has fulfilled primarily inner-loop control functions (Fig. 5.7). Control automation assists or supplants a human pilot in guiding an airplane through the maneuvers necessary for mission accomplishment. In this document, the term also includes devices devoted to the operation of aircraft subsystems, which are complex in modern aircraft.

Aircraft Attitude Control

Maxim's 1891 device maintained pitch attitude, but other early automated controllers maintained attitude in the roll axis (Fig. 5.2). Later generations of such single-axis stability augmentation devices have been called "wing levelers," and they continue to be used in general aviation aircraft today. Later autopilots added other axes of control; the device used in the world flight of the *Winnie Mae* maintained the aircraft attitude in pitch, roll, and yaw by controlling the positions of the elevators, the ailerons, and the rudder (Fig. 5.3).

Flight Path Control

In early generations of autopilots, the gyroscope that controlled roll was also used as a heading, or directional, gyro in the cockpit. Some autopilots of this period also incorporated a barometric altitude sensor, which could be used to hold altitude as well, once the proper altitude was attained and set into the sensor. In these developments, we see the beginnings of *intermediate loop* control, in which the pilot could specify a goal: a heading and altitude to be maintained, rather than roll and pitch attitude.

As aircraft performance increased, air mass data became necessary for precise control of aircraft speed and height. Central air data computers were provided when jet-powered transport aircraft entered service in the 1950s; these devices provided integrated precision sensing of static and dynamic air pressures. The analog computers likewise incorporated rate sensors that enabled precise climbs and descents.

Swept-wing jet aircraft are susceptible to Dutch roll, a roll–yaw interaction that can be suppressed only by well-coordinated pilot or machine inputs. Early jet transport control required very precise coordination to counter this tendency. When the 707 was introduced, yaw dampers were provided to counter the problem. Although nominally under control of the pilot (they can be turned off), yaw dampers in fact operate autonomously in all swept-wing jet aircraft. Pitch trim compensators, used to counter the tendency of jet aircraft to pitch down at high Mach numbers, likewise operate essentially autonomously.

Spoilers or wing "fences" were installed on jet aircraft to increase control authority and reduce adverse yaw, to assist in slowing these aerodynamically clean aircraft, to permit steeper descents, and to decrease aerodynamic lift during and after landings. Early jets had manually controlled spoilers; later aircraft had spoilers that were activated either manually, in flight, or automatically after main wheel spin-up during landings. The Lockheed L-1011 introduced direct lift control by means of automatically modulated spoiler deflection during precision approaches.

Jet transports also required more precise control to compensate for decreased stability and higher speeds, particularly at high altitudes and during approaches to landing. Flight by reference to precision navigational data was made easier by the development of flight director displays, which provided pilots with computed pitch and roll commands displayed as shown in Fig. 5.8. The directors were much easier to fly than unmodified instrument landing system (ILS) localizer and glide slope

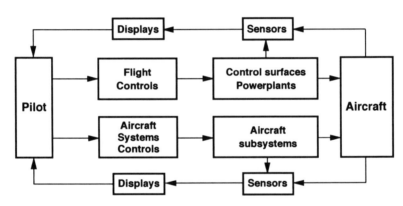

FIG. 5.7. Flight and aircraft systems inner control loops.

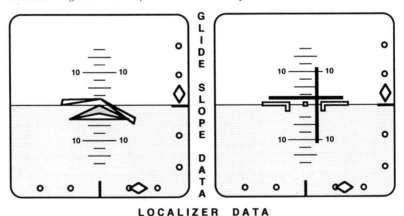

LOCALIZER DATA

FIG. 5.8. Single-cue (left) and dual-cue (right) flight director displays. Deviation data are at right and bottom of each display. The airplane is left of localizer centerline and slightly low on glide slope; director symbols are commanding a right turn and climb to regain the ILS center-line.

deviation data. Such displays rapidly became a mainstay of transport aviation. They provided roll and pitch commands to the autopilot; they also made it possible for pilots of average ability to fly approaches manually with high precision, although concern was expressed about "losing sight of the 'raw' data" while relying on the directors for guidance. A Delta DC-9 impacted a seawall short of a runway at Boston; its crew is believed to have followed the flight director, which was misset in "attitude" rather than "approach" mode, without adequate cross-checking of localizer and glide slope data (Boston, 1973).

Navigation Systems

Precision radio navigation systems capable of providing both azimuthal and distance information were introduced during the late 1940s and early 1950s. Very high frequency (VHF) navigational radios were developed during World War II. When introduced in civil aviation beginning in 1946, they eliminated problems due to low-frequency interference from thunderstorms, although they were limited to line-of-sight coverage.

VHF omnidirectional range (VOR) transmitters became the foundation of overland aerial radio navigation in the United States; ICAO soon adopted a similar standard. Distance measuring equipment (DME), consisting of airborne interrogators and ground transponders, was colocated with VORs and provided range data.

To provide guidance during approaches to landings, VHF directional localizer transmitters and ultra-high-frequency glide slope transmitters were located on airport runways; together they formed the basis for the instrument landing systems (ILS) that are still the standard of the current system (Fig. 5.9). Later, DME units were colocated with ILS to improve precision.

These devices provided aircraft with positional information of high precision. Their signals provided azimuthal and distance information that could be used either by pilots or by autopilots to provide intermediate loop control of aircraft paths. ILS signals, which provided glide slope guidance as well, were used to permit both manual and automatic ("coupled") precision approaches to runways. They enabled the design and implementation of autopilots with a range of capabilities (modes) including control of pitch, roll, and yaw, maintenance of a track to or from a surface navigational aid, and the capture of localizer and glide slope centerlines followed by the conduct of automatic approaches.

To improve schedule reliability, carriers began to study automatic landings (autoland). After automatic landing demonstrations in 1965, the British Aircraft Corporation *Trident III* (a three-engine medium-range transport) was the first production-series transport to be approved for automatic landings in category III

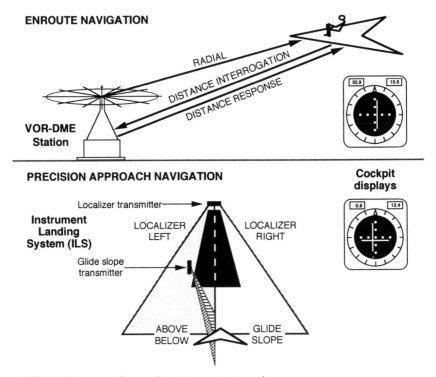

FIG. 5.9.   En route and approach precision navigation aids

weather (Fig. 5.10). The airplane utilized three autopilots with flare[1] capability and roll-out guidance, and a voting system to ensure concordance of the control outputs from the three analog autopilot computers. This equipment enabled the Tridents, operated by British European Airways from 1965 to the mid-1980s, to continue flying their routes when nearly all other aircraft were grounded.

Many newer transport aircraft have autoland capability, although pilots as well as aircraft and navigation facilities must be certified for such bad-weather approaches. In recent years, some carriers have utilized head-up display (HUD) equipment[2] to provide pilots with a better means to transition to a visual landing during extremely low visibility.

---

[1]The "flare" maneuver slows the rate of an airplane's descent just prior to its landing; this is accomplished by raising the airplane's nose slightly while reducing engine power to dissipate kinetic energy (air speed).

[2]Head-up displays utilize a semitransparent combiner plate mounted between the pilot and the windscreen, on which can be projected symbology and data depicting the airplane's flight path (see Fig. 6.7). When it becomes visible, the pilot can also see the runway through the plate. The projected display is collimated to optical infinity so the pilot's eyes can remain accommodated at a distant focus while using the HUD.

Integrated Flight Control Systems

Aircraft control automation was well advanced by 1970. Capable analog computers were the basis for autopilots that performed all inner-loop and many intermediate-loop functions (see Fig. 5.14), although pilots were still responsible for providing the devices with tactical instructions and monitoring the performance of the automation. Because the outputs of the autopilot and autothrottles were reflected both in control movements and airplane behavior, the pilots' monitoring task required only the displays also used by them for manual flight. A few new instruments provided surveillance of autopilot functions and indications of autopilot modes when automatic navigation was in use.

Two wide-body airplanes introduced during the 1960s and early 1970s, the Douglas DC-10 and Lockheed L-1011, introduced more complex autopilots with comprehensive mode annunciation and a broader range of options for both lateral and vertical aircraft control. Mode control panels (Fig. 5.11), located in the center of the instrument panel glareshield, commanded autoflight and autothrottle functions and the flight directors whose computers provided flight path commands to the integrated autoflight systems. The L-1011, which entered service in 1973, was the first commercial type to incorporate direct lift control, which modulated lift automatically during landing approaches by means of partial spoiler deployment and thus improved landing precision (Gorham, 1973). This feature, the forerunner

PRECISION APPROACH CATEGORIES

Category	Decision Height	Visibility or RVR*
I	200 ft.	2400 ft. (1/2 mile)
II	100 ft.	1200 ft.«
IIIa	50 ft.	700 ft.«
IIIb	†	150 ft.«
IIIc	†	††

* Runway visual range (RVR).
† No decision height specified. Visibility is the only limiting factor.
« No fractions of miles authorized when determining visibility. The runway served by the ILS must have operable RVR equipment.
†† No ceiling or visibility specified. Aircraft must be equipped with autoland.

FIG. 5.10.   Ceiling and visibility limits for ILS approaches

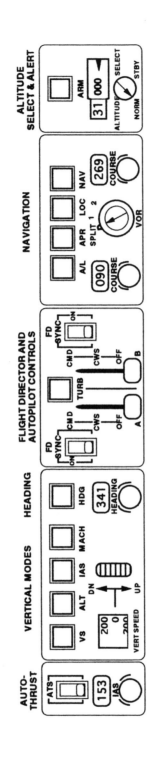

FIG. 5.11. Lockheed L-1011 Avionic Flight Control System mode control panel. From *Automatic Flight Control and Navigation Systems on the L-1011: Capabilities and Experiences*, by J. A. Gorham, 1973, paper presented at the USSR/US Aeronautical Technological symposium, Moscow. Copyright 1974 by J. A. Gorham. Adapted with permission.

of gust alleviation and lift control seen in some of the most modern transports, was an integral part of a category III fail-operational autoland system incorporated when the airplane was initially certified—a first in transport aviation.

The 1011 flight control system was more highly integrated than any other in service at the time and provided a number of autoflight modes (Fig. 5.12). These systems provided pilots with more sophisticated tools than had previously been available (at the cost of considerably more complexity). Training officers noted that some pilots and flight engineers had difficulty in learning the new systems, as their forebears had when first-generation jet transports entered service during the late 1950s.

Advanced Flight Control Systems

Until 1988, control of large aircraft, whether manual or automatic, was carried out through hydraulic actuators. Large, centrally located control columns ("yokes") and rudder pedals controlled the hydraulic actuators; they moved when actuated by the pilots or autoflight systems and thus provided both visual and tactile (touch) feedback of flight control inputs. Throttles (actually thrust levers: they were connected to electronic engine control systems rather than fuel valves) were electrically driven; they likewise moved when actuated by the pilots or the autothrust system.

The 1988 Airbus A320, whose flight controls are unconventional (fly-by-wire)[3] represented a departure from previous civil designs. Attitude control in the A320 is by hand controllers (sidesticks) located outboard of each pilot. The two sidesticks are not coupled to each other; they do not move to provide tactile feedback during autopilot control inputs or when the other pilot is making manual inputs. Likewise, the thrust levers in the center console do not move during autothrust inputs, although they can be moved by the pilots to provide instructions to the full-authority digital engine controllers (FADECs), which control the power systems (Fig. 5.13). Electronic Centralized Aircraft Monitor (ECAM) visual displays indicate both power commanded and power delivered, but ancillary tactile or visible feedback is not provided by the levers themselves.

The introduction of fly-by-wire systems in the A320/330/340 and B-777 has provided control system engineers with more flexibility to tailor aircraft control responses to match desired characteristics through software in the flight control computers. An inherently unstable airplane can be made to feel, to the pilot, like an extremely stable platform. Indeed, some modern aircraft (such as the MD-11)

---

[3]*Fly-by-wire* denotes a flight control system in which pilot or autopilot inputs are conveyed to the control surface actuators by electrical impulses rather than by hydraulic lines or cables. The surfaces are usually positioned by hydraulic power, as in conventional aircraft.

PITCH AXIS	ROLL AXIS
• Control wheel steering	• Control wheel steering
• Altitude hold	• Heading select
• Vertical speed hold	• VOR hold
• IAS hold	• R-nav coupling
• Mach hold	• Localizer hold
• Altitude capture	

DUAL AXIS

- Approach
- Approach/land
- Go-around
- Take-off
- Turbulence

FIG. 5.12. Lockheed L-1011 Avionic Flight Control System functions. From *Automatic Flight Control and Navigation Systems on the L-1011: Capabilities and Experiences*, by J. A. Gorham, 1973, paper presented at the USSR/US Aeronautical Technological Symposium, Moscow. Copyright 1971 by J. A. Gorham. Adapted with permission.

incorporate reduced longitudinal stability, which is compensated for by a stability augmentation system. Even manually controlled flight in such aircraft is actually accomplished by one or more computers interposed between the pilot and the machine. This control architecture offers other opportunities to the designer, who may now limit the flight envelope by making it impossible for the pilot to exceed certain boundaries, or provide precisely tempered degradation of flying qualities as safe operating limits are approached. This is called *envelope protection*.

Effects of Control Automation on Human Operators

Figure 5.14, an expansion of Fig. 5.5, suggests some of the effects of adding control automation to the pilot's resources. It indicates that the pilot has an additional means of controlling his aircraft attitude and flight path. In this sense, it relieves the pilot of inner-loop control tasks, which require a relatively high level of skilled psychomotor activity and considerable attention on a more or less continuous basis. Providing an alternative means of accomplishing the control functions gives the pilot considerably more time to devote to other functions and tasks essential to safe flight.

On the other hand, note that the pilot must now give at least intermittent attention to additional equipment and displays. The pilot must understand the functioning (and peculiarities) of an additional aircraft subsystem, remember how to operate it, and decide when to use it and which of its capabilities to utilize in a given set of circumstances. When it is in use, its operation must be monitored to ensure that it is functioning properly. If it begins to malfunction, the pilot must be aware of what it is supposed to be doing so he or she can take over its functions. Finally, the pilot must consider whether the failure impacts in any way the accomplishment of the mission and whether replanning is necessary; if so, the replanning must be done either alone or in communication with company resources.

Issues Raised by Integrated Flight Control Systems

Although the considerable psychomotor workload of the pilot is reduced by an autopilot, the cognitive workload is usually increased by the introduction of automated devices, and the pilot's tasks are always changed by the provision of such devices. The addition of an autopilot provides the pilot with additional resources that can offload high-bandwidth, flight-critical tasks, but the addition is not without cost in terms of the attentional, knowledge, and information-processing requirements placed on the flight crew. Note also that the pilot's management tasks increase.

The decrease in pilot workload when an autopilot is in use can be dramatic, but even this benefit is a two-edged sword. As an example, there have been several instances in which early-series Boeing 707 and 747 autopilots have malfunctioned subtly by disconnecting or introducing a gradual uncommanded roll input to the airplane controls. In at least some of these, pilots have first noticed the uncom-

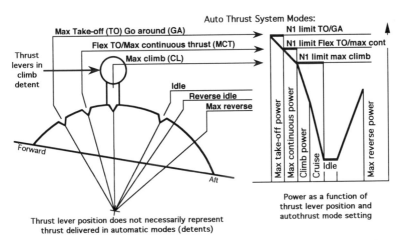

FIG. 5.13.   Diagram of dual-function thrust levers on Airbus A320/330/340 aircraft, showing detents for autothrust modes. Thrust levers may also be moved to intermediate positions for manual power control. From *A320 Flight Deck and systems Briefing for Pilots, Issue 4* (Rep. No. A1/EV-O 473 774189) (p. 24) by Airbus Industrie, 1989, Toulouse, France: Author. Copyright 1989 by Airbus Industrie. Adapted with permission.

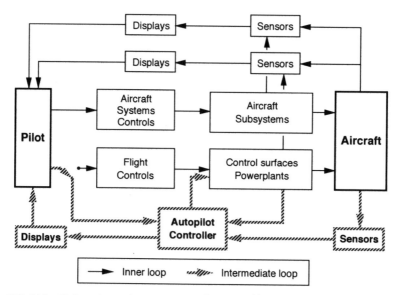

FIG. 5.14.   Flight and aircraft systems intermediate control loops.

manded maneuver only when the airplane was in a steep bank and dive from which level flight was regained eventually only after severe maneuvers (Atlantic Ocean, 1959; Nakina, Ontario, 1991). It has long been known that humans are not very good monitors of uncommon events, especially when they are tired, bored, or distracted by other tasks (Mackworth, 1950; Broadbent, 1971). Autopilot functioning is annunciated in the cockpit, but a subtle system interaction such as this, with or without a failure, may produce little in the way of obvious visual signals aside from evidence of the gradual attitude change on the attitude indicator, and the human vestibular system is unable to perceive a gradual roll acceleration.

These systems are more complex and tightly coupled than their predecessors. They require that pilots know more about system behavior under both normal and abnormal circumstances. Because the systems are very reliable, most system anomalies will occur only rarely, presenting the pilots with behavior that may not be understood when it occurs. It is difficult for pilots to keep all knowledge about uncommon system states and failures available in memory, and equally difficult for them to access such information when it is needed. This is sometimes called *inert knowledge.*

Moreover, even if the systems exceed the limits of their design envelopes, there may be little information provided, aside from an alerting message if the pilot is expected to take action. If the fault is one for which corrective action is not thought necessary by the designer, the system may provide no explanation of its behavior. Although this approach serves to keep pilots from improvising solutions to problems that may not require them, it does little to increase their confidence in the automated systems.

Issues Raised by Advanced Flight Control Systems

The major differences between previous aircraft control systems and those in recent Airbus aircraft have evoked fairly widespread concern in the operational and human factors communities (see, e.g., Folkerts & Jorna, 1994), although it should be said that the concern does not appear to be serious in airlines operating this aircraft type. After a survey of pilots operating these aircraft, British Airways concluded "that the A320 [thrust lever] design provides advantages in respect to engagement and selection of power settings, [but] that [thrust lever] movement provides better disengagement and information on system function…from a Flight Operations perspective a future system should consider providing movement between the idle and climb power positions, retaining the A320 thrust setting and engagement 'detents' technique" (Last & Alder, 1991, p. 51).

The lack of tactile feedback to the sidestick controllers either from autopilot inputs or between the two pilots' controls in the A320/330/340 has also been a matter of concern to human factors engineers because these airplanes differ from all other civil aircraft in this respect. Reports indicate that there have been a few situations in which opposing inputs from the two pilots have summed to produce no change in airplane flight path (e.g., incident at Sydney, Australia, 1991), although a button on each sidestick permits either pilot to remove the other from the control loop. This control arrangement would be likely to present problems only if a nonflying pilot were to initiate a go-around or an evasive maneuver because of an emergency before being able to tell the flying pilot that he or she was assuming control. To cover such a case, it is possible that procedures and training should be modified to include using the lockout when the nonflying pilot assumes control, to insure that only one pilot is flying the airplane. Simultaneous inputs from both sticks are not annunciated in the cockpit except when the airplane is on the ground.

Based on operating experience to date, it appears that pilots are usually able to obtain all needed information concerning flight and power control either with or without tactile feedback of control movements initiated by the automatic systems. This may be a case in which there is not one best way, based on empirical or analytical knowledge, to automate a system, and in which, therefore, any of several methods of providing feedback may be effective provided that pilots are given sufficient information to permit them to monitor the systems effectively. Unfortunately, information concerning the rare cases in which a particular innovation is *not* effective in providing adequate feedback may not come to light until a mishap occurs. How much feedback is enough? It depends on the context, as discussed in later chapters.

Power Control

Reciprocating-engine aircraft had only limited inner-loop automation of power control systems. Automatic mixture controls that utilized barometric altitude data to adjust fuel–air ratios were installed in the DC-3 and later transports. Automated control of propeller pitch (by means of governors) was also introduced during the 1930s, not long after controllable-pitch propellers. In later multiengine aircraft, propeller autosynchronizers were installed to match the propeller speeds of all engines. Some superchargers, installed in high-altitude aircraft, had automatic sensing devices that controlled the amount of air pressure or "boost" provided to the engine air inlet. Throttles, propeller and mixture controls were not integrated, however.

Following World War II, surplus military aircraft were purchased in considerable numbers by civil operators. Some of these aircraft were extremely demanding to fly after an engine failure at low speed during or shortly following takeoff. To lessen the asymmetric drag caused by a windmilling propeller and assist pilots in maintaining control during the critical moments after takeoff, automatic propeller feathering systems were introduced in some aircraft. These devices sensed a loss of thrust in a malfunctioning engine and rapidly aligned its propeller blades with the airstream to reduce drag.

Autofeathering devices provided critical assistance when they functioned properly, but several accidents occurred after functional engines were shut down autonomously. Autofeathering systems, once armed by pilots just before takeoff, are independent of pilot control and do not notify the pilot before taking action. To that extent, they remove a portion of the pilot's authority while leaving him or her with the responsibility for the outcome, a topic on which more is said in chapter 10.

Control of Aircraft Subsystems

In early generations of jet aircraft, the many aircraft subsystems were operated in the conventional way, with switches in the cockpit controlling most aspects of system operation. The flight engineer's primary task was the operation and surveillance of power, electrical, fuel, hydraulic, and pneumatic systems. Discrete controls for every system were located on the flight engineer's panel (Fig. 5.15).

FIG. 5.15.   Boeing 727-200 Flight Engineer workstation. Photos courtesy of NASA-Ames Research Center.

In aircraft designed for a crew of two pilots, attempts were made to simplify system operations somewhat to decrease flight crew workload. Passenger alerting signs were activated automatically. Automatic load shedding was introduced to simplify electrical system reconfiguration following a generator failure; air conditioning pack deactivation was automatic following an engine failure on takeoff, and so on. These and other measures represented a piecemeal approach to the problem, however; subsystems were still considered in isolation by designers, and until recently, manual systems operation during failures was still complex.

Automated flight path control systems usually provide immediate feedback to pilots concerning their continued functioning. Feedback concerning aircraft subsystem status may be much less obvious. Older three-person aircraft incorporated a multiplicity of lights and gages to provide the flight engineer and pilots with such information. Cockpit automation and simplification efforts have attempted (with considerable success) to minimize the amount of system information that the crew must monitor. The provision of simpler interfaces, however, has not always been due to the design of simpler aircraft subsystems. On the contrary, system complexity in some cases has increased greatly.

Cockpit simplification has included drastic reductions in the number of subsystem controls and also standardization of those controls, nearly all of which are now lighted pushbuttons with legends located on the overhead systems panel. Critical buttons may be guarded. The switches are usually located in subsystem diagrams (Fig. 5.16). The use of pushbuttons of identical shape and size in place of a variety of toggle switches has cleaned up the overhead panel, but it has made more difficult the location by feel of a given switch. Manufacturers state that their "dark cockpit" concept, in which buttons are lighted only if they require attention, indicates those that must be used, and that buttons should be actuated only after visual confirmation of which button to press.

In the MD-80 and B767/757, airplane subsystem control was considerably simplified wherever possible to reduce flight crew workload, though the systems remained conventional. The only alerts that were permitted to appear were those that required pilot decisions or actions, which were carried out largely by actuating lighted push-button switches on the overhead panel. Legends on the buttons showed switch position and, where necessary, related system state. Failure to differentiate switch position from system state has led to problems for operators; this ambiguity was a contributory factor in the nuclear power station accident at Three Mile Island (Woods, personal communication, 1994).

Douglas has taken a different approach to subsystem management in the MD-11. Many of its subsystems are automatically reconfigured by an Automated System Controller (ASC) if a fault occurs. The Douglas design philosophy, motivated by a desire to decrease flight crew workload, was stated by its Chief of MD-11 operations:

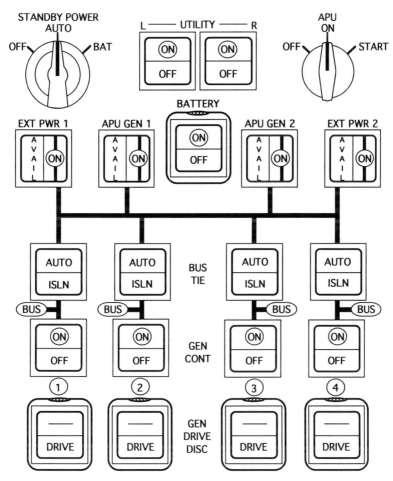

FIG. 5.16.   Boeing 747-400 overhead AC electrical system control panel. Reproduced by permission of the Boeing Company.

"One of our fundamental strategies has been, if you know what you want the pilot to do, don't tell him, do it" (Hopkins, 1990). Many normal subsystem functions formerly performed by the flight crew have also been automated. Douglas has made no attempt to automate any function which can irreversibly degrade aircraft capability.

The failure to display the basic or root causes of the faults in the MD-11 implementation of this philosophy presents the potential for pilot confusion or surprises, particularly in the case of a very complex system. This has been of concern to FAA certification pilots, and Douglas has found it necessary to provide ASC "task lists" in its abnormal/emergency checklists. These lists enable pilots to determine possible malfunctions and the actions the ASC takes when such malfunctions are detected, to clarify possibly ambiguous system states following ASC rectification of faults. This is another example of the use of automation to alleviate pilot workload when dealing with anomalies, at the expense of a possible increase in cognitive demand caused by the opacity of the automation.

INFORMATION AUTOMATION

Alhough control automation followed a generally evolutionary path until the introduction of the A320, information automation is largely a product of the digital revolution. The period from 1970 to the present has been marked by major changes in the appearance of the flight deck due to the introduction of electronic display units (EDUs) in the 767/757, the Airbus A310, and following types.

For those unfamiliar with glass cockpit terminology, Fig. 5.17 is a generic flight deck layout showing the panels that are discussed here. Up to 6 electronic display units, together with backup flight instruments (liquid crystal displays or electromechanical instruments) and a few critical systems indicators, are found on the main instrument panel. Aircraft systems controls are located on the overhead systems panel. A mode control panel (also called flight control unit) is located centrally on the glare shield below the windscreens. Other flight management system control units and communications controls are located on the pedestal between the pilots, together with power and configuration controls. These displays, together with paper documents, verbal communications, aural signals, and the pilots' own knowledge, provide all real-time information to the pilots. The flexibility of glass cockpit displays has made it possible to provide any sort of information in new and different formats, and to modify that information in any way desired by designers to fit any need.

Attitude and Flight Path Displays

Electronic primary flight displays (PFDs) have generally shown aircraft attitude and state data in much the same ways it was earlier presented on electromechanical displays, although in the latest generation of aircraft all of the formerly available information is shown on a single large display directly in front of the pilots (Fig. 5.18). This representation adds additional information, in particular trend information, but few attempts have been made to alter radically the format of the data aside from the use of linear "tape" presentations of altitude, airspeed, and vertical speed in place of the former round dial displays. There is still some argument about whether linear or circular displays are preferable, although linear tapes are now the rule.

Over the years, human factors researchers and design engineers have brought forth a variety of concepts for the simplification and integration of the data

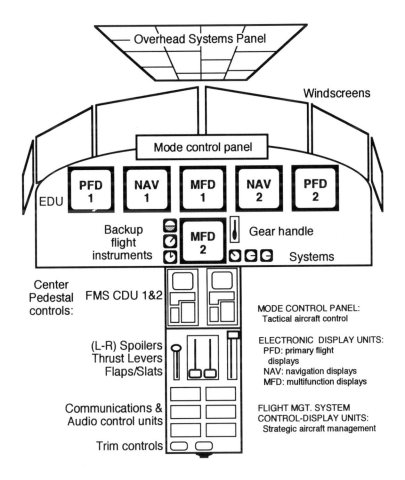

FIG. 5.17.   Generic "glass cockpit" showing control and display locations.

presented on the primary flight displays. Most of these have involved some sort of pathway (or tunnel) through the sky concept (Fig. 5.19). Such displays, based on concepts developed in Germany during the 1950s, have been tested in simulation and flight, and are still under development (Grunwald, Robertson, & Hatfield, 1980). Other displays with the same objectives are under development by Langley Research Center, the Air Force Armstrong Laboratory, and various airframe manufacturers. In all cases, the intent is to provide integrated information concerning attitude and flight path, similar to the integrated navigation displays that have been so successful in glass cockpits.

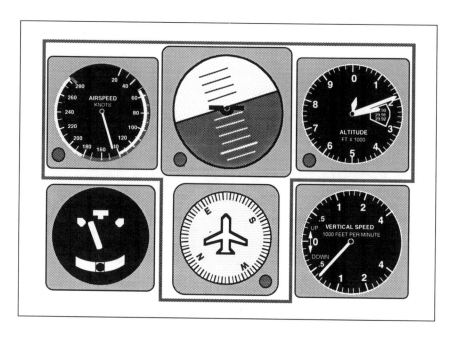

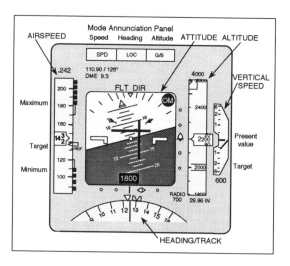

FIG. 5.18. (Above) Primary flight display on electromechanical instruments; standard "T" arrangement of primary instruments is boxed. Upper row: airspeed indicator, attitude indicator, altimeter. Lower row: turn and slip indicator, heading indicator, vertical speed indicator. (Below) Electronic primary flight display showing the same data (Boeing 747-400). Note that in general, the "T" arrangement of the most essential information has been preserved in this electronic display. Reproduced by permission of The Boeing Company.

Air Force human factors experts have conducted an intensive search for simpler, more intuitive means by which to convey primary flight information, navigation information, and threat alerting (Stein, 1986). Civil airframe manufacturers have shown interest in such concepts but have been inhibited in bringing them to service use, in part by financial constraints and in part by the mix of aircraft in nearly all airline fleets. Pilots fly a variety of aircraft during their careers, some with advanced cockpits, some with conventional electromechanical instruments. There has been considerable concern among operators that transitioning back and forth between the older displays and advanced, more integrated, primary flight displays could increase training requirements and perhaps compromise safety. At least two U.S. air carriers, each operating various B-737 models, have gone so far as to install electromechanical instruments rather than EDUs in their -300 and -400 aircraft to insure uniformity of displays across their fleets. (The new 737-700 and -800 models will offer electronic analogs of these older displays (Fig. 5.20) in order to permit pilots to operate both the older and newer models of the 737 under a single FAA type rating.)

Navigation Displays

Nowhere has information automation been used more effectively than in aircraft navigation displays. Glass cockpit navigation displays are a radical departure from their electromechanical forebears. All aircraft manufacturers have integrated the information formerly presented on electromechanical instruments into a single plan view map display to which has been added other features derived from the flight management system database. Terrain detail, explicit location of ground navigation

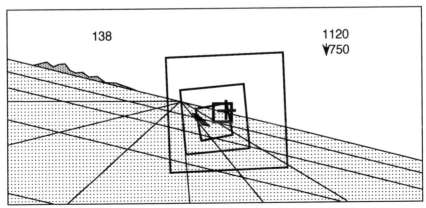

FIG. 5.19.   Generic "tunnel in the sky" flight path display. Flight director guidance is often incorporated in the displays, as shown here; alphanumeric speed, altitude, and other data are often displayed.

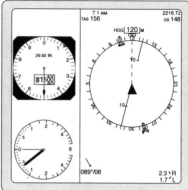

FIG. 5.20.   Optional primary flight and navigation displays in the Boeing 737-700 and -800.
These displays are similar to electromechanical and electronic displays in earlier series 737s, and
are offered to permit the earlier and latest models to be operated under a common type rating.
Reproduced by permission of The Boeing Company.

aids and pilot-constructed waypoints, airport locations, and other data can be
portrayed on a large EDU, together with the programmed route, alternative routes
if they are under consideration, and other data. Figure 5.21 shows how such
information was formerly presented to pilots, and a contemporary navigation display.

Navigation displays are the feature most liked by pilots transitioning to the glass
cockpit (Wiener, 1985b, 1989), and with good reason. No single feature has
mitigated flight crew cognitive workload as much as these new displays, and it is
probable that no technological advance has done as much to make the modern
airplane more error-resistant than its predecessors. In several advanced aircraft,
these displays also permit the pilots to preview the flight plans they have entered in
the flight management system on a large-scale map display, to assist in detecting
errors in FMS waypoint insertion.

Current navigation displays integrate a variety of complex data into a clear,
precise, and intuitive representation of aircraft position with reference to a preplanned course. As such, they are an information management tool of considerable
power. In the Map mode, they also assist in flight path management by displaying the
results of FMS entries. They are extremely compelling, although they hide a great
deal of data. An example of a serious problem that can be created by the nonob-
servability of source data occurred for a period of time a few years ago at Kai Tak
Airport in Hong Kong, an exceptionally difficult airport because of very close high
terrain and man-made obstacles. The problem was caused by a navigation transmitter
in nearby mainland China that caused spurious location data to be input to aircraft
flight management systems and thus induced map shifts on navigation displays.

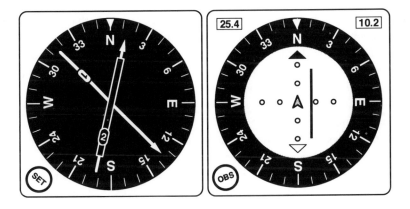

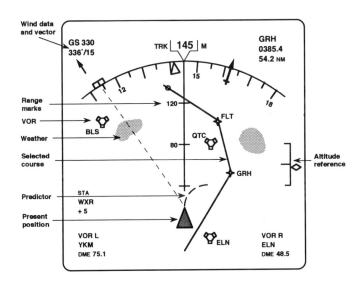

FIG. 5.21. (Above) Electromechanical navigation displays: radio magnetic indicator (RMI) at left, horizontal situation display at right. HSI also shows glide slope deviations when an ILS is tuned, VOR to–from indications, and DME range from two stations. (Below) Electronic navigation display (Boeing 747-400). Note superimposed waypoints, overlaid weather radar images and other data on periphery of screen. Reproduced by permission of The Boeing Company.

Issues Raised by Advanced Flight Path Displays

The principal issue raised by flight display innovations is that of feedback of automation actions and intent to pilots. Complex data become informative only when they are transformed in an effective representation, and the representation is effective only if it tells the operator what is required to be known at a certain point in an operation. Navigation displays are an excellent representation of what the pilot needs to know. They are not perfect, because the data used in their generation are not infallible.

The complex process that produces the map display is not observable unless the pilot becomes suspicious and utilizes displays that also show raw data (see Fig. 6.8). Given the effectiveness of the map displays and their reliability under all usual circumstances, however, the tendency of pilots to rely on usually reliable information may weigh against the likelihood of checking the raw data when they are busy preparing for an approach to landing. There may be no entirely satisfactory answer to this automation conundrum, but pilots, like operators in other domains, must be taught to be suspicious of all of the very capable automation under their control. This is not an easy sell in today's aircraft and environment, and it will only become harder in the future.

Power Displays

System performance displays generally have two objectives: to inform the pilots of the state or status of the system on an ongoing basis, and to aid the pilots to detect anomalous system performance. The first objective links system performance to some value or state having external significance. The second objective links performance to some value or state having internal significance (D. M. Fadden, personal communication, November 1995).

In older aircraft, power displays, by their arrangement, permitted pilots to compare the performance of one engine to the other(s). It was easy for pilots to compare needle positions on the electromechanical analog instruments to determine whether all engines were behaving the same, although an implicit weakness in this approach is that it may fail to show the effects of a problem that affects both or all engines equally. In an Air Florida takeoff from Washington National Airport (1982) with engine inlet icing that affected the engine pressure probes, both engines were developing only a fraction of takeoff power, but (incorrect) exhaust pressure ratio (EPR) indications were the same from both engines and the pilot flying failed to detect the problem during the takeoff roll. One way of circumventing this problem is to compare measured performance with a model of expected performance; an example is shown in Fig. 5.22.

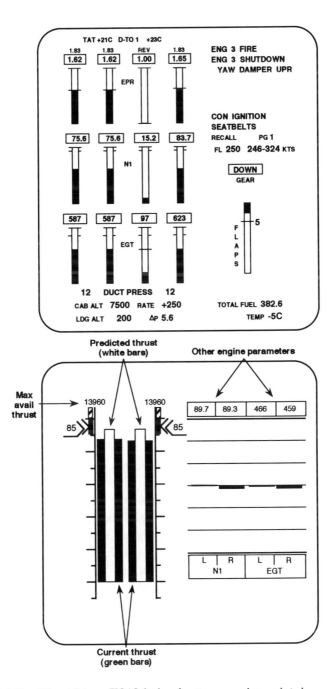

FIG. 5.22. (Above) Primary EICAS display of engine power, alerts and airplane configuration (Boeing 747-400). (Below) Experimental engine monitoring and control system display (E-MACS). From *A Simulation Evaluation of the Engine Monitoring and Control System Display* (NASA Tech. Mem. 101685) (p. 7) by T. S. Abbott, 1990, Hampton, VA: NASA Langley Research Center. Adapted with permission.

The Boeing 757/767 and A310 introduced electronic engine status displays. They depicted information that had previously been available on electromechanical instruments, together with adaptive EGT[4] limits, data on commanded versus actual thrust for autothrust operation, and so on. The later Airbus A320 provided a similar set of electronic displays and alphanumeric information. The Boeing 747-400 power displays were the first to utilize a simplified tape format on a primary and secondary display (Fig. 5.22). The format eases the task of comparing engine parameters. The MD-11 primary and secondary power displays are again CRT representations of the earlier electromechanical displays.

T. Abbott and coworkers proposed and evaluated a concept for a simplified set of power displays using bar graphs that show relative data versus expected values for engine parameters (T. Abbott, 1989, 1990). This display is similar to the 747-400 displays, but it compares engine power with a mathematical model of expected power (Fig. 5.22).

The engine monitoring and control system (E-MACS) display, like the navigation display, is an example of a general trend toward more integrated, pattern-oriented representations to help human operators deal with increasing volumes of data. The E-MACS concept represents an attempt to reduce cognitive workload by providing a simplified, more integrated representation of power being delivered, using simple dynamic bar charts. The processed information is based on a simplified functional model of the monitored engines, derived from the engine parameters shown in the Boeing display. A glance at the graphics is sufficient to inform a pilot about engine condition and whether the requested thrust is being supplied. The concept has been tested against round-dial representations in simulation and yielded a decrease in operator errors.

The E-MACS concept performs the several cognitive steps necessary to translate raw data into a concept of engine condition and thrust available, and presents summary results to the pilot. It should be noted, however, that its usefulness depends on the adequacy of the system's internal engine model. If actual engine behavior were to differ from the model's predictions, the result could be more confusing than that from a display having a simpler design concept.

The concern about overloading operators with information is also the motivating factor for attempts to provide more integrated representations of flight attitude, as noted earlier with respect to the primary flight display (although a practical display that incorporates all necessary information for this application has not yet been implemented). Similar concerns exist with regard to aircraft subsystem displays, the topic of the next section.

---

[4]EGT: exhaust gas temperature, related to engine thrust.

Aircraft Subsystem Displays

Although there is still a philosophical controversy about the necessity or desirability of providing synoptic (summarized diagrammatic) subsystem information in the cockpit, many pilots and operators clearly find it desirable to have such displays and they are provided in many glass cockpit aircraft. Some designers believe that simplified synoptics of complex systems may increase the risk of misinterpretation. Part of the controversy relates to certification issues; manufacturers wish to incorporate as few essential systems as possible to avoid grounding airplanes when they fail, and the overhead panels on these aircraft permit full manual operation of all subsystems without the aid of the synoptics. On the other hand, pilots are not normally required to operate in this manner and do not practice it; flight crew workload could increase considerably during the time required to reconfigure the affected systems. The Boeing 757/767 cockpit does not provide subsystem synoptics, although the engine indication and crew alerting system (EICAS) messages provide information on aircraft system status. Subsequent Boeing aircraft (B-747-400, B-777) do incorporate synoptics, but their designers, and FAA in its certification of the -400, did not consider them essential and the aircraft can be dispatched without them.

As noted earlier, the Douglas Aircraft Company has automated most normal and abnormal actions in the MD-11 subsystems. The synoptics in the MD-11 are simplified diagrams of each subsystem. When an abnormal condition is detected, the appropriate automatic system controller takes action autonomously; an alerting message is displayed on the engine and alert display. The appropriate subsystem pushbutton on the systems control panel is also lighted. When actuated, this pushbutton brings up the synoptic, which will show the system diagram with altered icons indicating the fault, what action has been taken, and a list of the consequences for the conduct of the remainder of the flight. Fig. 5.23 shows an example of a level 2 alert (number 1 hydraulic system fluid loss) that has been resolved automatically by inactivation of the two system 1 hydraulic pumps (at the left of the synoptic diagram) after low system 1 hydraulic quantity was detected. The depleted system 1 reservoir is also shown.

Issues Raised by Automated Subsystem Displays

Most manufacturers have tried to make subsystem synoptic displays as simple as possible. Multiple faults, however, still require careful pilot attention to several screens to understand fully the nature of the problems: More information is available, but more "navigating" through the menus and representations is necessary to access it. Herein lies another facet of the controversy over what the pilot

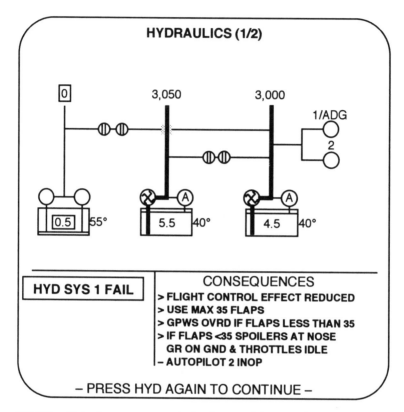

FIG. 5.23.   Hydraulic system synoptic display showing a failure (Douglas MD-11). Reproduced by permission of Douglas Aircraft Corporation.

needs to know. Modern airplanes are designed to require specific actions (usually as few as possible) in response to any fault or combination of them. The required actions are spelled out in checklists, which are designed to be followed precisely. These aircraft are also designed to require no more than checklist adherence for safe flight completion.

There is continuing concern among designers that providing too much detailed information on subsystem configuration may lead some pilots to adopt more innovative approaches to solve complex problems, and thereby negate the care the manufacturer has taken to simplify fault rectification. Such behavior has caused serious incidents in the past, among them the destruction of a engine in flight, and will probably continue to do so in the future despite the best efforts of designers to achieve simplicity and clarity in their designs and procedures.

Pilots argue, however, based on experience, that faults not contemplated by the manufacturer may well occur in line operations. They point, as one instance, to an L-1011 that landed safely at Los Angeles (1977) after its crew was faced with a compound set of faults (an elevator jammed in the full nose-up position) for which no "book" solution existed (McMahon, 1978). They do not wish to be deprived of any information that could assist them in understanding and coping with such problems. The problem for the system designer is to strike the right balance between too little information and too much, recognizing that the pilot's actual needs may not be clear in advance.

Proponents of each approach argue vigorously for their positions regarding display of synoptic information, but because not all information can be presented, the question that must be answered is at what point an appropriate compromise can be found. Better models both of system behavior and of cognitive responses to malfunction information are needed to answer this question. The answers will not be easy to find because designers cannot always predict the contexts in which the information will be required in line operations. In these and other areas, an important issue is the increasing complexity and coupling of automated systems and the potential for surprises (for both the designer and the pilots) due to the opacity of such systems (Perrow, 1984; see also chapter 9).

Practices with respect to the provision of information regarding subsystems have varied, from tightly coupled linking of systems, procedures, and alphanumeric displays in the Boeing 767/757, to the provision of synoptics simply for pilot information in the 747-400 (Fig. 5.24), or synoptics that are the primary means of subsystem feedback in the MD-11 and A-320. The A320 and MD-11 also present a limited number of normal checklists on their ECAM screens; a broader implementation of electronic checklists with automatic sensing of skipped actions is implemented in the Boeing 777 and will likely be seen in other future transport aircraft. Such automation will permit the flight crew to alternate among several checklists when necessary to resolve compound faults. Automated prioritization schemes for such faults are under consideration by NASA and other human factors researchers.

Alerting and Warning Systems

Configuration Displays

Landing gear and other configuration warning systems have been used since it was first discovered by a hapless pilot that retractable-gear aircraft could be landed with the landing gear retracted. Even with these systems, gear-up landings continue to occur occasionally and incidents involving gear-up near-landings occur more commonly, usually due to distractions or interruption of routine cockpit task flow. Early warning systems simply provided an aural alert if throttles were pulled back to idle during the flare maneuver. The use of idle power routinely during descents

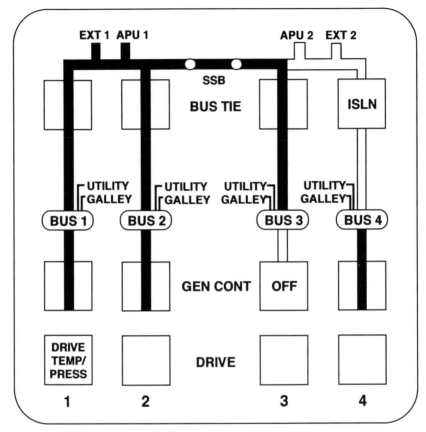

FIG. 5.24. Synoptic display of AC electrical system (Boeing 747-400). Compare with overhead AC system control panel, same airplane, Fig. 5.16. Reproduced by permission of The Boeing Company.

in jet aircraft required that the landing gear warning system be modified to take account of barometric altitude or other factors that could indicate that landing was not contemplated at the time. Aircraft without such modifications provided large numbers of nuisance warnings to pilots and therefore tended to desensitize them to the importance of the warnings.

Configuration warning systems probably represented the first information automation of any consequence. They date from the early 1930s. In later aircraft, additional surveillance was performed by these systems. In all jet aircraft, a configuration warning system operates prior to takeoff (inferred by landing gear on ground and throttles set at high power) if the airplane wing's high lift devices (leading edge slats and trailing edge flaps) are not in appropriate positions for takeoff, or the elevator trim is not positioned within limits determined in flight test to be appropriate for the takeoff maneuver. Before landing (as inferred from throttle and flaps positions), configuration warning systems operate if either gear is up or slats and flaps are in positions other than those permitted for landing.

Nearly all current-generation aircraft have configuration displays that provide aircraft status information in graphic form. The complexity of these displays can be high because of the number of items that are pertinent and the ease with which complex graphics can be generated. Although color can help to direct a pilot's attention to parameters that are abnormal, a good deal of data must still be scanned (Fig. 5.25). Cockpit designers have done an excellent job of eliminating large numbers of discrete "lights, bells, and whistles," within limits imposed by certification regulations, but they have substituted large amounts of discrete data integrated into a smaller number of displays. Operational constraints often require pilots to review, by whatever means, a great deal of important status information prior to takeoff and during approach, periods that are already busy.

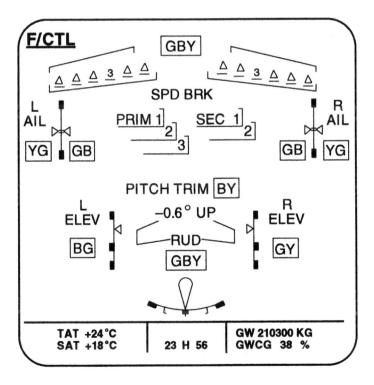

FIG. 5.25.   Flight control configuration synoptic display (Airbus A340). The letters B, G, and Y on this synoptic refer to the hydraulic systems that power the indicated control surfaces. From *A340 Flight Deck and Systems Briefing for Pilots*, Issue 5 (Rep. No. AI/ST-F 472 502/90) (p. 5-24) by Airbus Industrie, 1995, Toulouse, France. Copyright 1995 by Airbus Industrie. Adapted with permission.

Altitude Alerting Systems

In the early 1970s, it was noted by regulatory authorities that the high rates of climb and descent of jet aircraft were causing substantial numbers of altitude deviations, in which aircraft either exceeded the altitude to which they were cleared or failed to reach it. A backup altitude alerting system was mandated for transport aircraft. The altitude alert system consisted of a window on the panel in which the altitude cleared to could be set (Fig. 5.26), and sensors to detect actual altitude. When in a climb or descent, visual and momentary aural signals were actuated approximately 900 ft before reaching the set altitude; the visual signal remained illuminated until 200–300 ft before reaching the new altitude, then was extinguished. If the airplane thereafter strayed from the assigned altitude by more than 200–300 ft, the aural and visual signals again appeared.

Malfunction Alerting Systems

The central multifunction displays in glass cockpit aircraft accommodate alerting and warning information (as shown in an amber boxed legend "HYD SYS 1 FAIL" in Fig. 5.23). Older transports had warning and alerting message lights in so many locations that centrally located master warning (red) and master caution (amber) lights were placed on the glare shield in the pilots' direct field of view. Later, dedicated alerting and warning panels with lighted segments containing alphanumeric legends were incorporated wherever there was room for them. As aircraft became more complex, the number of discrete warning signals became progressively greater, reaching several hundred in the early B-747s (Randle, Larsen, & Williams, 1980). In the analysis following a fatal Trident takeoff accident at Heathrow Airport (London, 1972), investigators cited "a plethora of warnings" that overwhelmed the

FIG. 5.26.   Early altitude alerter control and display panel (Boeing 727).

remaining flight crew after the captain's incapacitation and a serious airplane configuration error occurred almost simultaneously (R. Smith, personal communication, May 1974). Newer aircraft have eliminated nearly all of these discrete warning indicators, although the master warning and master caution lights have remained. Alerting information is now presented on the central cockpit screens, usually in the form of alphanumeric messages in a dedicated location.

Although the number of discrete alerting devices has decreased markedly, the number of discrete alerting messages that may be displayed and may require action is still large. Nonessential warnings and alerts are routinely inhibited during takeoff and final approach. Nonetheless, fault management may still be complex, and newer aircraft are operated by a crew of two instead of the former three persons, so there may be more for each crew member to do during busy periods. It is largely for this reason that Douglas has automated many MD-11 subsystems management tasks.

Other Displays

Aircraft equipped with flight management systems but electromechanical instruments utilize a small, usually monochromatic, display in the flight management system for the presentation of alphanumeric information (see Fig. 5.29 and 5.30). These control-display unit (CDU) screens will also be used in the future for ATC messages received by data-link units in such aircraft.

The traffic alert and collision avoidance system (TCAS) incorporates a planform display of traffic in the vicinity of one's own aircraft. In some installations a dedicated EDU is used. In others, TCAS information may be shown on a color radar screen, and in still others, a new color LCD display combines a presentation of the instantaneous vertical speed indicator (IVSI) with a small planform display of traffic (Fig. 5.27). This instrument replaces the conventional 3¼-inch IVSI. In nearly all future glass cockpit aircraft, it is expected that traffic information will be shown on integrated primary flight and navigation displays.

Wind shear advisories are aural and visual, as are GPWS alerts. In glass-cockpit aircraft, wind shear advisories may be displayed on the primary flight display, as are TCAS alerts; in these aircraft, the permitted maneuvering range during a TCAS resolution advisory is shown on the IVSI tape.

Issues Raised by Automated Alerting and Warning Systems

Configuration Alerting Systems.   Ways of summarizing and representing aircraft configuration and subsystem data that can quickly alert pilots to a potential problem are highly desirable. Indeed, in many newer aircraft, pilots have no alternative means of accessing this information. As an example, preflight exterior

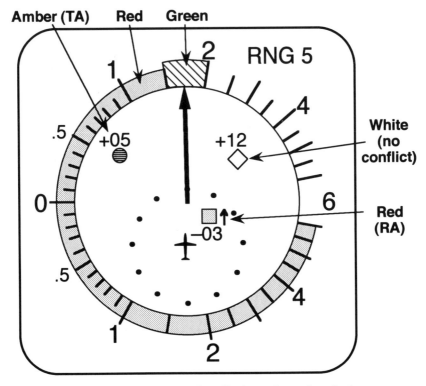

FIG. 5.27.   AlliedSignal liquid crystal display for traffic alerts and vertical speed indications (TA/VSI). This instrument is designed to replace the conventional IVSI in older aircraft. (Copyright 1995, AlliedSignal, Inc., used by permission).

inspection will not show abnormal control surface positions if the hydraulic systems are not powered, because unpowered surfaces drift. There have been cases in which extended wing spoilers on one wing, not indicated on the control surfaces position indicator in the cockpit, were detected before takeoff only by an alert pilot in a following airplane. In at least one case reported to the NASA Aviation Safety Reporting System (NASA ASRS, 1986), an airplane actually took off with two spoilers fully extended and locked on one wing. Fortunately, the airplane was light in weight and the pilot managed to maintain control while returning for an emergency landing.

Alerting messages and aural signals are still used in newer aircraft for critical items prior to takeoff and approaching landing, as noted earlier. These takeoff and landing configuration warning systems have prevented many accidents, but their occasional failure, and their ability to generate spurious or nuisance warnings, raise a problem of a more general nature: Devices that are extremely reliable will come, over time, to be relied on by pilots. In the rare cases when they fail, or are disabled, pilots may not be sufficiently alert to detect the condition for which the device was originally provided.

Altitude Alerting Systems.   Reports to ASRS indicated that many pilots, after they became accustomed to the automatic altitude alerter, tended to relax their previously required altitude awareness and to rely on the alerter to warn them that they were approaching a new assigned altitude. If the ordinarily reliable system malfunctioned, or if the pilots were distracted by other tasks and failed to attend to its signals, they busted the new altitude. These reports were dramatic evidence that devices installed as a secondary or backup alerting system had become instead the primary means by which pilots derived information.

Pilots also complained about aural alerts that did not represent an anomalous condition (the signal approaching altitude). They considered these alerts as "nuisance warnings," like the frequent inappropriate configuration warnings referred to earlier. In response to these complaints, FAA modified the regulations to require only a visual signal before reaching the assigned altitude. From those airlines that thereafter modified the systems to remove the aural alert approaching altitude, it was noted that reports began to be received indicating that some pilots, accustomed to the unmodified systems, began to report altitude "busts" because of the absence of the aural alert approaching altitude! These reports further reinforced the hypothesis that at least some pilots had come to rely on the alerter as a substitute for altitude awareness.

A similar phenomenon has been observed with respect to configuration warning devices. Although they were intended as backup systems, at least some pilots came to depend on them. This was demonstrated in two mishaps in which takeoff configuration warning devices malfunctioned or had been disabled; in both cases, flight crews failed to detect that flaps and slats had not been deployed prior to takeoff. Both aircraft crashed immediately after leaving the ground, with substantial loss of life (Detroit, 1987; Dallas-Fort Worth, 1988).

Hazard and Malfunction Alerting System.   Devices that produce too many "false alarms" will be mistrusted by flight crews. In the extreme case, they will simply be ignored after pilots have become accustomed to them. The earliest models of the ground proximity warning system (GPWS) were prone to nuisance warnings; several accidents have occurred because crewmembers ignored, disabled, or were slow to respond to warnings that were appropriate (e.g., Kaysville, Utah, 1977; Pensacola, Florida, 1978). Later GPWS models incorporated more complex algorithms and the number of nuisance warnings dropped dramatically, although the false alarm problem is still very real at certain locations.

We are now seeing similar problems with large-scale implementation of TCAS-II. This collision avoidance system, mandated by Congress after many years of development by FAA, has unquestionably prevented a number of collisions, but it is an extremely complex device whose control algorithm was not flexible enough to cope with new ATC procedures to speed the flow of traffic in high-density terminal areas, nor with the large number of aircraft in certain airport areas. As a result, pilots have been burdened with large numbers of nuisance warnings in the vicinity of certain airports such as Orange County, California, and during departures from certain terminals, among them Dallas-Fort Worth, Texas. The result has been erosion of confidence in the system, and concern that in certain cases, TCAS may actually worsen the situation (Mellone, 1993).

TCAS was mandated by the Congress, as were GPWS and WSAS. They were designed as self-sufficient, add-on systems; in older aircraft, they are not integrated with other cockpit systems, nor with each other. We are already seeing the emergence of new traffic surveillance requirements for TCAS, particularly in over-water navigation where radar surveillance is not available. The TCAS displays were not designed for these purposes and they may or may not provide information in a form that assists flight crews to perform the additional functions implicit in the new requirements. This is likely to become a more serious problem if pilots are required to take over more functions now carried out by air traffic controllers (see discussion of *free flight*, chapter 8).

The use of TCAS in line operations has caused considerable concern among air traffic controllers, who are faced with sudden pilot deviations from cleared altitudes without advance warning under circumstances they cannot control. Their ability to maintain effective command of air traffic rests on the assumption that pilots will do what they are told to do, and the interjection of this source of uncontrollable variance has caused them great discomfort.

There are fundamental tensions between systems capabilities and limitations and human characteristics. False warnings always diminish human trust of warning systems, yet the danger of a missed potentially catastrophic situation requires that conservative warning limits be embodied in such systems. Such situations can arise suddenly and can require immediate action, yet controllers, without an understanding of the immediate problem, cannot function effectively without knowledge of pilot intent (nor, for that matter, can pilots function fully effectively without knowledge of controller intent). Communication of intent in advance of action by both humans and machines is an important issue in any real-time dynamic system if all players, or agents, are to remain informed of system status and progress.

MANAGEMENT AUTOMATION

During the 1960s, area navigation (RNAV) systems independent of surface radio aids began to be introduced into aviation. The earliest such system made use of Doppler radar to determine relative movement over the earth's surface. The system provided considerable assistance during the long overwater portions of intercontinental flights and did not require of its operators the highly developed skills required for celestial navigation. During this period also, inertial navigation systems (INS), first developed for long-range missiles, began to be adapted to air navigation. As with Doppler, all required equipment was carried onboard the aircraft.

INS systems used highly accurate gyroscopes and accelerometers to determine the movement of the system (and the airplane that carried it) after being given a very accurate statement of its initial position prior to flight. INS, like Doppler, was totally reliant on this initialization. If an inaccurate initial position was input, it could not be corrected after the aircraft took off. Several transoceanic flights had to be aborted after it was determined that the initial position entry was incorrect.

Both of these early area navigation (RNAV) systems permitted pilots to enter a series of latitude and longitude waypoint specifications, after which the systems would provide navigation data to the autoflight systems. To this extent, these systems represented the beginnings of flight, or at least navigation, *management automation*. The systems provided pilots with much greater flexibility, but at the expense of greater complexity. They also enabled new types of human error associated with manual entry of waypoint data into navigation computers, a cumbersome and error-intolerant process.

The most revolutionary changes brought about by the introduction of digital computers into aircraft automation have been in the area of flight management. Flight management systems (FMS) in the contemporary sense have been in service for little more than a decade, but they have transformed the pilot's tasks during that time. This section contains a brief description of the modern flight management system, the functions it performs, and its interfaces with the flight crew.

Flight Management Systems

The introduction of the MD-80 and the Boeing 767/757 marked a fundamental shift in aircraft automation, as noted earlier. In these machines, the first systematic attempts were made to integrate a variety of automated devices into a seamless automation capability designed for routine use in line operations. Although pilots had been able to program overwater flight paths using inertial navigation systems in older aircraft, the new flight management systems were designed to be the primary means of navigation under all conditions.

Figure 5.28 again shows the control loops diagrammed earlier, but with the addition of an outer loop that represents management functions. Once again, automation has relieved the pilot of certain tasks, but has added other tasks involving additional cognitive workload. These tasks are the product of the complexity and self-sufficiency of the new flight management systems. They impose additional knowledge requirements, even while they relieve the pilot of tactical

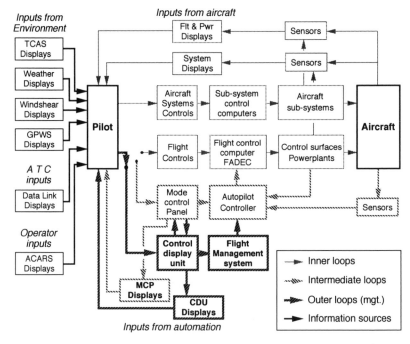

FIG. 5.28. Flight and aircraft systems outer control loops involved in strategic aircraft management; external information sources are also shown.

management chores. Most important, there is more information to be gathered and processed to ascertain the state of the aircraft and its more complex automation.

Flight Management System Functions

Contemporary flight management systems are complex computational devices linked to and communicating with a great many other aircraft systems as well as with the pilots (Honeywell, 1990). Figure 5.29 shows this diagrammatically for the MD-11, and the following discussion describes this system, although others have similar capabilities.

FMS software, resident in a flight management computer (FMC), includes an operational program (containing, in this case, over 1,400 software modules), a navigation database, and a performance database for the aircraft in which it is installed.

The FMC navigation database includes most of the data the pilot would normally access by referring to navigation charts. This information can be displayed alphanumerically on the CDU or symbolically on the navigation display. The geographic area covered includes all areas where the airplane is normally flown. The database, tailored to specific airline customers, contains 32,500 navigation points and airway route structure data. The stored data include the location of VHF navigation aids, airports, runways, geographical reference points, and other airline-selected information such as standard instrument departures, standard arrival routes, approaches and company routes. Up to 40 additional waypoints can be entered into the database by the pilots.

The FMS software executes these functions:

Navigation	Determination of position, velocity, and wind; management of navigation data sources.
Performance	Trajectory determination, definition of guidance and control targets, flight path predictions. Time and fuel at destination.
Guidance	Error determination, steering, and control command generation.
Electronic instrument system	Computation of map and situation data for display.
Control-display unit	Processing of keystrokes, flight plan construction, presentation of performance and flight plan data.
Input/output	Processing of received and transmitted data.
Built-in test	System monitoring, self-testing, and record keeping.

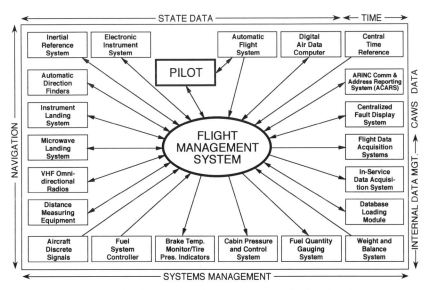

FIG. 5.29.   Interaction of flight management system computer with other aircraft systems and avionics (Honeywell FMS for MD-11). Reproduced by permission of Honeywell, Inc.

Operating system   Executive control of the operational program, memory management, and stored routines.

The FMC performance database reduces the need for the pilot to refer to performance manuals during flight; it provides speed targets and altitude guidance with which the flight control computer develops pitch and thrust commands. The performance database is also used by the FMC to provide detailed predictions along the entire aircraft trajectory. The data stored in the database include accurate airplane drag and engine model data, maximum altitudes, and maximum and minimum speeds. Functions performed by the FMS include navigation using data from inertial reference units aboard the airplane, updated by a combination of surface and/or satellite navigation aids when available. It provides lateral guidance based on a stored or manually entered flight plan, and vertical guidance and navigation during climb and descent based on airplane gross weight, cost index, predicted winds at cruise altitudes, and specific ATC constraints.

Flight Management System Controls

Interaction with all flight management systems is through a control and display unit (CDU) that combines a monochromatic or color CRT or LCD screen with a keyboard. An example of a CDU is shown in Fig. 5.30. The unit contains a CRT display screen, line select keys on each side of the CRT, 15 mode select keys, a numeric keypad, and an alphabetic keypad. The mode select keys provide access to FMS function pages and data; the alphanumeric keypads permit entry of data into the computer.

Newer FMSs provide modes and functions to reduce pilot workload. Among them are the engine failure (ENG OUT) function, which provides automatic or manual access to the flight plan (F-PLN) or performance (PERF) pages to assist in evaluating and handling an engine failure condition. Entry of data is accomplished by using the keypads. The entered data are shown on a scratch-pad line (discussed later); when a line select key is pushed, the data are transferred to the indicated line if they are in a format acceptable to the computer.

Flight Management System Displays

The CDU display consists of a large number of pages, each containing up to 14 lines of alphanumeric information as shown in Fig. 5.31.

The CDU screen shown here appears when the initialize (INIT) mode select key is actuated. The title line shows that this is the first of three flight plan screens; others may be accessed with the PAGE key. The scratch-pad line is at the bottom

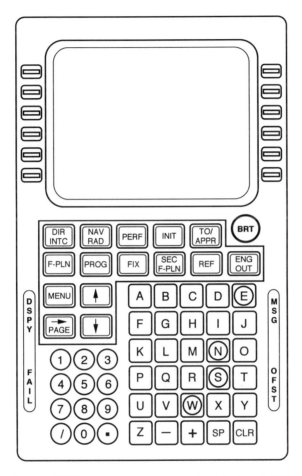

FIG. 5.30.   Honeywell flight management system control and display unit (CDU). Reproduced by permission of Douglas Aircraft Corporation.

of the display. Vertical arrows indicate that the arrow keys may be used to increment values. The small font displays are predicted, default, or FMC-calculated values, and labels. The 50 CDU pages are arranged in a tree architecture. A portion of this logical, but complex, architecture is shown in Fig. 5.32.

These diagrams show the tree structure for two modes of this FMS. There are 12 such structures, but in a study of another FMS of the same generation, it was found that the number of sequences utilized was several times the number planned for by the manufacturer (Corker & Reinhardt, 1990). These data structures, as well as the displays, vary greatly among aircraft types and avionics manufacturers. This large number of potential trees involves a considerable attentional demand on the

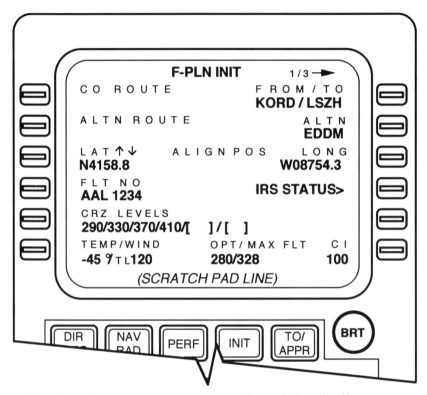

FIG. 5.31.   MD-11 control and display unit screen (Honeywell). Reproduced by permission of the Douglas Aircraft Corporation.

pilot even if he or she is fully proficient in the use of the FMS. Because flight plan changes are most commonly required during departure and arrival, reprogramming the FMS can divert a significant amount of attention that may be needed for outside scan and for cross-cockpit monitoring.

Flight Management System Operation

The two CDUs are redundant. In the MD-11, both pilots may interact with the FMS simultaneously; however, the system will accept flight plan modifications only one at a time. There are two FMCs, each of which may accept data from either CDU; one FMC is designated as master, and both must confirm data entry before new data will be accepted. The two computers communicate with each other through a private data bus.

Effects of Management Automation

Programming of INS and Doppler units is an exacting task, requiring precise and accurate entry of many alphanumeric characters. Slips (Norman, 1981) were not uncommon and, once made, sometimes went undetected. Air carriers instituted a variety of procedural requirements to detect such errors both during data entry and thereafter during overwater flight, utilizing various crewmembers to read, enter, and confirm data and special progress charts to be filled out en route, in the hope of trapping undetected errors before they affected traffic separation over water where no other means of position evaluation was available.

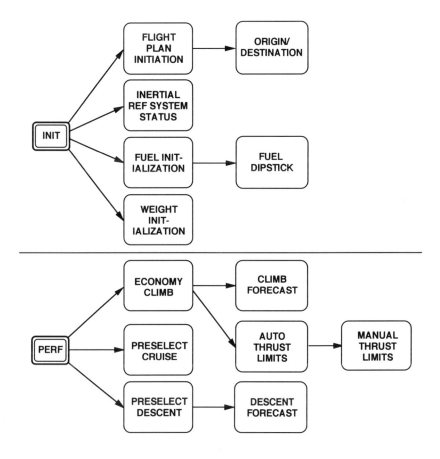

FIG. 5.32.   Flight management system mode and display screens (MD-11).

A few serious errors still crept through the procedural barriers, however, and some led to near collisions many hours after the initial programming was accomplished, as in an incident between a Delta Airlines L-1011 and a Continental 747 over the North Atlantic Ocean (1987) (Preble, 1987). Also, autopilots had to be properly coupled to the navigation systems; if this was not done, the aircraft could fly a long distance in heading mode[5] rather than in the intended navigation mode. Based on data made available by Russia in recent years, it is now thought that this may have been the error that led to the destruction by Soviet fighters of a Korean Air Lines B-747 after it flew many miles over Soviet territory enroute to Seoul from Anchorage, Alaska (Sakhalin Island, 1983). It was also the cause of a more recent near-collision between an El Al 747 and a British Airways 747 south of Iceland (Atlantic Ocean, 1990).

Fundamental issues posed by management automation are discussed more fully in chapter 10, but it should be noted that even the early attempts at management automation sometimes distanced pilots from the tactical details of their operations. This, of course, depended on whether the human operators maintained a high degree of alertness concerning the progress of their missions. The safety record indicates clearly that most did, regardless of whether automation was in use. What the increasingly capable automation provided, however, was the *opportunity* to become somewhat less involved, an opportunity that could easily permit tired, fatigued, or preoccupied pilots to lose track of their situation if they were not on guard against it.

Issues Raised by Flight Management Automation

In all FMSs, the complexity of the mode and display architecture poses substantial operational issues. Much has been done to simplify routine data entry, but recovery from errors in programming (e.g., an acceptable but incorrect entry) can be difficult. Entry of certain types of data remains cumbersome and diverts attention from other flying tasks, as discussed later. If an unacceptable entry is attempted, it is rejected, but without explanation of the error that led to the rejection, as one instance.

Interaction with the FMS is through one of two or three identical CDUs mounted on the center console. Even with color to assist, operation of the FMS requires close visual attention to the screen, and precision in entering data on the keypads. Alphanumeric data entry is known to be subject to human errors: Numbers may be recalled incorrectly from short-term memory, they may be input incorrectly,

---

[5]In the autopilot Heading mode, the airplane follows a heading inserted into the mode control panel (see Fig. 5.11), whereas in the Navigation mode, the trajectory is a great circle route calculated by the navigation management system. The difference may not be obvious for some period of time.

or they may be misread when the entries are verified in the scratch pad before entry into the computer. Some data must be entered in a specific sequence, which imposes an additional memory load on the operator; screen prompts are not always clear even when they are available.

Avionics and aircraft manufacturers have made many efforts to make interaction with the FMS more error resistant. Standard or frequently used routes are stored in the navigation database and may be recalled by number. SIDs and STARs[6] are also in the database; if a change is required by ATC, only the name of the procedure need be entered. Changing the arrival runway automatically changes the route of flight. Appropriate navigation radio frequencies are autotuned as required. Perhaps most important, newer FMSs interact directly with navigation displays; pilots are shown the effect of a change of flight plan in graphic form. They can thus verify that an alternative flight plan is reasonable (although not necessarily what was requested by ATC) before putting it into effect.

In most newer aircraft, entry of tactical flight plan modifications (speed, altitude, heading, vertical speed) can be done through the mode control panel (MCP; see Fig. 5.11) rather than the CDU. These entries either may supersede FMS data temporarily, or may be entered into the FMS directly from the panel.

It is likely that these improvements may resolve some problems with tactical data entry, though pilots must keep track of still more potential mode interactions. Mode control panels now contain numerous multifunction control knobs (turn to set; pull to activate, push to transfer data to FMC; see Fig. 5.33), which has posed problems of a different sort when pilots have inadvertently activated a mode other than that intended.

Vertical navigation profiles generated by the FMS take account of standard ATC altitude constraints as well as airplane performance constraints, although the air traffic control system is not, at this time, able to take full advantage of the capabilities of management automation, which calculates profiles based on actual rather than average aircraft weight. These optimal descent profiles therefore differ enough to cause sequencing problems for ATC.

In some newer aircraft, manual tuning of navigation radios is possible only by interacting with the CDU. Many pilots have complained that alphanumeric entry of frequency data is more time-consuming and requires more prolonged attention inside the cockpit than setting the rotary selector knobs in older aircraft.

Flight management systems truly permit pilots to manage, rather than control, their aircraft. The systems are extremely effective and have enabled many improve-

---

[6]Standard Instrument Departure (SID); Standard Arrival Route (STAR). These routes are promulgated by the air traffic control system.

Mode Control Panel Knob Operation

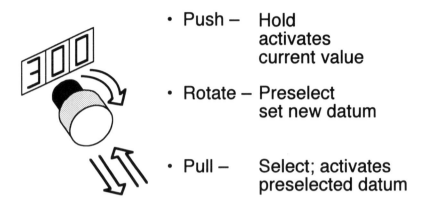

- Push – Hold activates current value

- Rotate – Preselect set new datum

- Pull – Select; activates preselected datum

FIG. 5.33.   Mode control panel (MCP) operation (Fokker F28-100).

ments in operational efficiency and economy. The greatest improvement in FMS display capability has been its integration with aircraft navigation displays, improving visualization and freeing the systems from some of the constraints imposed by small alphanumeric CRTs. The addition of colors, matched with those used on the navigation displays, to the CDU display may help (early displays were invariably monochromatic), although the resolution of the color displays is somewhat less and the usefulness of color in this application has not received much systematic study. The design of pages, however, still represents a compromise between the amount of alphanumeric data per page and the number of pages necessary to enable a particular function. Pilots must look at a very large amount of data through a relatively small keyhole (Woods, in press; see chapter 13).

Initial FMS designs were based on the notion of FMS use at high altitude in cruise flight. The success of the concept resulted in extension of the cruise concept into use throughout flight, without redesign of the interfaces, a common problem with successful automation (D. M. Fadden, personal communication, November 1995). The attention required for reprogramming has led to undesirable ad hoc procedures in the cockpit; appreciable numbers of pilots prefer not to interact with the systems below 10,000 ft during descent, in order not to compromise aircraft control and scan for other traffic (Curry, 1985; Wiener, 1989).

This approach permits human resources to be devoted to more important tasks, but at the cost of losing some of the benefits of the FMS during flight in the terminal area (such as its knowledge of altitude restrictions). As noted by Fadden, this is a problem of human–system interface design, rather than a problem in the functionality of the systems themselves. Research and development efforts are underway to improve these interfaces and specifically to make them less totally dependent on cumbersome alphanumeric data entry, but considerable attention to the CDU displays is also warranted. There remain important questions about the integration of these systems into the overall cockpit and automation design, and it is these integration issues that most need to be resolved.

COMMENT

Aircraft automation's major benefits, among them improved fuel economy and operating efficiencies, have been accompanied by certain costs, including an increased cognitive burden on pilots, new information requirements that have required additional training, and more complex, tightly coupled, less observable systems. To some extent, both benefits and costs are inherent in these highly automated systems. Other costs have accrued because today's automated systems are not optimally designed to work cooperatively with their human operators. Finally, some costs have accrued because the automation was designed to operate in an ATC system that is constantly evolving, forcing human operators to adapt and tailor their uses of and responses to the automation and the changed requirements.

Plus ça change, plus la même chose—the system changes (as usual); the pilots and controllers adapt (as usual). This is not new, but as elements of the system become more complex and less transparent, the task of adapting becomes more complex and more difficult. The further changes likely to be seen in future aircraft, and their likely effects on operators, are the subject of the next chapter.

CHAPTER 6

AIRCRAFT AUTOMATION
IN THE FUTURE

INTRODUCTION

This chapter considers aircraft automation proposed or already developed for use in future aircraft. Airframe and avionics manufacturers and operators alike are constantly on the watch for emerging technology that can expand their scope of operations. Satellite navigation, as an instance, offers the promise of freeing aircraft from constraints imposed by ground navigation facilities, especially if those same satellites can enable landing at any suitable airport. On the other hand, new technology is expensive, and air carriers have only recently emerged from a period in which they have suffered the greatest losses in the history of commercial aviation.

Given this economic climate, new aircraft will have to be more efficient, more reliable, and less expensive to maintain than those presently on the market. It will not be easy for airframe and powerplant manufacturers to meet these goals. If new technology can improve efficiency or productivity (as is the case with satellite navigation systems), it will be embraced. If not, it is not likely to find its way onto future aircraft, at least in the near term. Let us look for a moment at some of the enhancements that have been proposed for near-term (1996–2015) implementation (Fig. 6.1).

Control automation is already highly sophisticated; its future applications will probably extend its capabilities rather than making new functions available. An exception to this generalization is the possible requirement for automatic flight during approaches to closely spaced parallel runways. Navigation functions will be revolutionized during the next decade by the use of global navigation satellite systems (GNSS) for guidance and automatic dependent surveillance (ADS) for flight following. *Information automation* is an area in which many new functions have been proposed; some are now in test or demonstration. In the area of *management automation*, efforts will be directed toward the improvement of the human–machine interfaces and (hopefully) toward new functions or enhancement of existing functions to improve the error tolerance of the aviation system.

ENHANCEMENTS PROPOSED FOR
FUTURE TRANSPORT AIRCRAFT

• Control Automation

>Low-visibility taxiing assistance or guidance
>High-precision in-trail operations in terminal areas
>Automated collision avoidance maneuvers
>Automated wind shear avoidance maneuvers

• Information Automation

>Electronic library system–"paperless cockpit"
>Satellite navigation and flight following
>Digital data link–high bandwidth communications
>Satellite communications world-wide
>Automatic dependent surveillance
>"Big picture" integrated cockpit displays
>Enhanced head-up displays
>Enhanced or synthetic vision systems

• Management Automation

>Easier, more intuitive FMS interfaces
> – Cursor modification of flight plans
> – Improved error-checking
>Direct FMC-ATC computer communication
>Improved error tolerance
> – Enhanced error monitoring and trapping
>Improved electronic checklists
>Improved mental activities models for design

FIG. 6.1.   Proposals for future aircraft automation.

In *Trends in Advanced Avionics*, Curran (1992) provided a review of avionics evolution throughout the history of aviation and discussed present and future trends in avionics. Curran stated (p. 160) that "Past avionics advances have permitted the elimination of the radio operator, flight navigator, and the flight engineer positions in the cockpit. Future improvements should result in better avionics functional capability, integrity, and availability for the remaining crewmembers." He concluded (p. 172):

Avionics designers must find ways of keeping flight crews more involved as the need for automation increases. Avionics designers must become more aware that there is a kind of automation that improves situation awareness and there is a kind that diminishes this awareness. The challenges for avionics designers are many.... These improvements must be accomplished without creating unacceptable workload and information overload.

Curran was quite correct that automation can either enhance or diminish situation awareness, and that there is a real need for designers to understand the difference between them. Unfortunately, neither Curran nor most other authors have stated how this understanding comes about, or even what the critical differences are. This is a question of some gravity, for without such understanding we cannot improve the design of future automation or the performance of the human–machine systems in which it will be embedded. It is addressed at several points in this book.

AIRCRAFT AUTOMATION TODAY

If we wish to examine future aircraft automation, the Boeing 777 and Airbus A330 are convenient benchmarks. They demonstrate the state of the art in transport aircraft and presage the future of aircraft automation. The A330's cockpit, however, is as nearly identical as possible to that of the A320 and the four-engine A340 to minimize problems in transitioning among these aircraft. The 777 is a new aircraft type and its cockpit (Fig. 6.2) is not a derivative, although it has much in common with the slightly older 747-400. Because I have discussed the Airbus aircraft in chapter 5, I spend some time here in an examination of the 777, using various Boeing materials as primary sources.

The Boeing 777

The Boeing 777 is the world's largest twin-engine transport airplane (Fig. 6.3). It was designed for extremely long-haul routes ("B" market version), although a shorter range "A" market variant was the first to enter production. The A330 is slightly smaller than the 777; the Airbus consortium's A340 is its longer-legged companion. The B777 will cover and exceed the range spectrum of the 747-400, although with a smaller capacity.

The overall philosophy espoused by the 777 flight deck design team was "crew-centered design and automation" (Kelly, Graeber, & Fadden, 1992, p. 1). This philosophy had been under development for some time before the program was launched (Braune & Fadden, 1987). It is based on the principles set forth in Fig. 6.4.

Kelly et al. pointed out the similarity of these principles to those presented in chapter 3 of this book. They are a distillation of experience—what has worked well and what has not—in earlier aircraft. They also pointed out, however, the difficul-

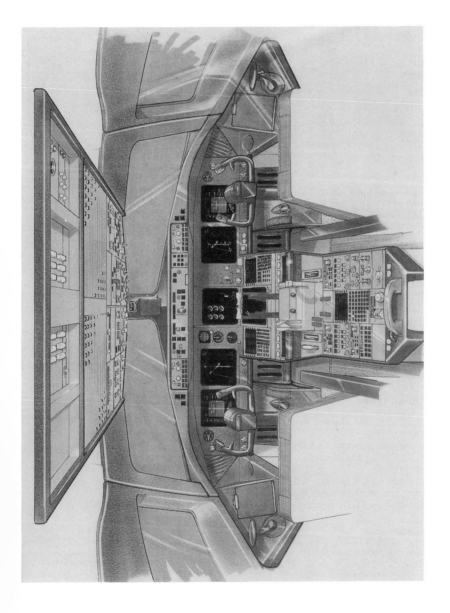

FIG. 6.2. Boeing 777 cockpit. Courtesy of Boeing Commercial Airplane Group.

121

A340, A330 & B777-200 SPECIFICATIONS AND PERFORMANCE

	A340-300	A330-300	B777-200 A-Market	B777-200 B-Market
Size				
Wingspan	198 ft	198 ft	200 ft	200 ft
Length	209 ft	209 ft	209 ft	209 ft
Tail height	55 ft	55 ft	61 ft	61 ft
Cabin width	17'4"	17'4"	19'3"	19'3"
Max. TO weight	558,900 lb	458,600 lb	515,000 lb (1)	632,500 lb (2)
Performance & capacity				
Range	6,750 n.m.	4,550 n.m.	4,240 n.m.	7,380 n.m.
Maximum speed	M 0.86	M 0.86	M 0.87	M 0.87
Fuel capacity	35,660 gal	24,700 gal	31,000 gal	44,700 gal
Passenger capacity	295	335	305-440 (3)	305-440 (3)
Cargo volume	5,751 cu ft	5,751 cu ft	5,056 cu ft	5,056 cu ft

Notes:
(1): A 535,000 lb variant with a range of 4,820 n.m. carrying 305 passengers will also be offered.
(2): Data shown are for the largest of three variants planned.
(3): Three-class seating, 305-328; two-class seating, 375-400; single class economy seating, 418-440.

FIG. 6.3. Comparative specifications of advanced transport aircraft. Data from Airbus Industrie (1990), *Aviation Week and Space Technology* (1994), and Boeing (1995).

Pilot's Role and Responsibility

- The pilot is the final authority for the operation of the airplane.
- Both crewmembers are ultimately responsible for the safe conduct of the flight.
- Decision making on the flight deck is based on a goal hierarchy.

Pilot's Limitations

- Expected pilot performance must recognize fundamental human performance limitations.
- Individual differences in pilot performance capabilities must be accommodated.
- Flight deck design must apply error tolerance and avoidance techniques to enhance safety.
- Flight decks should be designed for crew operations and training based on past practices and intuitive operations.
- Workload should be balanced appropriately to avoid overload and underload.

Pilot's needs

- When used, automation should aid the pilot.
- Flight deck automation should be managed to support the pilots' goal hierarchy.
- Comfortable working environment.

FIG. 6.4. Boeing 777 flight deck design philosophy. From *Applying Crew-Centered Concepts to Flight Deck Technology: The Boeing 777*, by B. D. Kelley, R. C. Graeber, and D. M. Fadden, 1992, paper presented at the Flight Safety Foundation 45th International Air Safety Seminar, Long Beach, CA. Reprinted with permission of the Boeing Company.

ties inherent in translating these principles into the specifics of a particular flight deck design, in part because of their lack of specificity and because economic and market issues heavily impact the operational features that will actually appear on a new flight deck. "Recently, for example, head-up displays, electronic library systems, and some improvements to flight management functions have been difficult to justify because they did not appear to provide new capabilities which would result in return on investment" (p. 2).

The 777 is Boeing's first commercial fly-by-wire airplane; it entered service in June 1995. Large control columns were retained; the two columns are cross-linked and are back-driven by the autopilots to provide tactile feedback to pilots of control inputs either by the other pilot or the automation. Similarly, the thrust levers are back-driven by the autothrust management system. Control laws provide speed stability; manual trimming[1] is required when speed or pitch is changed. This approach also provides more feedback to pilots, although at the expense of somewhat greater workload during manual flight (actually assisted: all flight control is through the electronic systems; see chapter 10).

Perhaps the most obvious innovation in the 777 cockpit is a cursor control used to respond to electronic checklist items, to navigate through menus, and to interact with data link functions. An electronic checklist function is provided. The system senses many checklist items and indicates their completion during checklist execution; other functions are marked through the cursor control when completed. The checklist system also keeps track of checklist items not completed and indicates these on command. Both normal and abnormal–emergency checklists are incorporated in the system, to minimize the number of memory items required to be performed by the pilots.

Other significant innovations have been provided for but cannot be implemented until industry standards are developed. There is a data link interface, for instance, but its final form will depend on the standards set by the FAA in the future for its Automated Telecommunications Network. Similarly, the airplane is equipped for satellite navigation using the U.S. global positioning system (GPS), but primary reliance on GPS depends on development and implementation of a navigation satellite integrity monitoring and alerting system, as well as the installation of differential GPS stations at or covering airports to be served by GPS precision approaches.

The flight management system (FMS) on the 777 is not new, although certain aspects of its operation have been simplified to ease programming (and particularly

---

[1]Control surfaces are adjusted to a neutral, or zero-force, position by trimming. The small control tabs or other devices that accomplish this are adjusted by the pilots separately from their primary surface position controls.

reprogramming) workload, a design effort that has been in progress for several years. The flight management computer (FMC) automatically detects certain anomalies such as an engine failure and recalculates aircraft capabilities. It also constructs a flight path back to a departure airport if an engine fails during the initial climbout. When instructed by a single keystroke, the FMC builds a transition from a selected runway approach course to another at the same airport and retunes the navigation radios automatically to those appropriate for the new runway, relieving pilots of significant distraction when ATC requires a runway change.

The 777 control-display unit is the first airline unit to use a color screen in its FMS; colors correspond to those used to highlight specific data on the navigation display, which is generated from the FMC when routes are programmed. This is an excellent use of color and another effective step in the integration of cockpit interfaces.

Both the 777 design philosophy and its implementation are more conservative than in some other new aircraft. Although Boeing has always been a conservative company, this may in part be due to an unprecedented effort by the firm to involve customers (as well as human factors experts) in the design process from the outset. United Air Lines, British Airways, and All-Nippon Airways had operations and maintenance engineering staff on the Boeing premises throughout the design and development of the airplane. A full-time human factors group was a part of the flight deck design team, and Boeing also consulted human factors specialists from government and universities at intervals during the design phase.

Perhaps most important, and unique in civil transport development programs, engineering simulators were available from early in the process. The first simulator, although not then complete, became available at the beginning of 1991; it was fully functional before the flight-deck functional definition was complete and was utilized heavily for familiarization and testing by consultants as well as by the design team. A second simulator was operational before the end of 1993. These devices made it possible to evaluate not only individual devices and functions, but how they were integrated, before the first cockpit was actually built.

Beyond the 777

What lies beyond this point in transport aircraft automation? As noted earlier, many features thought desirable for tomorrow's flight decks have not been implemented in the 777 because of economic factors. Nevertheless, several innovations are under active development at this time, either by airframe or avionics manufacturers or in a few cases by air carriers. Some nearly made their way into the 777 design, such as the Electronic Library System. Others were prepared for in that design, to save the expense of later retrofit: data link modules are an example. Still others are being tested in aircraft now flying the line; satellite navigation, communication, and automatic dependent surveillance fall in this category.

Other innovations now under development include synthetic vision systems designed to provide pilots with an adequate view of the runway and airport environment during conditions of extremely poor visibility. Finally, there is a set of innovations under consideration or early development whose future use is uncertain. Among them are very-large-screen integrated displays incorporating synthetic or enhanced views of the aircraft surround and also information concerning aircraft state and status. These devices are sometimes referred to as *big picture displays*; originally considered for military aircraft, they are also under serious consideration for future civil aircraft in which outside visibility will be limited, notably a future supersonic (Mach 2+) civil transport without a forward view from the cockpit (discussed later).

FUTURE AIRCRAFT AUTOMATION

Much of this chapter is devoted to technology trends, but the reader must not lose sight of the real issue: the relationship of humans and machines in a complex human–machine system. Each new technology element discussed here, if introduced, will shape human operator behavior. Will the technology *and* the humans who operate it significantly improve the safety, reliability, or economy of the overall operation? New devices must have that potential, or they would not be introduced, but it is sometimes a far cry from what should happen to what will happen. This tension must be at the forefront of our minds as we consider new technology for the future system.

I continue to categorize technology as control, information, and management automation, although the separation among these categories becomes blurred by the increasing integration of various functional elements. It is a comparatively short step, as an instance, from the provision of a wind shear advisory system (information automation) to the provision of an autoflight module that responds autonomously to such an alert with a predetermined avoidance maneuver (control automation).

CONTROL AUTOMATION IN THE FUTURE

Because control automation is already so advanced in the newest aircraft, one would expect less further innovation in this area of aircraft automation. Rather, I would anticipate that near-term efforts will be directed toward making existing functionality yet more self-sufficient and autonomous, a trend that could further bound pilot authority with the intention of avoiding execution errors under difficult circumstances. Such a trend, of course, would also increase automation complexity and would probably increase the opportunity for surprises.

Minimizing Separation Requirements in Terminal Areas

The need for increased system capacity and throughput has already stimulated the FAA, with NASA, to begin an intensive examination of how technology may be used to increase airport acceptance rates by enabling parallel independent approaches to converging or closely spaced parallel runways (Fig. 6.5). At present, independent approaches to parallel runways at the same airport are not normally permitted under instrument meteorological conditions unless the runway centerlines are 4,300 ft apart. Parallel runways at many major airports are spaced more closely than this because of land restrictions; at San Francisco, as an example, runways 28 left and 28 right, the major landing runways, are only 750 ft apart. The overall acceptance rate for this airport is roughly halved when instrument conditions exist, as they often do because of fog or low cloud. Although it is not likely that independent operations will be permitted to runways this close together, FAA and NASA have conducted simulation and flight experiments to determine whether the 4,300-ft limitation can be reduced by the use of surveillance radar with a 1-sec rather than the present 3-sec scan rate.

These studies indicated that with improved ATC and radar surveillance, pilots can consistently conduct approaches to two or even three parallel runways whose centerlines are separated by less than 3,500 ft, although tight control is required and the problem of blunders or slow turns to the final approach course becomes more serious as separation decreases. A difficult aspect of such operations is the "belly-to-belly" cockpit visual restriction when aircraft bank for the turn to the final approach courses from opposite sides of the extended runway centerlines. The FAA's Aviation Capacity Enhancement (ACE) plan discussed many ongoing and planned efforts to increase airspace and airport capacity (FAA, 1994).

Issues Raised by Reduced Traffic Separation Concepts

If these types of operations are approved for general use at equipped airports, they will be combined with minimum safe longitudinal spacing to make maximum use of the capacity of each runway, possibly using TCAS displays or radar for station-keeping. The dual tasks of very precise station-keeping and lateral control will be demanding, and control automation may be introduced and even required to obtain maximum flight path precision under these circumstances. I would also expect that automated alarm systems will be developed to augment controller surveillance of such operations.

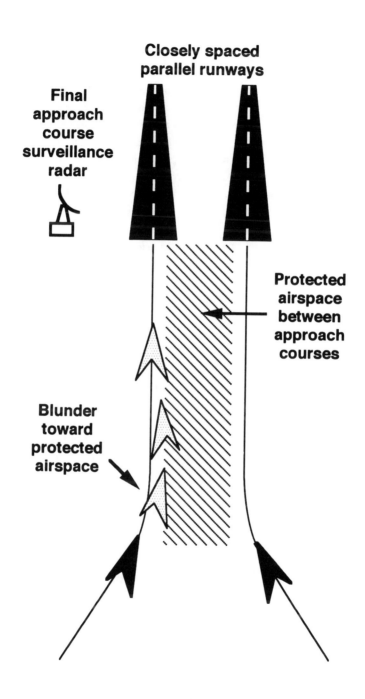

FIG. 6.5. Independent parallel approaches to closely spaced runways.

Whether manually or automatically flown, tightly spaced final approaches under IMC will impose considerable cognitive workload both on pilots and on controllers, especially when unforeseen circumstances force departures from nominal flow. If a leading aircraft is slow to clear the runway, following aircraft will have to execute missed approaches, possibly toward aircraft taking off on crossing runways. Design of procedures must ensure that one or several aircraft have safe escape paths from any point on the approach. With closely spaced aircraft under IMC, this may not be easy.

The temptation to use such innovative technology to the fullest will be difficult to resist; indeed, increased throughput is the motivation for this technology and these procedures. *Human operators must not be placed, however, in a situation from which they cannot safely and reliably extricate themselves and their aircraft if some element of the automation fails,* which may mean that the full potential benefits of the technology cannot be realized without eroding safety margins. This dilemma will become commonplace as we attempt to squeeze every possible increase in capacity from our finite airspace.

Protection Against Environmental Threats

Three types of automated environmental alerting and warning functions are now implemented in transport and some corporate aircraft. They are ground proximity warning systems, traffic alert and collision avoidance systems, and wind shear advisory systems. Each is designed to detect threats that may not be obvious to pilots, especially under instrument meteorological conditions. At this time, each is an information system; pilots must respond manually to the warnings. Each requires a pitch mode response; for TCAS-II warnings, the pilot may be required to descend or climb, whereas GPWS and WSAS advisories require a maximum rate climb to a safe altitude.

Each of the older systems (GPWS, TCAS) has appreciably enhanced safety. GPWS, although not universally effective, has a documented safety record (e.g., Porter & Loomis, 1981). Even at its present state of development, TCAS-II is perceived by pilots and the FAA to have prevented at least several midair collisions—just how many is impossible to tell. WSAS is too new to have accumulated such a record, and newer devices (active sensors in aircraft to improve detection capability, as compared with passive inertial sensors, and Doppler radar at airports, the first of which was commissioned at Houston in July 1994; Hughes, 1994) are under development, but there is good reason to believe that some form of WSAS will be helpful to pilots, especially during takeoff and approach in the vicinity of convective turbulence. During tests of the Doppler system at Denver in 1993, a considerable number of airplanes were helped to avoid severe wind shears.

Issues Raised by Environmental Protection Systems Automation

Both GPWS and TCAS have produced variable numbers of false and nuisance alarms, particularly early in their periods of line service. Although ground proximity warning systems have been greatly improved since they were mandated in 1976, they still give rise to nuisance warnings. In the case of one large international carrier, 247 of 339 GPWS warnings during a recent 12-month period were false or nuisance alarms (73%). Like all new technology, TCAS has also caused new problems, most importantly for air traffic controllers. Inadequacies in the TCAS software have also burdened pilots with nuisance alerts in considerable numbers, and with a few resolution advisories that if followed would have put the aircraft in danger.

The problem of false/nuisance warnings is not trivial. If a substantial fraction of the warnings received are evaluated by pilots in hindsight as false or unnecessary, they will not trust these systems, even if some of the warnings are correct and could save the aircraft. Pilots' (or controllers') perceptions (whether correct or not) about the inaccuracy of warning systems will always shape their behavior toward trying to verify whether the warnings are correct—yet delays in responding to appropriate or true warnings may negate their effectiveness. Airlines have mandated full responses to GPWS warnings, but have had to backtrack on these procedures in the face of numerous nuisance warnings at specific locations. Procedures may be required in the short run, but they are not the best answer.

To my knowledge, no aircraft now flying in line service responds automatically to these warnings, although autonomous responses could be implemented and would have some theoretical advantages. Manually flown TCAS responses, for example, often exceed the vertical plane separation boundaries established by ATC, and this has been a source of intense discomfort to controllers who are faced with sudden altitude excursions without advance warning. The initial operational simulation evaluation (Chappell et al., 1988) indicated the likelihood that such excursions would occur, and operational evaluations have confirmed it. It is likely that automated resolution advisory responses could minimize such excursions.

The great danger of an inadequate response to a true GPWS warning has motivated nearly all air carriers to require a full procedural response unless it is visually obvious to the crew that no danger of controlled flight into terrain exists (see, for example, Kaysville, Utah, 1977). Cases continue to crop up, however, in which an inadequate crew response failed to avoid the terrain that motivated the warning. Like TCAS avoidance maneuvers, GPWS responses could easily be automated, and it has been suggested that this be done. On the other hand, false or nuisance GPWS alerts occur under a variety of circumstances, among them in holding patterns[2] when an aircraft passes directly over another below in the same pattern. A GPWS response under these conditions could cause the maneuvering airplane to climb into the path of yet another aircraft holding 1000 ft above. (It is worth noting that ATC, which has only a planform view of traffic, cannot detect such an excursion in a holding pattern, so an important element of redundancy is not available. TCAS should warn of a potential conflict, but it is not infallible either.)

Severe wind shears, often caused by microbursts associated with thunderstorms, have been responsible for many aircraft accidents over the years (Caracena, Holle, & Doswell, 1989; Boeing, 1994). The most recent occurred at Charlotte, North Carolina (1994). They are particularly dangerous to aircraft flying slowly in a relatively high-drag configuration; such configurations occur routinely during approach to landing. Wind shear advisories, like GPWS alerts, require an immediate maximum-performance climb, trading kinetic energy (airspeed) for potential energy (altitude) if necessary. There is little doubt that this escape maneuver could be more precisely performed by automation than by the human operator, because not all of the inertial and air data information necessary for the performance of the maneuver is available in the cockpit and a very rapid, precise response is required. This would seem, therefore, to be an ideal candidate for control automation.

In each of these cases, however, the false alarm problem, together with the many other variables not known to or accounted for by the logic in these systems, suggests a considerable measure of caution with respect to automating escape maneuvers. Leaving aside issues of passenger comfort, a secondary consideration when safety is threatened, the record to date suggests that very substantial numbers of unnecessary and sometimes dangerous escape maneuvers would occur if pilots were not in the loop, and given the time-criticality of these threats, it is likely that pilots would not be able to moderate or inhibit automated response maneuvers.

Further, the initiation of an unannounced escape maneuver by the autopilot when a pilot was flying manually would almost certainly be countered (at least initially) by the pilot, who would consider the maneuver initiation to be a turbulence or other input which required corrective action. At least one recent accident involved pilots unsuccessfully attempting to counter automation inputs (Nagoya, 1994; see also Paris, 1994). If escape maneuvers were to be automated, highly salient displays to inform the pilot of the intervention would be required, and pilots, as well as ATC, would have to have special procedures available to protect against conflicts that might be introduced by the performance of the maneuver.

Automated warning systems have saved lives and aircraft, but they are good example of the dictum stated earlier: What *should* happen and what *will* happen

---

[2]A racetrack pattern flown by aircraft as directed by ATC to delay their progress when ATC handling capabilities cannot accommodate all traffic.

when new technologies are introduced are often at variance. If new systems are introduced without considering the full range of behaviors they may evoke and the new problems they may create, they are liable to do more harm than good.

Ground Maneuvering Assistance

A third area in which control automation (together with information automation) may be introduced is on the ground at airports, to assist pilots in guiding their aircraft between parking gates and active runways under conditions of poor visibility. Today's aircraft can land automatically, or even manually, under visibility conditions that are inadequate to permit them to taxi safely from the runway to a gate. This fact and the serious problem of incursions of aircraft into airport movement areas without clearance (Billings & O'Hara, 1978; Tarrell, 1985; Detroit, 1990) have stimulated a serious search into how aircraft may be assisted in surface navigation on airports when unaided visibility is inadequate. Some proposals have included either manual or automatic steering with reference to taxiway centerline guidance devices, usually cables buried just beneath taxiway surfaces. Steering guidance during takeoffs has also been considered. (The incursion issue is more serious than just getting lost on the airport; avoidance of conflicts between aircraft and other aircraft or surface vehicles is another vexing facet of the problem.) Most such proposals have assumed that pilot vision will also be aided by devices that can produce enhanced or synthetic views of their immediate surround (discussed later), although some simulator experiments have been conducted using only airport maps and enhanced GPS navigation aiding.

Advanced Navigation Systems

Satellite-based position determination systems are rapidly reaching a level of maturity that can permit them to serve as the primary basis for aircraft navigation. Such systems are in wide use, although they are not yet approved as a sole means of navigation because adequate monitoring systems (for satellite signal integrity) are not yet available.

As described by Paulson (1994), two satellite navigation systems are now in place. The U.S. Department of Defense Global Positioning System (GPS) is essentially complete, with 28 Navstar satellites in 55° orbits. The Russian Glonass will encompass 24 satellites in 65° orbits; 19 satellites were functional in early 1995 and the remainder were due to be launched in early 1996 (Simpson, 1995). The Glonass system, like GPS, is under the control of military authorities, and this fact has caused considerable apprehension among civil operators who are concerned about reliability and guaranteed access.

Both the United States and Russia have declared their systems available for civil use. The Inmarsat organization, recognizing the need for a health warning system for satellite signals, agreed to include navigation transponders on its four third-generation geostationary communications satellites; signals from these transponders would provide both wide-area differential capability and an integrity monitoring service, broadcasting warnings to aircraft and ATC in the event of a satellite failure or malfunction. These satellites (or another means of accomplishing this function) were scheduled to be deployed in 1995–1996.

From a technical viewpoint, either or both systems could be made available for precise en route navigation. GPS antennas and decoders are widely available at reasonable cost, and several newer aircraft have made provisions for GPS navigation in their flight management systems. ICAO final standards are not yet in place, but FAA has given its permission for use of GPS provided it is not the sole means of navigation, and Europe's Joint Airworthiness Authority has certified the A330 and A340 avionics suites for satellite navigation.

The use of GPS, augmented by inertial data, for precision approaches is under test at this time. Although it is not yet clear whether such a system, enhanced either by differential signals or by other means, can routinely meet the standards for category II or III approaches, there is general agreement that it can provide at least category I accuracy (see Fig. 5.10). The FAA, which spent many years developing microwave landing systems (MLS; Federal Aviation Administration, 1987), has recently cancelled its MLS production contracts, although ICAO has adopted the U.S. MLS standard and Europe is committed to MLS as its future landing system. The wide availability of GPS technology, however, has led to much uncertainty about the landing systems of the future (Butterworth-Hayes, 1995). Economics will be important; MLS is an expensive system, but some means of conducting category III approaches will be an absolute necessity, and GPS has not met this need thus far.

Issues Raised by Advanced Navigation Systems

As far as pilots are concerned, the source of their guidance signals is of less importance than the accuracy and reliability of those signals. They will continue to require a way of monitoring signal integrity, but they will accept whatever guidance brings them dependably to a position from which a landing can be assured. It is believed that GPS, like MLS, can be used for more complex approaches than the long straight-in approach paths required with ILS. The FAA has experimented with very complex curved-path approaches for use in noise-sensitive and confined areas (Scott, Goka, & Gates, 1987), but it is not clear whether such approaches will be widely used except in very difficult locations such as the New York (LaGuardia–Kennedy–Newark) area.

I mentioned earlier that although pilots of advanced aircraft are able to evaluate the sources of the information on their map displays, what raw data are being used to synthesize the integrated information is not obvious. When GPS alone is used, it is impossible for the pilot to determine either the source or the accuracy of the data because of the complexity of the calculations used to derive instantaneous position from four to six satellites. If both GPS and Glonass are integrated into the future navigation system, the positions derived from the two independent satellite systems can provide new redundancy; each will have about equal precision. If the ability to compare their data is made available, this redundancy will be available almost anywhere over the earth's surface.

INFORMATION AUTOMATION IN THE FUTURE

This is an aspect of automation in which many innovations will be offered in the near future. Some are already in test; others await technology advances such as large flat panel displays. All will be able to make still more information available in the cockpit at a time when there may already be more than many pilots can attend to in the time available. But the new technology, if properly implemented, can simplify rather than complicate the pilot's task. I next review some of the new functionality that has been proposed and then examine the likely effects on flight crews.

Digital Data Link

Digital data link, combined with satellite communication, has been under evaluation in civil aviation for several years. At present, ACARS transceivers are used; in the future, mode S ATCRBS transponders may become the preferred medium for exchange of ATC data. At this time, the usefulness of automatic dependent surveillance (ADS) on overwater routes seems assured. Several carriers have participated in tests over the Pacific Ocean. Russian authorities are also considering ADS for primary use over large portions of its huge land mass, where radar air traffic surveillance is not available.

ADS involves the frequent reporting, without crew intervention, of position, altitude, velocity vector, and often wind speed and direction. These data are received by a ground communication service and retransmitted to air traffic control facilities, where they are automatically plotted and can be used by controllers to survey traffic under their control. At present, voice communication with aircraft in oceanic airspace still depends largely on high-frequency (HF) radio equipment, but all parties hope that satellite communications, already available for passenger telephonic communications on a few air carriers, will soon become available for the pilots of those aircraft as well. As one pilot remarked, "I hate to be using a lousy HF channel when the passenger behind me is talking to his wife on the phone!" (T. Demosthenes, personal communication, November 1995).

Data link provides the capability for high-bandwidth data communication; the issues relate not to the technology but to its uses. The FAA is working on standards for integrated data and voice communications services for the future, the Aeronautical Telecommunications Network, which will tie the entire aviation communications system together. A host of issues concerning communications architecture, protocols, vocabularies, standards, policies, and procedures remains to be enunciated, however, and equipment manufacturers cannot provide equipment without these details. This is a major reason why ATC data link is not yet implemented in the 777 and other new aircraft, and why aviation communications technology is still a patchwork.

Data link may eventually enable nearly all routine communication between ATC and aircraft to be carried out without recourse to voice contact, leaving voice for urgent messages and nonroutine transactions between pilot and controller. Weather en route and terminal airport information are among the types of data that will be sent in this way. Through ACARS, two-way data link is already used for much company communication, and the ACARS system is now used for preflight clearance delivery and other non-time-critical data transfers. Many new aircraft have printers in the cockpit, so that ATC messages can be saved as hard copy when desired. It is likely that such devices will be needed to spare crews the need to page forward and backward through many stored messages when a particular datum is needed, and to give them the ability to refer to such information quickly. The flexibility of the display systems should permit pilots with differing cognitive styles to adapt information handling to their own preferences.

Thus far, I have not discussed new functions that may be enabled by data link. The high-bandwidth capability of digital data link permits it, at least in theory, to be used to downlink a considerable amount of aircraft data not now made available to ground facilities or airline operations centers. This offers the potential for error checking by ATC computers of clearances that have been uplinked, accepted, and executed by pilots, as well as the exchange of more aircraft data with the ground, as was done in United Kingdom CAA trials in 1991 (see page 213).

Among the functions that are routinely performed by ACARS data link are the transmission of "out-off-on-in" data (times of departure from gate, takeoff, landing, and gate arrival), diversion or delay information, engine performance data, arrival gate data, and airplane malfunction information to assist ground maintenance personnel in planning for repairs or parts replacement without causing delays. Passenger needs on arrival are also communicated. Other data could also be transmitted, including performance data and nonroutine events, although the transmission in real time of such data is of concern to pilots. Transmission of sensitive data over broadcast channels also brings up questions of data security, especially if the data concern identifiable flights or persons.

Issues Raised by Data Link

The routine use of data link for controller–pilot communications will change in fundamental ways the interaction processes between these two groups of human operators. Where they now work together in direct person-to-person conversational contact, most of these contacts will be by alphanumeric messages that must pass through two computers. Further, unlike voice messages today, which are primarily broadcast on a party line, data link messages to aircraft will be selectively addressed; others in the air will not have access to them. The implementation schemes for data link all envision the availability of a voice communications channel for urgent messages, but the potential for decreased *team* (pilot–controller) involvement in cooperative problem solving is worrisome, and the possibility that both human operators will be further distanced from a sense of immediate involvement in the air traffic management task could pose a potential safety hazard.

Electronic Library Systems

An electronic library system (ELS) was planned for the 777, but most airplane customers did not feel it to be financially viable at this time. At least one airline and an avionics manufacturer have actively explored this concept, however. With today's computer technology, it would be possible to store virtually all of the information required by pilots (and now carried in their capacious flight bags) on CD ROM disks or other electronic media, and to make it instantly—or nearly instantly—available on a dedicated screen in the cockpit. Approach and en route navigation charts as well as the flight and airplane operating manuals could be encoded in such a database.

Issues Raised by Electronic Library Systems

I say "nearly instantly" because instantly *available* and instantly *accessible* are not quite synonymous. Admittedly, pilots must now thumb through hard-copy manuals to find a desired bit of information. (Quick-reference handbooks assist in emergency and anomaly checklist retrieval.) With an electronic library system, they would have to navigate serially through numerous menus to find the same bit of information. With an electronic system, however, they would also have to learn the data archi-

tecture, preserve it in memory, associate the structure with the abbreviated identi-
fiers on the screen, learn economical ways of accessing what they needed, and then
perform the on-screen manipulation necessary to bring the desired data to hand.

If pilots find it necessary to print material stored in an ELS (such as an approach
chart) in order to scrutinize it more carefully or to move it to where they want it,
little purpose will have been served by the provision of yet more expensive technology in the cockpit (although the cost of updating the information may be decreased,
and pilots may be spared the recurrent task of updating their manuals and charts.
On the other hand, most pilots review changes in the data when they are performing
this task, and the benefits of that review may be lost.).

Most of the material in the flight bag is alphanumeric, and simply transferring it
to an electronic medium seems a clumsy way to use this technology. Because the
ELS will be a single (nonredundant) system, it is unlikely that certification authorities will permit it to be interconnected with flight-critical systems such as the FMS,
and without such connectivity more of its potential usefulness will be compromised.
With connectivity, the automation becomes yet more complex, coupled and susceptible to unwanted surprises.

A capable expert system might be helpful to assist in navigating through ELS
information, and some research has been done toward that end. Lacking such a system,
one must consider whether a "paperless" cockpit represents a substantial improvement
on what we now have. Things have improved since Ruffell Smith (1979) pointed out
the 20-m^2 "blizzard of paper" required for a trans-Atlantic flight (Fig. 6.6).

Much of the flight path navigational data that pilots need is now available in the
large FMS database; few pilots using FMS find it necessary to refer more than
occasionally to their navigation charts, although all pilots still use approach plates,
even for familiar airports, as memory aids. (The Cali, Colombia Boeing 757 accident
on December 20, 1995, presently under investigation, is thought to have involved
the disappearance of a critical navigation fix from the navigation display when a
revised approach route was entered into the FMS. This was probably a major factor
in loss of position awareness by the flight crew, operating at low altitude in
mountainous terrain.) Charts are another area in which the printed page is a
substantial improvement over electronic data. The best resolution available on
monochrome CRTs is substantially less than can be achieved on printed charts;
simple reproduction of such charts would not provide adequate spatial resolution
of the data now provided, and navigation and approach charts would have to be
redesigned for effective electronic presentation. Nevertheless, it is likely that at
some time in the future, electronic libraries will become available in transport
aircraft, especially if the computer equipment used to enable them is also found to
have a commercially profitable purpose such as providing services for which
passengers will pay.

FIG. 6.6.   Paperwork required for a 747-100 flight from Washington via New York to London in 1976. From Ruffell Smith (1979, Courtesy of NASA-Ames Research Center).

Enhanced Vision Systems for Pilots

Although air transportation is now highly reliable, visibility restrictions due to fog can still shut down airports completely for an indefinite period. If this occurs at a major airport such as Chicago's O'Hare, air traffic over a large part of the United States will be affected within a few hours. Category III autoland can enable safe landings at suitably equipped airports in very bad visibility, but taxiing may be impossible. To provide independent monitoring capability in the cockpit during such operations, the government, avionics firms, and some air carriers have studied how pilot vision might be improved by sensors operating in portions of the electromagnetic (EM) spectrum less attenuated by these weather phenomena than the visible spectrum.

Two portions of the EM spectrum have been explored in depth. One is the infrared (IR) band; the other is in the millimeter-wave (MMW) portion of the microwave spectrum. The studies aimed at providing pilots with a synthetic visual image, either projected on a head-up display or on a head-down screen on the instrument panel, that would assure them of the location and orientation of a runway with respect to their airplane. Other studies have been carried out to determine whether images derived from more than one portion of the EM spectrum could be fused to provide such imagery (see Cooper, 1994b).

These programs have been variously called *synthetic vision, enhanced vision,* or *image fusion.* Such technology could permit pilots to land without assistance from the ground on any appropriate surface if they were guided to the proximity of that surface by appropriate on-board navigation equipment. Thus, one major benefit of such devices could be a decrease in the number, and therefore cost, of ground navigation aids, a significant factor in less developed nations.

The technical difficulties lying in the way of such technology are formidable. Infrared images are different from visible images in that they reflect temperature differences among objects in the environment rather than brightness or chromatic differences; even though outlines may be clearly detected, they may not be the outlines expected. Also, although IR imagery can detect objects either colder or warmer than their surround, there are times of day when objects are at essentially the same temperature as their surround as they are being either heated by solar radiation or cooled in its absence. Nearly all infrared radiation is attenuated by airborne moisture, dust, or smoke between the sensor and the objects of interest; for this reason, IR sensors may be useful only at fairly short ranges.

Millimeter-wave radar relies on EM impulses generated in and propagated from the airplane toward the earth ahead. A fraction of this radiation is reflected from solid objects in the path of the radar beam, and a small fraction of the reflected radiation returns to the transmitting and receiving antenna. Metal objects are highly reflective, paved surfaces less reflective, and earth absorbs most microwave radiation impinging on it. Large objects such as a runway can be visualized. Much smaller objects made of metal, such as surface vehicles, are easily detected; such obstructions on a runway can be detected easily.

MMW equipment has been demonstrated in aircraft and has been shown to provide sufficient information to permit an approach to be completed under at least test circumstances. Forward-looking (passive) infrared equipment (FLIR) is in wide use for target detection by the armed forces, often in combination with other sensors such as low-light-level TV or synthetic-aperture radar.

Finally, it has been proposed that enhanced terrain maps stored in aircraft and correlated with precise geographic position information from GNSS could be used to generate entirely synthetic imagery for pilots landing at airports. This technology could in theory free pilots entirely from environmental constraints to vision (but it would not be able to show runway obstacles unless it were augmented by forward-looking sensors of some type, operating in real time).

Issues Raised by Enhanced Vision Systems

The human factors issues associated with the use of this technology are likewise formidable. Because the images are qualitatively different from visual images, questions arise as to whether synthetic imagery should be transformed in order to make it more obvious what the pilot is seeing, or whether pilots should be taught the differences and required to use the processed imagery to decrease the likelihood that they will form a false or misleading impression of what the sensor "sees." Although much research has been done over many years (e.g., Kraft & Elworth, 1969; Roscoe, 1979; Stout & Stephens, 1975) to elucidate what visual cues pilots require for landing in impoverished visual environments, none of it has been able to specify an exact minimum set of required cues, and given the number of human variables, there may not be such a set.

If synthetic or enhanced imagery is projected on a wide-angle head-up display in the cockpit, questions arise as to whether pilots will be able to attend both to the display and to the external environment behind it. Most synthetic runway representations on head-up displays have been outline forms to make it easier for pilots to transition to outside visual cues during the landing maneuver. The problem may be that the head-up display symbology that is used during the approach is more salient than the external scene, especially when viewed through fog by an inexperienced pilot (Lauber, Bray, Harrison, Hemingway, & Scott, 1982).

Another problem is the relatively slow scan rate of radar. Most jet aircraft are traveling over 200 ft/sec when they enter the landing flare; the environment is changing very rapidly and rapid updating of visual cues is necessary. We do not know exactly what image update rate is required for fully effective inner-loop control, although studies of this are underway.

Several air carriers have installed head-up displays to provide category II and III landing capability without the expense of triplex autopilots and other equipment. From a perceptual and cognitive viewpoint, these devices can provide cueing information of considerable help (Fig. 6.7). The danger of such devices is that pilots may be misled by what they think they see through the device, or that they may not see what they need to complete a safe landing. Some have proposed that synthetic or enhanced imagery should be presented "head-down" on the instrument panel to obviate the latter problem, but this poses a new problem: the time required to transition from head-down to head-up visual orientation, a process that requires at least a few seconds and may take longer if external cues are minimal.

The decision to land is one of relatively few in civil aviation that must be made very quickly (within very few seconds) under poor visibility conditions. If GNSS and enhanced vision technology are used to permit landings at airports without surface precision navigation aids, pilots must be provided with unequivocal information as to their precise location and the suitability of the runway ahead before they can commit to landing and throughout the landing process, including rollout and taxi.

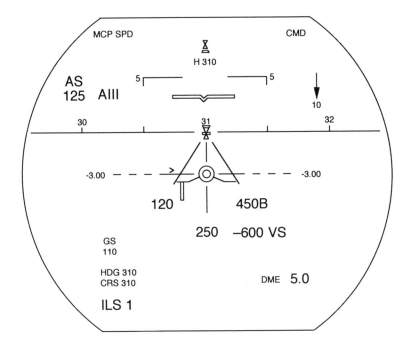

FIG. 6.7.   Head-up display for precision approaches. From *Aviation Week & Space Technology*, 12/12/94, p. 50. Reprinted with permission.

Advanced Integrated Displays

Recognizing the extreme perceptual and cognitive demands placed on military pilots during combat operations, the armed forces for many years have been investigating large flat-panel display technology in the hope of being able to provide pilots with highly integrated intuitive situation displays. These "big picture" displays, coupled with adaptive automation (see chapter 12), would provide pictorial and analogical representations of terrain, threats, targets, predetermined course, and aircraft and weapons status. The technology is not yet available to provide flightworthy displays of the size desired, but in laboratory simulations, the representations appear to integrate much of the information required by pilots under such circumstances.

Big picture displays have been proposed for use in civil aircraft, although the costs have been perceived thus far to outweigh possible benefits. This situation may change, however, if a new high-speed (supersonic) transport reaches the development stage (Fig. 6.8). One desired feature of such a transport is the ability to provide pilots with sufficient forward visibility without the considerable structural weight penalty associated with a movable visor and nose assembly that covers the windscreens during high-speed flight. Such a visor apparatus is used on the Concorde to permit a view over the nose of the aircraft during takeoff and approach, when pitch angle is high compared with that of conventional aircraft.

A supersonic transport without a visor assembly would have cockpit side windows but none oriented directly forward, for aerodynamic reasons. Some sort of forward visual display would be necessary both for maneuvering at low altitude and for taxiing. It would probably be driven by a combination of television and other sensors, though some have proposed an entirely synthetic (virtual) computer-generated display for this purpose.

An additional problem for ground maneuvering in a supersonic transport would be the position of the pilots, far forward of the steerable nose gear as well as the main landing gear position much further aft. Even if they had forward vision, additional views, perhaps from the nose gear position, might be necessary to enable them to remain within the confines of narrow taxiways and to negotiate turns with variable radii on airports. These technologies would qualitatively change the ways

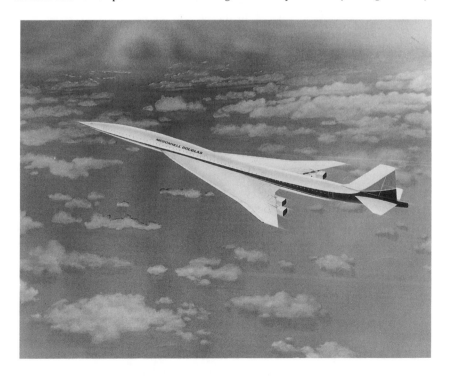

FIG. 6.8.   Artist's sketch of a proposed high-speed civil transport. McDonnell-Douglas Corporation conceptual design, reproduced by courtesy of NASA Langley Research Center.

in which pilots maintain contact with their external environment. They pose both perceptual and cognitive questions related to reliability, trust, automation complexity, and transparency (literally!) that will require much further research, not only on the technologies themselves but on the human's ability to remain in command in the range of situations in which he or she would be dependent on them.

Issues Related to Information Management

It is important to keep in mind the need for independent sources of data in a real-time, highly dynamic system. Although a pilot may have access to several apparently different types of information concerning a single topic in a highly automated airplane, he or she must always consider whether the redundant information was collected by independent systems, or whether it is merely two ways of representing data from the same source. If the former, it can be used for cross-checking; if the latter, a single sensor could corrupt both representations. In tightly coupled systems, the difference may not always be obvious. For this reason, among others, pilots need to know the sources of the processed information that reaches them.

We have reached a point at which multiple sources of similar data are usually available to pilots. As just noted, in the near future pilots may have access to representations of the airplane environment derived from the visual, infrared, and microwave portions of the electromagnetic spectrum. Even if a way can be found to synthesize congruent imagery from each source, or to fuse disparate imagery into a consistent representation, it will be important that the pilot understand what data sources are being used, and the limitations of those data. The training burden imposed by such technology will not be trivial unless pilots are given the opportunity through simulation and flight experience to become thoroughly familiar with what can be trusted and what cannot be under specific circumstances.

Another case in which disparate data sources are used is data from surface navigation aids, satellite data, and inertial sensors within the aircraft. In the past, the data derived from various sources have been presented in a common manner, or the data have been reconciled within the flight management computer prior to their presentation on the navigation display. Pilots can gain access to the sources of this information on their navigation displays (see Fig. 6.9 for an example), but they must select an alternative display mode to obtain it.

MANAGEMENT AUTOMATION IN THE FUTURE

Kelly et al. (1992, p. 2) indicated that "some improvements to flight management systems have been difficult to justify because they did not appear to provide new capabilities which would result in return on investment." It is probably fair to say

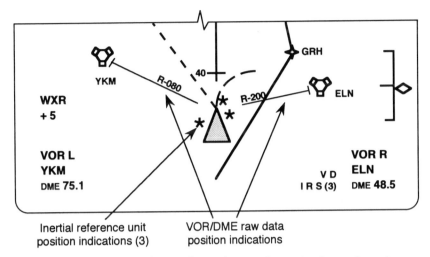

Inertial reference unit   VOR/DME raw data
position indications (3)   position indications

FIG. 6.9.   Simultaneous visualization of raw and processed navigation data on electronic Navigation Display (Boeing 747-400). Reproduced by permission of the Boeing Company.

that many pilots now flying FMS-equipped aircraft would welcome a simpler, more intuitive flight management system with which they could interact more easily than is possible with today's CDUs. It is also likely that most designers and human factors specialists, given the benefit of hindsight, would welcome the opportunity to redesign this interface, and that members of the avionics community have the knowledge necessary to do it better.

The fact remains, however, that today's flight management systems work, and replacing them would be extremely expensive. A very considerable investment in training has already been made, and the vast majority of pilots have found it possible to adapt successfully to present FMS idiosyncrasies. Many steps have been taken in newer systems to simplify reprogramming with the intent of reducing the inherent clumsiness of the system, to speed FMC response times (which were very slow in early devices), and to improve the legibility of the CDU screen. Any attempt to radically revise the FMS interface will require extensive retraining of operators at considerable expense. Unless carefully done, a redesign may impose training transfer problems, and thus safety issues, for pilots moving from the older to the new devices. These factors, leaving aside the return-on-investment issue, make it likely that airline flight management systems and their interfaces will continue to look and operate much as they now do for a considerable time to come.

Having said this, are no improvements possible without starting over with a clean sheet of paper? The answer to this question, if there is one, lies in looking carefully at problems known to be associated with FMS use in line service. Several investigators, prominent among them Sarter and Woods, have conducted such inquiries.

Their data, gathered in flight observation and simulation experiments, indicate two principal sources of FMS interaction problems. The first class of problems involves mode errors or lack of mode awareness. As Sarter and Woods (1996) pointed out, today's flight management systems are "mode-rich" and it is often difficult for pilots to keep track of them (see Fig. 9.2). The second problem, which is related to the first, involves lack of understanding by pilots of the system's internal architecture and logic, and therefore a lack of understanding of what the machine is doing, and why, and of what it is going to do next. (See, for example, the mode problem at Manchester, United Kingdom, 1994.)

Simply saying that improvement of the flight management system is not economically justifiable rationalizes away the many lessons learned from operational experience with these systems. If the box itself cannot be redesigned, it is possible and likely that redesign of some of the displays associated with it (the CDU format, the map display, and particularly the mode annunciation panel) might accomplish some of the same purposes. There is, for instance, no true vertical navigation display in transport cockpits at present, yet it is during the climb and descent phases of operation that a majority of the problems in the interaction between operators and the automation arises. Reworking of mode annunciation panels to make mode data, and particularly mode changes, more salient could improve pilot understanding of what the automation is doing (E. Hutchins, cited in Automated cockpits, 1995b).

Although modifying procedures is a poor substitute for fixing the basic problems that motivated the modifications, there are at least three possible approaches to these problems in addition to the display improvements just mentioned, each of which has both advantages and drawbacks. Each, however, seems worthy of consideration by designers and operators. They are briefly discussed here.

• *System software revisions could be made to simplify complex FMS functions with the intent of making them more understandable and/or transparent to operators.* In newer aircraft, flight management systems are flight-essential equipment. Any changes in the systems or in how they function are subject to rigorous configuration management and certification criteria. Software changes, however small, are extremely expensive. Further, given the tight coupling among software modules, changes in one module may have cascading effects on other software elements. Any proposal for software modifications is subject to even greater cost constraints than are hardware modifications. Further, cosmetic changes in software are unlikely to have appreciable effects on system complexity, the real issue underlying present problems with the FMS. A wholesale redesign of the system would probably be required to simplify it in a useful way.

Despite these negative comments, however, further research should be undertaken to learn which of a large number of approaches to FMS architecture would convey the greatest benefits in terms of real system simplicity and transparency. I am unwilling to accept the thesis that advanced flight planning, management, and guidance systems cannot be made easier for human operators to understand and to operate. Several research groups are now working on various aspects of this problem, although none, to my knowledge, has looked at the overarching question of FMS architecture and functionality.

• *Procedural modifications for the use of the FMS could simplify the use of the systems. Such modifications would involve the use of only a subset of FMS functions. Excluded functions could either be disabled or simply not used.* Procedures have always been used to make up for deficiencies in equipment and technology, but it is also true that uses of technology are often suboptimal because proper procedures for its utilization have not been developed and applied. We have FMS technology in being; it is unlikely to be fundamentally modified, and we know that human operators are having some difficulties in using it effectively. Here, I am not suggesting ways to get around specific problems; rather, I suggest that a systematic look should be taken at those FMS functions that are widely used, necessary for safe and effective mission accomplishment, and least likely to be misunderstood or misused. Functions that do not meet these criteria should be considered for abandonment.

Is it really necessary to have four distinct descent modes, or would two suffice? Is it really necessary that pilots be able to demonstrate their ability to use all FMS functions, or is it necessary only that they be able to use a limited subset of the available modes to accomplish their mission under all foreseeable circumstances? Any reduction in FMS complexity would pay dividends during training, would decrease the cognitive burden imposed on pilots by the equipment, and would simplify flight procedures. Automation complexity is the fundamental problem in this domain; reducing that complexity offers the greatest hope of a successful resolution of that problem, even if system redesign is not possible.

Simplifying procedures for the use of the FMS would also permit us to avoid those corners of the FMS functional envelope that have posed the most serious problems in the past. The open descent issue in current Airbus airplanes is one example; climbs using vertical rate rather than speed modes is another. A third is operations that may cause pilots inadvertently to disable an airplane's altitude capture function. The recent A300 accident at Nagoya (1994) suggests that mode interactions that permit simultaneous manual and automated control should be avoided. United Airlines is focusing its FMS training on preferred modes of operation of the FMS to simplify its pilots' tasks in managing their airplanes (H. A. Langer, personal communication, June 5, 1995).

These are only examples designed to provoke thinking about whether we can simplify the use of this very complex tool by avoiding some of its less important capabilities. Pilots could point to several other possibilities, perhaps more important than some mentioned here, if they were asked to. Though a number of pilot opinion surveys has been conducted, to my knowledge none of them has asked, "What modes or functions do you *never* use, and why?"

• *Pilot training should be examined and revised with the intent of providing operators with a better understanding of system logic and behavior under the full range of conditions likely to be encountered in line operations.* To paraphrase Sarter and Woods (1994), "What is needed is better understanding of how the machine operates, not just of how to operate the machine." It has also been said that, "If you can't see what you need to know, then you've got to *know* what you need to know" (to which T. Demosthenes [personal communication, November 1994] added, "And if you don't know, you've got to be told!"). Given the flexibility and complexity of the current FMS, some of the mistakes pilots make in its operation suggest that they simply do not understand how it operates, and why it does things that way. There are good reasons in most cases, and they are known to the designers of the equipment. Some are imposed by certification requirements, others by the system architecture, and still others by the range of FMS interactions with the airplane and with other automation. But the problem of inadequate user understanding persists.

Explanations of these interacting requirements during training would be costly in terms of training time. They would certainly be less expensive, however, than the loss of an aircraft and its passengers because of the lack of such knowledge. Accidents to date (Fig. 3.1) and growing experimental evidence (Sarter, 1994) does indicate inadequate understanding of FMS behavior and operating constraints, and pilots responding to surveys indicate that they have not been satisfied with the thoroughness of their computer-based training (Uchtdorf & Heldt, 1989).

An adequate internal model of an automated system is vital to a pilot's ability to predict how that system will function under novel circumstances. I believe that research in progress will point toward a better understanding of what pilots require to build correct and adequate models of the systems they operate. It is hoped that air carriers can find ways to assist them in forming such mental models during their training.

MANAGEMENT OF HUMAN ERROR

The alternatives already presented are not mutually exclusive. Experience with advanced automated systems indicates the need for simplicity, transparency, and comprehensibility in the systems we use, as well as predictability in the behavior of those systems. Even though today's flight management systems fall short of human-centered principles in certain respects, it will be difficult to generate much interest in radical rework of any systems that are functional, let alone systems as capable as our present FMSs. Yet we must find ways to improve the error resistance and error tolerance of both our current systems and those of the future. I end this chapter with a short discussion of these all-important concepts.

An error-resistant system is one in which it is very difficult for a human operator to make an error in system operation. Simplicity in system design and the provision of clear, unambiguous information on display interfaces are important tools with which to improve error resistance. Error-tolerant systems are those that are able either to trap errors or to mitigate their effects; an example is the autonomous landing gear-lowering system in Piper *Arrow* general aviation aircraft; it operates when it senses a combination of low engine power and low air speed, both of which occur during the last stages of an approach to landing.

The aviation system has been plagued by the problem of human error since it began. One of many reasons for this has been that our investigations of accidents have tended to focus rather narrowly on the specifics of individual cases, wherein a specific set of often unlikely circumstances, including erroneous actions by humans, has led to an undesired outcome. Points of commonality among accidents have been discerned and often corrected, but on the whole, our remedial measures have been specific and focused on the "sharp end" of the system.

In recent years, several investigators looked farther in an attempt to discover more generic factors involved in accidents, among them Perrow (1984), Reason (1990), Lauber (1989; 1993), and Woods et al. (1994). Reason's latent failure model has been influential; in oversimplified form, it suggests that a variety of latent factors that can potentially degrade operating safety are present in most organizations and endeavors. Under certain circumstances, they may act to affect the course of an operation or production process in such a way that an untoward outcome ensues: an accident. Woods et al. carried this construct further and have explored the variety of circumstances that can potentiate the effects on the operators at the sharp end of such enterprises. Lauber stimulated systematic searches for such factors in the background of transportation accidents.

The Dryden, Ontario (1989) accident briefly summarized in appendix A is a classic example of latent organizational factors in an operator, its owner, and the government regulatory and oversight organizations (Moshansky, 1992), but they have been major contributors to many other mishaps as well. Without full information concerning the context and environment(s) in which accidents occur, it is not possible to understand their genesis and how to take rational steps to prevent future accidents. Accidents are not only human failures; they are also failures of design, operation, management and often oversight. In short, they are *system* failures. They must be looked at as such if they are to be fully understood.

Error Resistance

Ideally, aircraft automation should prevent the occurrence of all errors, both its own and those of its human operators. This is unrealistic, but it is necessary to design systems to be relatively error resistant. *Resistance* is an opposing or retarding force, a definition that recognizes the relative nature of the phenomenon. Resistance to errors in automation itself involves internal testing to determine that the system is operating within its design and software guidelines. Resistance to human error is more subtle; it may involve comparison of human actions with a template of permitted actions (reasonability checks), a software proscription against certain forbidden actions under specified conditions (envelope limitation or protection is an example), or simply clear, intuitive displays and simple, uncomplicated procedures to minimize the likelihood of inadvertent human errors.

Automation of unavoidably complex procedures (such as fuel sequencing and transfer among a large number of widely separated tanks to maintain an optimal center of gravity) is necessary and entirely appropriate provided the human is kept apprised so he or she understands what is going on. The system must be able to be operated by the human if the automation fails; it must "fail safe" (in this case, it must be designed so a failure will not leave the airplane outside its operating limits), and it must provide unambiguous indications that it is (or is not) functioning properly. Guidance in performing complex tasks (and fuel balancing in some aircraft is such a task) is helpful, whether it is in a quick reference handbook or in the form of an electronic checklist. Prompting has not been used as effectively as it could be in aircraft human–system interfaces, although the newest electronic checklists attempt to assist in this task.

Questioning of critical procedures or crew instructions to the automation (those that irreversibly alter aircraft capabilities), or requiring that critical orders be confirmed by pilots before they are executed, can be additional safeguards against errors. These queries can also be automated, either by themselves or as part of a procedures monitoring module that compares human actions with a model of predicted actions under various circumstances. Such intent-inferencing models were developed in research settings (Palmer, Mitchell, & Govindaraj, 1990); some are now in use.

The human operator is known to commit apparently random, unpredictable errors with some frequency (Wiener, 1987; Norman, 1988); it is extremely unlikely that designers will ever be able to devise automation that will detect all of them. This being the case, it is essential to provide alternate means by which pilots can detect the fact that a human or an automation error has occurred. Such warnings must be provided in enough time to permit pilots to isolate the error, and a means must be provided by which to correct the error once it is detected. Where this is not possible, the consequences of an action should be queried before the action itself is allowed to proceed.

It must be noted here that automation also makes apparently random, unpredictable "errors" (e.g., the flap lockup at Hong Kong, 1994, second narrative in Fig. 6.10), and it is equally unlikely that designers will be able to devise the means to trap all of them. The human operator is the last (and best) line of defense against these failures, but operators must be given the means to deal with such failures. Figure 6.10 shows a few of these apparently random failures.

Error Tolerance

Because error resistance is relative rather than absolute, there needs to be a layered defense against human errors. Besides building systems to resist errors as much as possible, it is necessary and highly desirable to make systems tolerant of error. *Tolerance* refers to the act of allowing something; in this case, it covers the entire panoply of means that can be used to insure that when an error is committed, it is not allowed to jeopardize safety.

"As the nosewheel was about to touch down, the rudder moved, uncommanded, 16-17 degrees to the right. The airplane left the runway at about 130 knots..."

"As the aircraft banked, it encountered a wind shear...this buffeting triggered its automatic flap locking mechanism...the flaps locked at a full setting...the pilot aborted the landing. On the fourth try, he landed on runway 31...two passengers were slightly injured after the aircraft ran off the runway..."

"A V-2500 engine 'shut itself down' during a descent...because of a fault in the automatic fuel flow logic, which is being urgently investigated..."

"About half a mile from the runway threshold, the stick pusher activated while the airplane was slowing through 130 knots...the pilots estimated the pull required to overcome the forward yoke pressure at more than 250 pounds..."

"The airplane was operating at 31,000 ft, at night, with the autopilot engaged. The crew did not notice the initiation of the roll, and first noted a problem when the INS warning lights illuminated. They then noted...a roll to the right with a bank angle in excess of 90 degrees."

"After touching down, the pilots selected spoilers and reverse thrust, but there was a delay of 9 seconds before they deployed..."

FIG. 6.10. Examples of automation failures during transport aircraft operations.

Nagel (1988, p. 274) pointed out that "it is explicitly accepted that errors will occur; automation is used to monitor the human crew and to detect errors as they are made." The aviation system is already highly tolerant of errors, largely by virtue of monitoring by other crew members and by air traffic control. But certain errors possible with automated equipment become obvious only long after they are committed, such as data entry errors during preflight FMS programming (or even errors in the construction of the FMS database, a factor in the Mt. Erebus DC-10 accident). New monitoring software, displays, and devices may be required to trap these more covert errors.

As was suggested earlier, checks of actions against reasonableness criteria may be appropriate; for an aircraft in the eastern hemisphere, a west longitude waypoint between two east longitude entries is probably not appropriate. An attempted manual depressurization of an aircraft cabin could be an appropriate maneuver to rid the cabin of smoke, but it is more probably an error and should be confirmed before execution. Closing fuel valves on both engines of a twin-engine transport, an action that has occurred at least twice, is almost certainly an error if airborne (San Francisco, 1986; Los Angeles, 1987).

Given that it is impossible either to prevent or to trap all possible human errors, aircraft accident and especially incident data can be extremely useful in pointing out the kinds of errors that occur with some frequency. Formal system hazard analyses are appropriate to elucidate the most serious possible errors, those that could pose an imminent threat to safety. The latter should be guarded against regardless of their reported frequency (Hollnagel, 1993; see also Rouse, 1991).

Error Management

An epidemiological model of, and approach to, the problem of human error in aviation was suggested over two decades ago (Barnhart et al., 1975; Cheaney & Billings, 1981). In a recent comprehensive review, Wiener (1993) discussed intervention strategies for the management of human error. Wiener stated that "The aim of intervention is to strengthen the lines of defense at any barrier, or any combination of barriers, and to insert additional lines of defense where possible" (p. 13). He also proposed, however, that "Each proposed method of intervention ... should be examined with respect to its feasibility, applicability, costs, and possible shortcomings (e.g., creating a problem elsewhere in the system)"(p. 8). He offered guidelines for the design of error management strategies. This thoughtful study and the others already cited, deserve careful scrutiny by operators and managers, as well as designers, of complex equipment.

COMMENT

In this chapter, I presented a variety of automation innovations that I believe will be seen in, or at least proposed for application in, future aircraft. It is worth remembering again the criteria given by Kelly, Graeber, and Fadden (1992): Does a new system or function offer a reasonable likelihood of a return on investment? The return may be actual or potential, but it must be demonstrable in advance if the new system is to find its way onto a future airplane. It must be needed, not merely desired, in today's (and very probably tomorrow's) economic and competitive climate.

From a system viewpoint, prevention is a great deal less expensive than accidents. Two Boeing 737 accidents in recent years remain entirely unexplained at this time (Colorado Springs, 1992; Pittsburgh, 1994). Both airplanes had older digital flight data recorders that did not record control surface positions; that information might very well have led to an unambiguous finding of probable cause. In sharp contrast, the Aerospatiale ATR-72 that crashed after extended flight in icing conditions (Roselawn, Indiana, 1994), was equipped with a modern digital flight data recorder whose data enabled investigators to discover, literally within days of the accident, that icing had disturbed airflow over the ailerons beyond the pilots' ability to maintain control. It has been suggested that a substantial fleet could have been equipped with modern flight data recorders for less than the costs of the two 737 accidents.

Some of the innovations discussed here are clearly needed if the industry is to continue to expand its horizons; some form of enhanced or synthetic vision is an example. Improved error tolerance is imperative. Capacity must be increased, by whatever means. Global satellite navigation and satellite data and voice communication are certainties. The need for some of the other innovations discussed here is less certain, although the technology for them exists. Many could have been implemented in the Boeing 777 had there been sufficient demand for them—but there was not.

Other innovations not yet thought of will be proposed for aircraft still in the future, although most will be introduced in civil aviation only if they can meet the test proposed by Kelly and his coworkers. Even an entirely new supersonic transport, if one is built, will be subject to the demands of the marketplace, and our manufacturers cannot afford to take chances. They will build even a radically new airplane with the caution they have displayed throughout history—and that airplane is more likely to be both safe and economically viable because of that caution.

It is the task of the human factors community to make that aircraft and any other new models easier to manage, more error tolerant, and thus safer than those that have come before, despite the economic factors that militate against change if what we have is good enough. Accidents, even the few we have, are sufficient evidence that good enough is not satisfactory—that as long as preventable accidents occur, our job is not finished.

Chapter 7

Air Traffic Control
and Management Automation

INTRODUCTION

Aircraft automation has a very long history (chapters 5 and 6). In contrast, air traffic control (ATC) automation is of relatively recent vintage, dating from the 1960s, when the potential advantages of computer management of flight plan data were first recognized by the FAA, which manages essentially all air traffic control in the United States. This chapter discusses the evolution of air traffic control and management automation. The task of our complex ATC system is simple on its face: to provide safe separation among controlled aircraft and to expedite their passage to their destinations. Fulfilling the requirements of that tasking is less simple.

BACKGROUND

The U.S. National Airspace System (NAS) utilizes computers for a great part of its data management and information transfer, but the air traffic control process itself is still an almost entirely human operation conducted by highly skilled air traffic controllers whose information is derived from processed radar data, voice communication with pilots, and printed flight data strips. Although ATC system automation is primitive compared to the advanced technology in the aircraft that it controls, the system is a truly remarkable human–machine system that has accommodated itself to enormous demands on it. In recent years, the system has been called on to handle traffic volume well beyond what a few years ago was thought to be its capacity. It has done so because of the creativity and flexibility of its operators and managers.

During this same period, the air transport system itself been beset by constant change, totally unlike anything known during its 70-year history. In their former regulated (and stable) environment, air carriers were able to set operating standards at a level well above the minimums required by regulations. The same could be said of air traffic control; safety and conservatism were the overriding factors in its design and implementation.

This state of affairs changed dramatically during the 1980s for a number of reasons, including the air traffic controllers' strike in 1981 and an enormous increase in discretionary travel brought about by airline deregulation and the emergence of unfettered competition. The aviation system worked well despite these perturbations, but carriers found it necessary to adopt radically different ways of doing business. A major change was the introduction of "hub-and-spoke" flying, in which carriers selected "hub" airports, flew long segments between them, then shunted passengers onto shorter "spoke" flights to get them to their destinations.

This produced enormous concentrations of traffic that had formerly been more reasonably spaced, with consequent workload increases for controllers. The air traffic control system found itself handling considerable peak loads of traffic with outdated equipment, chronic understaffing, and less experienced controllers in many facilities. Since the early 1980s, the FAA has been working on plans for a radical upgrading of the ATC infrastructure involving major increases in automation to improve controller productivity, eliminate airspace bottlenecks, and increase traffic throughput. The first of the new equipment was scheduled to be installed in the Seattle Air Route Traffic Control Center (ARTCC) in late 1994, but the implementation schedule has slipped considerably and the costs have escalated by nearly $3 billion.

EVOLUTION OF THE AIR TRAFFIC CONTROL SYSTEM

Airport Air Traffic Control

Air traffic control began at airports during the late 1920s. The first controllers used flags and stood outside; later, control towers were built and controllers used light guns to provide one-way communication with airplanes. Radios began to be used during the middle 1930s, although most smaller aircraft did not carry them until after World War II, and light guns continued to be used well into the 1950s.

As all-weather transport flying increased and radar became available after the war, tower visual control of local aircraft was augmented by radar control of traffic in busier terminal areas. Terminal area controllers, attached to towers, were given separate radar facilities, which permitted them to provide departing air traffic with a transition to the en route environment and guide arrivals from that environment to a final approach to landing. Terminal radar approach control (TRACON) facilities were equipped with broadband radar, later augmented by data-processing equipment and automated data communication with en route centers. Full-performance-level controllers functioned as both tower and TRACON controllers.

Continuing increases in air traffic motivated the FAA to establish new categories of terminal airspace, in large part to separate fast jet traffic from slower, smaller (and harder to see) general aviation aircraft. Terminal control areas (TCAs) came into being; within these areas, generally shaped like an inverted wedding cake, all traffic, whether flying under visual (VFR) or instrument (IFR) flight rules, was required to submit to positive control by terminal area controllers. Beacon transponders and radio transceivers were required in order either to land at the primary airport or simply to transit the area. Other airspace reservations with less stringent requirements but also involving increased surveillance and control were put in effect around less busy airports (Fig. 7.1). The increasing requirements in these categories of airspace imposed a heavier workload on air traffic controllers. In theory, they lessened surveillance workload for pilots, although high levels of vigilance were still required, particularly at the vertical and horizontal margins of terminal airspace where many light aircraft flying just outside the controlled areas could still be encountered.

Effects of Increasing Terminal Airspace Complexity

These terminal areas assisted in air traffic segregation, but they imposed increased procedural and information processing burdens on pilots and controllers alike. When it became necessary to relax procedural separation standards and increase the use of visual separation procedures on approaches, the vigilance and information-processing requirements, especially on pilots, increased further. The total dependence of the system on low-bandwidth voice radio communications for real-time information transfer increased workload still more. These problems are made more pressing by the demand on air traffic control for still further increases in terminal area capacity, which has required innovative, complex procedures to stay ahead of (or even with) demand.

En Route Air Traffic Control

En route air traffic control began to be utilized in 1935 along airways marked by aeronautical beacon lights. En route air traffic control units (ATCUs, the first of which was established by TWA, United, Eastern and American Airlines at Newark because the government had no funds for it) communicated with air carrier dispatchers, who forwarded the information by radio to aircraft. Flight plans were made mandatory in 1936; ATCUs were taken over by the CAA in 1937–1938. In 1940, the CAA was reorganized to take account of its increasing responsibilities for

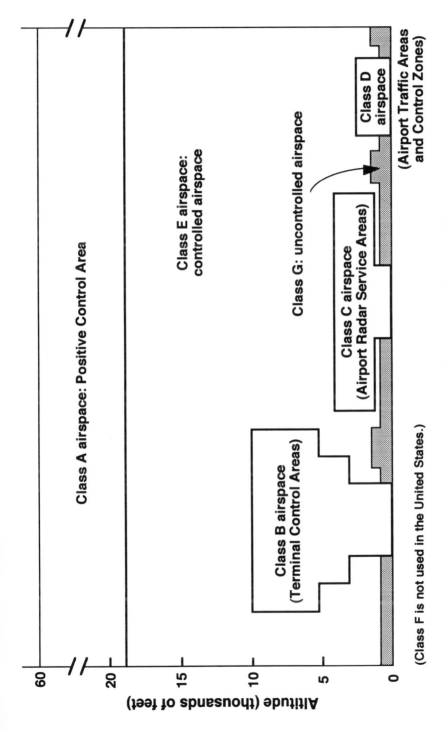

FIG. 7.1. Airspace categories in the United State. Reproduced by courtesy of FAA..

air traffic control; thereafter, it acquired control over traffic at all municipal airports and established 23 airway traffic control centers. During World War II, approach control facilities began to be established at some of the busiest airports (see Nolan, 1994, for an excellent succinct history of air traffic control).

Prior to the introduction of radar, en route control facilities kept track of traffic using flight progress strips. All information input was by voice radio; controllers kept a mental three-dimensional picture of traffic under their control and annotated the strips to correspond with reported positions and clearance modifications (see Fig. 7.3). Low-frequency radio navigation aids marked the various airway segments.

In 1956, radar became available and controllers began to be provided with a visual representation of traffic within their sectors of control. Primary radar provided only a display of aircraft locations; altitudes were still reported by voice, and procedural control was still required to ensure vertical separation. When the radar failed, controllers had to revert quickly to full procedural control based on flight strips and their mental picture of traffic, a function that required considerable skill. Communication was improved by the introduction of improved two-way VHF radio transceivers. Controllers still kept track of their traffic by annotating flight strips as instructions were given to each airplane.

The introduction of interrogation devices, the Air Traffic Control Radar Beacon System (ATCRBS) at radar sites, and transponders in aircraft that responded to queries from the interrogators made possible secondary surveillance radar (SSR) systems (also referred to as *narrowband radar*, as contrasted with *primary* or *broadband radar*). The transponders provided coded identification of specific aircraft, eliminated ambiguity as to location, and enhanced the radar returns from the responding aircraft. By the early 1960s, most en route centers were operating with SSR, to which was added over the next several years altitude reporting encoded in the transponder replies. This information, along with aircraft identification codes, provided controllers for the first time with positive three-dimensional location information for aircraft being controlled.

A disadvantage of the beacon system was that primary radar targets were poorly visualized, and not all aircraft operating within the system had transponders. The controller had to provide separation to these aircraft as well if they were operating under instrument flight rules, and this task became much more demanding as en route system controllers began to depend increasingly on direct representations of traffic. This is still a problem in many areas; although nearly all aircraft now carry transponders, some general aviation aircraft still do not have altitude encoding (mode C), and radar images of these airplanes consequently may be ambiguous as to altitude. (This is also a problem for TCAS, which cannot provide conflict resolution advisories without altitude data.)

Air Traffic Management

The development of the federal air traffic control system has often fallen seriously behind traffic demand. Even when Congress provided additional funding, usually following a major accident such as the midair collision of a TWA Constellation and a United Airlines DC-7 over the Grand Canyon (1956), it has been difficult to get ahead of the need for services. One result of this has been that controller morale has often been at a low ebb. This was evident following World War II, again during the Vietnam War, and more recently, during the late 1970s.

Manifest and latent labor–management problems got out of control in 1981, culminating in a disastrous walkout of union controllers in August. The President of the United States acted decisively to end the strike and 10,000 controllers who did not return to work within a few days were summarily discharged. The national airspace system was placed under draconian capacity controls but continued to function at a fraction of its former capacity, manned by supervisors and the relatively few controllers who had not participated in the strike or who had returned to work within the permitted window for such action.

It was at this time that strategic air traffic management, the foundations for which had emerged during the Arab fuel embargo of 1975, assumed critical importance in airspace management. Flow Control (now the ATC System Command Center; SCC) allocated airspace capacity to operators in accordance with system resources, established throughput targets at tolerable levels, and literally managed the entire system. The facilities, staff, and equipment available for this task were grossly inadequate, but the flow control system worked and provided ATC en route centers and terminal area control facilities with the buffer they required to continue to provide traffic services with whatever personnel were available. Pilots and operators cooperated in every way possible, and the system never broke down completely.

The activities of Flow Control during this period made startlingly obvious the need for a continuing strategic management function, supported by a communications and information management infrastructure that would provide it with a "big picture" of U.S. air traffic. After some years, the ATC SCC, located near Washington, DC, received an automated system visualization device, the aircraft situation display (ASD), which displays current aircraft positions and directions on a national scale, with superimposed maps of geographic and facility boundaries (Fig. 7.2). The system incorporates selective digital filtering to permit controllers to visualize special categories of aircraft or situations of interest; system software also enables SCC coordinators to project and visualize the effects of intended strategies for management of air traffic in response to weather or other contingencies. It is thus a strategic planning tool as well as a representation of the current traffic situation.

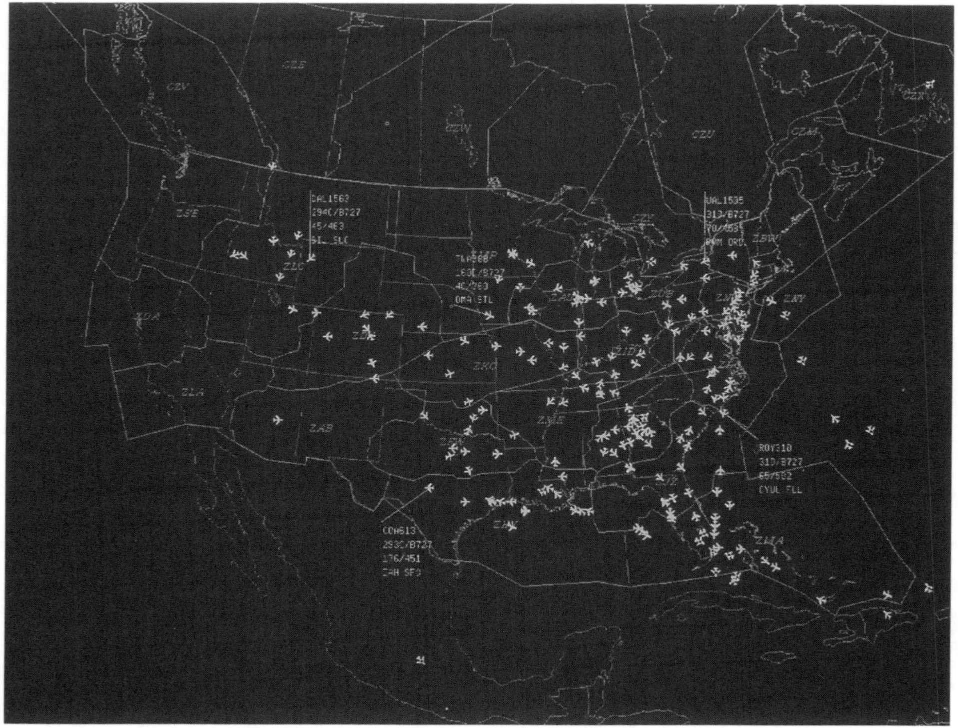

FIG. 7.2. ATC System Command Center aircraft situation display. Reproduced by courtesy of FAA.

The SCC's primary mission is to ensure that traffic demand does not exceed ATC system capacity. SCC personnel act as coordinators between users of the airspace (largely air carrier operations centers [AOCs], with whom the center has direct contact) and controllers in the various ATC facilities. During 1993, ASD displays also became available to AOCs, whose ability to manage their own traffic flow has been enhanced by access to the larger picture of air traffic activity. The SCC performs an extremely important integrating function for aviation, although it does not direct individual aircraft. This integration is a key to the efficient utilization of airspace whose capacity is strained by traffic demands.

Air Traffic Control Automation

Radar itself may be considered a form of automation, in that it integrates and provides a visual representation of a geographic or spatial phenomenon, and thus constitutes "a system in which a production process is automatically performed by a self-operating electronic device" (chapter 1, p. 6). Air traffic control radar incorporates a variety of electronic aids to reduce ground clutter, eliminate noise, overlay video maps on radar scopes, and so forth. In the early 1970s, the FAA began to install radar data processors (RDP) in en route centers, all of which make use of several remote radar sites to obtain full coverage of their airspace. Before radar data processing, sector controllers would utilize imagery from whatever individual radar provided acceptable coverage of their sector. RDP correlated the data from many radars to produce a composite synthetic image of all traffic using the best information available from its sensors. The result was a vastly improved visual representation of the best available data with less ambiguity and greater consistency, and thus decreased controller interpretive workload.

During the same time period, FAA installed flight data processors (FDP) that stored flight plan data, recognized the sectors through which flights would pass, and printed flight strips appropriate to each facility's responsibilities for flights. The FDPs were interconnected so that data on flights leaving a center's area would be passed automatically to the next center or terminal facility in line. FDPs also generated data for aircraft "tags" on controller plan view displays (PVD). Sector controllers continued to store the flight strips, annotate and move them to remind them of their flights' progress and requirements (Fig. 7.3). Hopkin (1994a) discussed the assistance that manual handling and marking of flight strips provides to controllers. He pointed out the information that adjacent sector controllers obtain simply by glancing at another sector's strip bay, the ability to re-sort the strips to take account of changes in traffic flow, and so on. Controllers have shaped this tool, as humans always do, to serve their needs. Some investigators believe that there is no longer a need for such tools (Vortac & Manning, 1994); others are less certain (Hughes, Randall, & Shapiro, 1992; Hopkin, 1994b).

N186MC	3465	OKK	OKK FWA MOTER DTW			
BE20/R	P2040		~~110~~			
979	170		YIP 090			

FIG. 7.3.   Flight data strip, showing annotations by controllers as the airplane's clearance is modified in the course of its flight.

During the past decade, despite severe limitations on data-processing capacity within aging ATC computers, several automated monitoring and alerting functions have been added to the ATC system. Conflict alert, designed to warn of a failure of separation minima, provides an audible alarm in the ATC facility if standards are transgressed. Unfortunately, violation of these separation minima subjects controllers to adverse action if they are found at fault. In response to controlled flight toward terrain incidents (and a small number of controlled flight into terrain accidents despite GPWS in aircraft), a minimum safe altitude warning (MSAW) module was developed. Later, an automated altitude monitoring function was added, which alerted controllers if pilots transgressed an altitude clearance limit (pilots violating their altitude clearances also face enforcement action by FAA).

Effects of Air Traffic Control Automation

As radar and ATC automation became more reliable in the 1970s, NASA Aviation Safety Reporting System reports began to contain comments from older controllers, highly expert in the art of procedural control, worrying that their younger colleagues had become dependent on radar representations of traffic and thus were less skilled in constructing a three-dimensional mental image of the traffic under their control. Although some ASRS reports did indicate that sudden radar failures produced short-term disruption of traffic control, reports of serious incidents were uncommon. A few really dangerous losses of control did occur (e.g., Atlanta, 1980); these were most often ascribed to training or proficiency problems, although the investigations usually did not delve into latent factors (Kinney, Spahn, & Amato, 1977). Air traffic controllers (and ATC system managers) cultivated the image of great mental strength and individuality; this image did not willingly admit to personal or system weaknesses that could compromise the performance of their critical tasks (Rose, Jenkins, & Hurst, 1978; Flight Safety Foundation, 1982).

Nonetheless, the ATC system was under strain when it met its greatest challenge in 1981, the sudden departure of the great majority of its experienced operators. That it survived this challenge, even under severe constraints, reflects the capability and dedication of the humans who remained to operate it after the strike. The men and women who continued to control traffic were severely tested, but the basic structure of the system survived and remains in place today, awaiting advanced equipment that hopefully will enable the system to meet still greater challenges ahead. The plans for the new advanced automation system (AAS) were drawn to provide greater flexibility, productivity, and capacity. Whether the system can meet its new demands will depend on whether its design provides controllers with the flexibility to meet the challenges of an environment many aspects of which are not under the control of the human operators in the system. (See chapters 8 and 11 for further discussion of these issues.)

COMMENT

One of the continuing problems in the aviation system has been that its two principal components, the aircraft and the air traffic control infrastructure, have usually been considered in isolation. Aircraft designers usually gave only passing consideration to the operational environment in which their machines must operate; ATC system designers have usually considered aircraft simply as point objects to be moved from place to place. Most controllers are not pilots, and virtually no pilots are, or have been, controllers. Designers in each sphere rarely have truly adequate knowledge of the other domain.

Although this has not created insuperable handicaps in the past, evolving automation in aircraft, unaccompanied by similar development of the ATC system, has led to increasing disparities between aircraft and ATC capabilities. These and increasing demands on the entire system are now manifest as delays, which are expensive both to operators and to airline passengers. Although future ATC automation, the subject of the next chapter, may help resolve some of these discrepancies, it is critical that the future system's architecture recognize that the aviation system is a *single* system (chapter 11). Only with this recognition will the system be sufficiently functional to meet the demands on it.

Chapter 8

Future Air Traffic Control and Management Automation

INTRODUCTION

The FAA is in the midst of the largest air traffic management system upgrade in its history. Its intended product is the Advanced Automation System (AAS). Europe is beginning the harmonization and integration of its multiple national air traffic control systems, and many members of the European Community are also undertaking massive equipment modernization programs. There is every reason to believe, therefore, that future air traffic control systems will look very different from the systems of today.

It is possible to suggest in some detail what these future systems may look like and how they will function. What cannot be stated with any clarity at this time is what human and machine roles will be in that system, because its designers have not approached the question except in general terms. Joseph Del Balzo's (1992) forecast of "an era where air travel is unhampered by ... the limitations of human decision-making" suggests the depth of the concern about human capability and reliability among senior system managers. If this concern is permitted to dominate the debate about the shape of the future aviation system, that system will not be a human-centered air traffic management system, and if ATC system automation is not human-centered, automation in the remainder of the system will not be either.

It is for these reasons that this book attempts to make a case for a human-centered automation system for air traffic control as well as for aircraft. There are not two systems (air and ground); there is one National Aviation System. Its elements must be designed and operated from a common philosophical base if the system is to be maximally effective. Forecasts of capacity demands indicate that even if the system operates optimally, its capacity will still be strained by early in the next century. We must, therefore, make the most of what we will have during that period.

162

FUTURE AIR TRAFFIC CONTROL
SYSTEM CHARACTERISTICS

During the past decade, planners and system designers have been pondering the shape and characteristics of the future air traffic management system. The many discussions on this topic have put forward alternatives ranging from full ground-based automation (the FAA's full automation system [FAS]; see scenario 3) to user-based separation assurance with minimal ground participation (the free flight concept recently advanced by air carrier managers; see scenario 4). As a result, the shape of the future system is less certain now than it has been at any previous time.

The concept of *free flight* appears to have been embraced, at least in the United States, by many of the stakeholders in the aviation system. The implementation of such a system, however, will require much research and development, and it may or may not be realizable. Nonetheless, the free flight concept must be described here; it will be seen as one extreme in a spectrum of possible approaches to air traffic management, with the other pole or extreme being the rigid central control of traffic by an automated ground system.

A GROUND-CONTROLLED AIR TRAFFIC SYSTEM

To consider first a future system in which ground-based air traffic facilities retain control of airspace and traffic, there are several possibilities with respect to human roles and tasks. These roles are discussed using terminology that is discussed in chapter 10, in which pilot and controller roles, responsibilities, and authority are described. I characterize these possibilities in the following scenarios.

Scenario 1: Management by Delegation

The first scenario, the least radical, involves a system in which the controller manages by delegation, as do pilots of present-day aircraft. It is an outgrowth of today's system in many respects: The controller is given an enhanced multicolor plan view of traffic in a sector of airspace, a data display that presents flight-strip analogs, rules governing the handling of that traffic, and a variety of automation tools that can be used to accomplish the task.

This scenario is roughly analogous to the environment of the pilot of a moderately automated airplane; a variety of aids, developed in consultation with controllers of widely differing experience and expertise, would be available for use as needed. If a controller wished to control traffic without such assistance, the machine would let him or her do so while monitoring the controller's actions for discrepancies (potential conflicts, actions not permitted by current procedures, potential incursions into special-use airspace, etc.). Most important, the computer would alert the controller to such anomalies before they resulted in transgression of permitted boundaries, to permit him or her to take corrective action early. In that sense, the computer would improve the error tolerance of the human–machine system.

Scenario 2: Management by Consent

This scenario assumes a higher degree of machine intelligence somewhat like that projected for the FAA advanced automation system's advanced en route air traffic system (AERA 2) automation. In this scenario, the ATC computer accepts requests for flight plans or flight-plan modifications. It examines the effects of these requested trajectories over a 20 to 30 minute period (and perhaps a longer period for initial requests), approves them if no conflict is detected, or suggests modifications to avoid a future conflict. (This may be done in automatic "negotiation" with the affected airplane's flight management computer.) The computer's output, when accepted by the controller, is an approved flight plan, either without or with modifications of the plan requested. This plan is shown (hopefully graphically, as on current aircraft navigation displays) to the controller. As in AERA 2, the controller may request another alternative or may input an alternative for evaluation. When a plan is acceptable to the controller, he or she gives consent, after which it is transmitted as a clearance to the affected aircraft. On acceptance by the pilots, it is executed in the FMS.

This scenario differs from the previous one in that the machine takes initial and primary responsibility for development of plans, subject to consent by the controller. The level of automation is fixed, but the human can bypass machine decisions by making an alternative plan acceptable to the error monitor. Scenarios 1 and 2 are not mutually exclusive; it would be possible to embody both architectures in a single machine, giving the controller the ability to select which automation option he or she wished to utilize.

Scenario 3: Management by Exception

This scenario involves a somewhat higher degree of automation than has thus far been suggested by air traffic management for near-term implementation. In view of increasing demands upon the ATC system, however, I think it likely that a more autonomous solution may be proposed as a "growth" version of the next-generation air traffic control system. In this scenario, computers would perform all of the functions listed under scenario 2, but they would select and exercise decision options autonomously. The human air traffic monitor (he or she would no longer actually control traffic) would be informed by some means of the present and future (intended) traffic situation and would manage by exception. Compliance with machine-generated clearances would likewise be monitored by the computer, which would alert the human monitor to any undesired behavior of aircraft. The human could intervene by reverting to a lower level of automation, or could instruct the computer to resolve potential conflicts or problems.

Here, the ATC computer, and pilots in flight, are controlling air traffic. The controller's role is to insure that the machine is behaving in accordance with predetermined rules and that its actions are in conformance with directives and procedures. Human monitors would be trained to evaluate and detect departures from permitted machine behavior and given means with which to limit machine authority as needed to maintain a safe operating environment.

Where would the responsibility lie in such a system? I believe it would have to vest in the operating organization and the system's designers. It could not remain with the system's operators; they would be too far removed from the details of system behavior to accept full responsibility for outcomes. This is a potentially troublesome problem, but a much more difficult problem would be to endow such an automated system with enough flexibility to encompass the range of environmental and other variables that can affect air traffic.

I recognize the exponentially increasing capability of computers and I will readily admit that such a capable system is at least thinkable. It would be extremely expensive, but it could convey enormous return on investment if it worked, and its throughput under normal circumstances might be as good as or better than that of other, less automated systems. The newest aircraft flight control and guidance systems are essentially capable of being managed by exception; it is at least conceivable that at some point, air traffic control automation will also be a highly intelligent machine agent, although it is far in the future.

Flow Control: Strategic Traffic Management

Each of these scenarios assumes the existence of a strategic management function. In the United States, the FAA's enhanced traffic management system (ETMS) will improve the ability of the ATC System Command Center to manage traffic in cooperation with traffic management units at each en route center and system operations center at airlines. As mentioned earlier, SCC aircraft situation displays are already available to air carrier dispatchers, and this has laid the foundation for increased cooperation between SCC and its customers. Recall that the basic purpose of the SCC is to insure that air traffic demand does not exceed ATC facilities' ability to handle it. If a runway, or an airport, becomes unusable, the SCC coordinates a reduction or diversion of traffic flow until the facility problem is resolved.

The ETMS will provide an enhanced aircraft situation display, a monitor-alert function, automated demand resolution, a strategy evaluation and recommendation module and other decision aids, and a directive distribution function, among other automated tools. Elements of these functions are already in use; the final product will provide personnel at the System Command Center with even more formidable strategic management capability.

A central flow management unit (CFMU) for western Europe, located in Brussels, came into operation in 1993. It is planned that there will be an equivalent facility in Moscow for eastern Europe. As in the United States, the CFMU will coordinate with flow management positions at each area control center.

Terminal Area Traffic Management

These scenarios also assume the existence of software modules that will extend some type of automated air traffic control from takeoff to landing. Oddly enough, terminal control elements, which are substantially more difficult than automated enroute control, are also farther along, in large part because of a research and development effort called the Center-TRACON Automation System (CTAS). CTAS, initiated by NASA's Ames Research Center in the mid-1980s, has developed as a decision aiding system; it does not operate autonomously. It provides displays and tools to help controllers secure maximum utilization of terminal airspace by flow planning and precise direction of descent and approach maneuvers (Erzberger & Nedell, 1989; Harwood & Sanford, 1993). Its functionality is generally similar to that proposed earlier in scenario 2.

CTAS is a time-based system (Tobias & Scoggins, 1986) that when fully implemented contains three modules: a traffic management advisor (TMA), a descent advisor (DA), and a final approach spacing tool (FAST). The TMA has been developed for use by traffic management units, which monitor the demand of arrival traffic and coordinate with ATC facilities to make decisions about balancing traffic flow so demand does not exceed capacity in center and terminal areas (Sanford et al., 1993).

CTAS has undergone a great deal of simulation testing in cooperation with the FAA's terminal air traffic control automation (TATCA) program. During 1994–1995, elements of the system were implemented at Denver for evaluation with live traffic (Harwood & Sanford, 1993); tests at Dallas-Fort Worth and other terminals are also planned. The system offers considerable promise with regard to maximizing terminal area traffic throughput.

En Route Air Traffic Management: The AERA Concept

Since the early 1980s, FAA and its contractors, notably the MITRE Corporation, have been developing an automated en route air traffic control system (AERA).

This proposed system has undergone many changes since it was initially proposed, but its outlines have remained. In brief, the AERA concept envisions automation similar to that described in scenario 2. Automation would maintain surveillance of air traffic movements, detect potential conflicts over a 20 to 30 min window, and provide revised clearances to mitigate detected conflicts. Controllers could accept these machine decisions or could propose alternatives to deal with the detected problems. Data link would communicate clearances to aircraft (Kerns, 1994); voice communications would be available for emergencies.

THE FREE FLIGHT CONCEPT

Prompted by the relative inflexibility of the present system of en route control of air traffic and growing understanding of potential economic benefits if more flexible routes can be approved, airline managements and their representative organizations, ATA and IATA, have considered more fundamental changes in air traffic management. They have recently proposed a "free flight" concept, in which operators would have the freedom to determine airplane trajectories and speeds in real time (IATA, 1994; RTCA, 1995). The radical nature of this proposal amounts, in essence, to a fourth scenario for future air traffic management. Relevant parts of the RTCA document are therefore extracted here.

Scenario 4: Free Flight

To quote from the Report of the RTCA[1] Board of Directors' Select Committee on Free Flight (1995):

"Free Flight" is defined as:

A safe and efficient flight operating capability under instrument flight rules (IFR) in which the operators have the freedom to select their path and speed in real time. Air traffic restrictions are only imposed to ensure separation, to preclude exceeding airport capacity, to prevent unauthorized flight through special use airspace, and to ensure safety of flight. Restrictions are limited in extent and duration to correct the identified problem. Any activity which removes restrictions represents a move toward free flight. (p. 3)

---

[1]The Radio Technical Commission on Aeronautics is a nonprofit organization sponsored largely by FAA. It proposes standards for technology and its uses in the national aviation system.

The Free Flight concept is described (p. 7, section 3.0) in this way:

Free flight can provide the needed flexibility and capacity for the foreseeable future (approximately the next 50 years). At its basis, the concept enables optimum (dynamic) flight paths for all airspace users through the application of communications, navigation, and surveillance and air traffic management (CNS/ATM) technologies and the establishment of air traffic management procedures that maximize flexibility while assuring positive separation of aircraft.

The primary difference between today's direct route clearance and free flight will be the pilot's ability to operate the flight without specific route, speed or altitude clearances. Restricting the flexibility of the pilot will only be necessary when (1) potential maneuvers may interfere with other aircraft operations [or] special use airspace, (2) traffic density at busy airports or in congested airspace precludes free flight operations, (3) unauthorized entry of a special use airspace is imminent or (4) safety of flight restrictions are considered necessary by the air traffic controllers.

In the free flight system, a flight plan will be available to the air traffic service provider to assist in flow management, *but will no longer be the basis for separation* [italics added]. It is possible, and highly desirable, to shift from a concept of strategic (flight path based) separation to one of tactical (position and velocity vector based) separation. There even may be instances included in the system's design where separation assurance shifts to the cockpit. When this occurs, there will be clearly defined responsibility (pilot or controller) for traffic separation.

Realization of free flight depends on construction (by ground and/or aircraft computers) of protected and alert zones around each airplane in the system, as shown in Fig. 8.1.

Position and short-term intent information is provided to the air traffic service provider who performs separation monitoring and prediction functions. The air traffic service provider intervenes to resolve any detected conflicts. Short-term restrictions are used only when two or more aircraft are in contention for the same airspace. In normal situations, aircraft maneuvering is unrestricted. Separation assurance may be enhanced by appropriate on-board systems [Author's note: Presumably an enhanced collision avoidance system]. (section 3.1; p. 7)

When alert zones touch...automation will determine the best position and velocity of each aircraft. The flight paths will be projected to determine if the minimum required separation will be violated.... If the projection shows the minimum required separation will be violated, the conflict resolution software suggests an appropriate maneuver. With controller concurrence, these instructions will be passed to the aircraft involved. The automation and the controller then verify the required separation is assured. (section 3.2; p. 8)

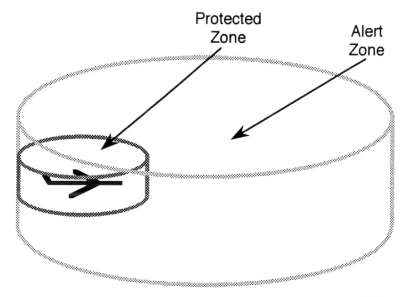

FIG. 8.1.   Computer-constructed protected and alert zones for aircraft in the proposed Free
Flight system. From *Report of the RTCA Board of Directors' Select Committee on Free Flight* (p.
7), by Radio Technical Commission on Aeronautics. Reproduced with permission.

Issues Raised by Future Air Traffic Management Concepts

Both the AERA 2 and free flight concepts raise substantial issues with respect to
human–machine cooperation in the aviation system. Both concepts envision radical changes in the architecture of airspace control, although the free flight concept
is more revolutionary than the AERA concept. Each concept implies major shifts
in human and machine responsibilities, and each could involve major shifts in the
locus of control of the aviation system. Some of the more important human–machine issues are discussed next.

Human and Machine Roles in the Future System: AERA-2

As implied earlier, air traffic control automation can go in several directions:
either toward a more autonomous machine system, or toward a system in which the
human operator remains in command. Each direction presents different theoretical
advantages and different potential problems.

There is substantial sentiment within FAA and Eurocontrol for a more automated system that reduces the probability of human error by reducing the span of

human control. Several studies of operational errors (defined as loss of prescribed separation between aircraft) have found a trend toward more errors under light or moderate, as opposed to heavy, workload (Kinney et al., 1977; Schroeder, 1982; Schroeder & Nye, 1993), although it has been hypothesized that more serious errors occur under heavy workload conditions (Rodgers & Nye, 1993). Deficient situation awareness has been implicated as a factor associated with the severity of operational errors. Not surprisingly, human operators are almost always found to be at fault.

Although our understanding of latent factors in the causation of human errors has progressed considerably, there is no doubt that many in the air traffic control community look to automation as the principal way to improve ATC reliability. This being the case, it will be necessary to make a compelling case for keeping the human controller in effective command of the system once advanced ATC automation is available. I say "effective command" because there seems little question about the controller's continuing responsibility for traffic separation regardless of the amount of automation interposed between the controller and his or her traffic. It is encoded in high-level operating guidelines for the AERA 2 system when it becomes operational (Celio, 1990):

- "Responsibility for safe operation of aircraft remains with the pilot in command."
- "Responsibility for separation between controlled aircraft remains with the controller."

I argue in this document (see chapter 10) that if the human remains responsible for safety, then the human must retain the authority with which to exercise that responsibility, by whatever means—automation must be a tool over which the human must have full authority. The operating guidelines offered by Celio do not give cause for comfort:

Since detecting conflicts for aircraft on random routes is more difficult than if the traffic were structured on airways, the controller *will have to rely* [italics added]on the [automated] system to detect problems and to provide resolutions that solve the problem.

Alerts may be given in situations where later information reveals that separation standards would not be violated.... This is due to uncertainty in trajectory estimation.... Therefore, alerts must be given when there is the possibility that separation may be violated, and the controller *must consider all alerts as valid* [italics added].

In its Executive Summary, the report states,

Machine-generated resolutions offered to a controller that are free of automation-identified objections *are assumed feasible* [italics added] and implementable as presented.

The controller *will use* [italics added] automation to the maximum extent possible.

Note that if a controller accepts a computer decision and it turns out to be faulty, the controller is responsible. If the controller rejects a computer decision and substitutes one that is faulty, the controller is also responsible. This sort of dilemma represents a classic double bind (Woods et al., 1994). Note also that in this sort of system, de-skilling (Cooley, 1987) is very likely to occur over time. Finally, because the AAS computer will resolve conflicts over a relatively long time window (20 min or so), the controller who issues a machine-recommended clearance may not be able to assess retrospectively whether that choice was correct, for the outcome of that clearance will often occur in a sector not under his or her control and not visible to him. The alternative, of course, is to revert to short-term controller-initiated conflict avoidance, as occurs routinely in the present system.

Pilot, Controller, and Machine Roles in a Free Flight System

Although the free flight proposal is new and has not yet undergone intensive scrutiny, it clearly represents a carefully considered proposal that expresses the frustration of operators with what is perceived as an increasingly outdated, cumbersome, and inflexible ground-centered concept of aviation operational control. The first stages of a free flight precursor are already under test at altitudes of 31,000 ft and above; lower altitudes will be included over the next year. Building on its previous experience with the aviation safety/automation program, CTAS, and other control strategies, NASA is planning a new aeronautics initiative called Advanced Air Transportation Technology (AATT) and has established interdisciplinary teams to explore this concept and the technology needed to bring it to fruition.

The AERA concept poses major questions concerning the role of the air traffic controller. I believe that the free flight concept, as set forth earlier, poses still more fundamental questions concerning human (both pilot and controller) and machine roles in the future air traffic management process. More important, it calls into question many of the fundamental assumptions on which the largely successful ATC system has been built. The architecture of a fully developed air traffic management system designed around this concept would have to be radically different from that presently proposed for the AAS because of its emphasis on short-term tactical, rather than strategic, management and its implication that management should be almost entirely by exception. Such a system would involve a qualitative change in the roles of the humans and machines that operated it. Some of the issues raised by the concept are set forth next.

The free flight concept envisions that flight paths would be selected by pilots, or more likely dispatchers working in air carrier operations centers (AOCs), and implemented by the pilots with notification to, but without a requirement for concurrence by, the air traffic management system. This concept envisions the entry of a third, more or less equal, authority into the control process—the AOC—and it thus raises many questions about diffusion of authority and responsibility for air traffic movements (see chapter 10).

The air traffic management subsystem would be relegated to an oversight role unless a conflict were detected. It appears that the ATM system would function in somewhat the way that airborne collision avoidance systems now function: by using aircraft data and extrapolated trajectories to develop time- or distance-based separation zones around aircraft, which are used to determine a need for alerting or conflict resolution in real time. The concept implies the existence within the ATM system of computer software that can accomplish the predictive functions planned for the AERA-2 system, but with additional uncertainty posed by random variations in flight trajectories.

These uncertainties would pose problems for controllers (and for the ATM system) similar to but more acute than those they now face when TCAS issues a resolution advisory to pilots, who respond prior to notifying ATC that they are doing so. Yet controllers are expected to intervene, or supervise the computer that will intervene, on a "by exception" basis, only when there is a high probability of conflict. The likelihood that controllers will be able to detect and diagnose probable conflicts under a high level of uncertainty is low, and the proposal recognizes this by stating that computers will accomplish this task and suggest appropriate resolution tactics. The computer, of course, will also have periods of uncertainty during aircraft maneuvers, before it is able to reestablish a stable trajectory prediction and project it forward in time to evaluate whether the maneuver has created a potential conflict. Yet the IATA (1994) proposal also states that the purpose of automation is to assist humans and not to replace human reasoning. (Regarding human factors, the RTCA [1995] report says only that "technology will not replace humans or replace their reasoning process but will allow them to do their job better" [p. 5]). The compressed times within which the humans would have to apply such reasoning and take action, perhaps given only retrospective notification of aircraft maneuvers, appear to have received rather less attention than they deserve in the development of the concept.

The free flight proposal does not explicitly mention significant additional requirements on pilots. Nonetheless, the lack of "assured separation" that is provided by the present, admittedly cumbersome system would actually require higher vigilance throughout flight operations, because maneuvers could be instituted by any pilot without advance knowledge of the locations of other aircraft whose own trajectories might be affected by such maneuvers.

The IATA and RTCA documents clearly envision TCAS, perhaps with lateral as well as vertical maneuver capability (the capability proposed for TCAS-3), as an additional means of conflict detection and resolution ("Separation assurance may be enhanced by appropriate on-board systems"; RTCA, 1995, section 3.1). Yet TCAS resolution maneuvers are a prominent source of problems for controllers in the ATC system today and would almost surely present more difficult problems for them and for ATC computers in a less constrained free flight system.

The requirement for knowledge of other aircraft positions, altitudes, and trajectories would require an enhanced cockpit display of traffic information. TCAS in its present form provides only a rough approximation of the information required for this task. As noted earlier, its representations of traffic are not entirely adequate even for its present tasks (and its software thus far has not been able to handle all of the situations in which it should or should not provide traffic or resolution advisories). It appears that free flight would impose substantially greater requirements on airborne collision avoidance equipment, as well as considerably greater separation assurance requirements ("see and avoid [by whatever means]") and therefore workload on pilots. Pilots are not presently required, or trained, to think in terms of the four-dimensional resolution of traffic conflicts, yet this new task would be required of them during climbs and descents. This proposal would certainly increase the *involvement* of pilots in enroute operations (see chapter 3), but it does not address how pilots would be kept adequately *informed* of the positions and paths of other aircraft that may become a future problem for them.

The IATA and RTCA documents differ somewhat with respect to the issue of timely notification of pilot-initiated maneuvers to the ATM system. The RTCA report states that "Position and short-term intent information is provided to the air traffic service provider" (p. 7), but the IATA plan says, "*Subsequent* [italics added] to any change, a revised plan will be data-linked to the ground system for planning purpose." The issue of intent communication is critical, as discussed next.

Perhaps the most worrisome aspect of the free flight proposal in its present form is its central assumption that automation can enable greater flight path flexibility without imposing greater order on the system it would control, and without providing human operators in the air and on the ground with the information they would require to maintain command over the system. Separation standards would no longer be constrained to provide time for prospective action to resolve *potential* conflicts; such conflicts would be resolved in real time as they occurred. *Intent* might no longer be required to be communicated in advance of an action.

It appears that little thought has been given to whether humans can operate and manage such a system. Rather, new technology will operate the system and humans will supervise its operation, but not necessarily with advance knowledge of how it is going to behave. As proposed, the airborne systems may not inform the ATM system of their intent in advance; the ATM system and its supervisory controllers will not predictably be involved in air traffic movements, and may not have advance knowledge of any individual airplane's future trajectory. For these reasons among others, the likelihood that pilots and controllers will be able to remain in command of such a system is very low, for the system will not be predictable. *The likelihood that those humans will not be held accountable for the results, however, is negligible.*

Wiener (1993, p. 4) pointed out (with respect to aircraft automation) that "Many in the aviation industry have assumed that automation would remove human error, replacing the fallible human with unerring devices. The research of Wiener and Curry ... suggests that this may be overly optimistic, and that automation merely changes the nature of error, and possibly increases the severity of its consequences." In an earlier paper (1987, p. 179), he said,

> The experience from commercial aviation shows that it is unwise to dream of automating human fallibility out of a system. Automation essentially relocates and changes the nature and consequences of human error, rather than removing it, and, on balance, the human operator provides an irreplaceable check on the system. The search should be directed toward the management of human caprice, not the elimination of its source. (See also pp. 146–150).

There is no reason to believe that automation in air traffic control and management will be a panacea, any more than it has been in the cockpit. As Woods has commented, any tool, including automation, shapes human behavior. The human errors expected in a highly automated system would be expected to be different, and indeed they are different. But automation does not, and cannot, eliminate human error (although if properly designed, it can sometimes mitigate the consequences of human error).

Automation, of course, is not infallible either. The literature abounds with failures of automation to perform as expected; a few examples in aircraft are shown in Fig. 6.9. These failures are among the reasons why humans must be an integral part of the system—they are there to compensate for the imperfections of the automation. They are also there, as noted earlier, to accept responsibility for system safety. If they are to remain in command, they must be involved in system operation—not only when the automation fails, but during normal operations as well, in order to be in the loop when the inevitable failures occur. The human operator is the final line of defense in automated systems, and the new systems proposed for air traffic management are no exception.

Implications of Future System Design Proposals

The complexity of either of these automated systems for air traffic management will be far greater than the complexity of a flight management system, and pilots' problems in understanding that system's behavior are discussed in chapters 5 and 6. It can be confidently predicted that similar problems will be encountered in the air traffic management domain if controllers are unable to form adequate mental models of the system's processes. Those processes must be both comprehensible and predictable, both so the controller can predict them and so failures of the automation can be detected. Consciously reducing the predictability of a highly integrated, cooperative human–machine system seems an unlikely way to achieve greater system safety.

A summary of comments regarding the human's future role in air traffic management, made during a recent conference on European air traffic management, is instructive but unsettling:

> "The present system of air traffic control has been in existence with few fundamental changes for 40 years. Even with technological improvements, the present system is likely to reach capacity limits by 2005," says Peter Whicher of Logica. The provision of automatic aids to assist the controller is only marginally likely to defer the problem, and the deadline [Whicher] sets for getting a new concept installed and running successfully, with a potential capacity of at least *five* times 1992 traffic, is 2010.

> The basic requirement *is to minimize human control involvement in routine events* [italics added] and concentrate skills on system and safety management, and on the resolution of exceptional situations.

> "To permit unrestricted ATC growth we should first determine how to eliminate one-to-one coupling between a proactive sector controller and every aircraft in flight—and so avoid him becoming reminiscent of the man with a red flag in front of early motor vehicles. With improved area navigation and flight management systems, pilots can and are willing to take direct responsibility for routine enroute track-keeping functions," Peter Whicher explains, "freeing controllers to concentrate on the key areas where human skills have most to offer—traffic management, system safety assurance, and dealing with the exceptional occurrence."

> Mr. Whicher foresees two possible concepts of control for the next century: one is *full aircraft autonomy*, [italics added] the other is its opposite—*full ground control automation* [italics added]. (Cooper, 1994a, p. 8)

The role of the future controller proposed here is that of a monitor, not a manager, except when exceptional situations are detected. If an automated ATC system works well, such situations should arise relatively rarely—and the human controller is back to the situation Mackworth (1950) described so effectively, searching over long periods of time for rare events that may not be particularly obvious when they arise. As with pilots of long-haul aircraft, some form of active involvement is required if controllers are to remain in command of the traffic situation. Further, active involvement in air traffic control is necessary to prevent skill degradation (Cooley, 1987; Rauner, Rasmussen, & Corbett, 1988).

If the controller is to remain in command, and if automation is responsible for conflict detection and resolution, it must inform the controller of what it is doing and how. We know from previous studies in aircraft, nuclear power plants, and elsewhere that complex automation tends to be opaque to its observers. Controllers must be informed, not only of the traffic situation, but of the processes that are being invoked to modify that situation, if they are to remain controllers rather than simply machine monitors. If the controller is *not* to remain in command, then system architects must state more clearly who or what is to replace the controller, and how. Responsibility for an adverse outcome *will* be placed at some human's door.

Whicher's concept of the future ATC system might be economical, but field observations and empirical research suggest that it is not likely to be effective. Are there alternatives that will still accomplish the objective of increased throughput? I believe that management by consent, as exemplified earlier in scenario 2, offers at least a greater likelihood of preserving controller involvement in the tactical management of air traffic. There is a problem with such an approach: It may be difficult to prevent situations in which consent is perfunctory rather than thoughtful, if a controller is tired or distracted. Nonetheless, it is preferable to management by exception (scenarios 3 or 4), in which the "controller" is not intimately involved in the control process.

Given that more controller operational errors occur during periods of lighter rather than heavy traffic, I would prefer from a human factors viewpoint to see a work environment in which the controller could adjust his or her workload as required, by invoking automation to offload some of the routine tasks while preserving authority over the more complex and challenging tasks such as planning and management of exceptional situations (as pilots do today). This approach to ATC automation resembles that available today in advanced aircraft, in which the pilot is able to preserve control skills by exercising them, but is also able to lighten routine workload when desired.

COMMENT

As Benjamin Franklin observed at the signing of the U.S. Declaration of Independence, "We must all hang together or we shall assuredly hang separately." Much the same can be said of human operators and automation in complex systems. What is required is a *cooperative* relationship between humans and machines, in which each intelligent agent augments the strengths and compensates for the deficiencies of the others. Can the basis for such a relationship be established in future automated ATC systems? I believe it can be, but that it must be a part of the fundamental architecture of such a system, which means its basis must be established very early in the design process.

Ongoing attempts to make air traffic management more effective can, and should, point the way to the shape of the future tactical air traffic control system. But we must not lose sight of the strengths of the current system, and of why it works as well as it does. Before new technology is designed for a future system, ATM concepts should be brought under intensive scrutiny to determine the ingredients of success in a complex, distributed human–machine system whose performance can be evaluated and measured quantitatively. The cognitive factors that make for success or failure in this team enterprise are beginning to be understood and can serve as a model for the design of the future tactical ATM system. Without this model to drive the architecture of the system, the technology is likely to fail.

The fundamental question raised by present proposals for the architecture of the future air traffic management system is simply whether future ATC automation should be designed to assist human controllers or to supplant them. As I have said earlier, a fully automatic ATC system may be thinkable, and might have important economic benefits. I do not believe its productivity would be appreciably higher than a cooperative human–machine system; it would be less flexible than a cooperative system by virtue of being unable to call on human creativity in dealing with unplanned contingencies, and there will always be such contingencies. Further, the difficulties that have already arisen in connection with the development of system software for AAS will be magnified many times by the enormous cost of developing a more fully autonomous system even if it is possible in theory.

Although the "free flight" concept envisions very important economic benefits for air carriers, and perhaps increased ATC system productivity (if fewer controllers were to be needed), it would require that a full ATM infrastructure remain in place to deal with exceptions. Further, much new and more complex ATM automation would be required to deal with conflict prediction in a less orderly system involving random, unpredictable flight paths. This factor would also decrease the amount of time available to human managers, who would be expected to exercise flexibility in the resolution of conflicts. The new automation could potentially bring with it more of the problems to be discussed in chapters 9, 10, and 11.

Dr. Hugh Patrick Ruffell Smith, a very wise human factors expert, observed in 1949 that "Man is not as good as a black box for certain specific things; however, he is more flexible and reliable. He is easily maintained and can be manufactured by relatively unskilled labour." We should think carefully about this observation as we contemplate the shape of the future ATC system.

PART III

THE ROLES OF HUMAN OPERATORS IN THE AVIATION SYSTEM

In part II, I discussed the developmental history of automation in industry and aviation. In part III, I try to encapsulate some of the benefits, and some of the costs, of aviation automation in terms of the human operator's ability to work cooperatively with highly automated systems. Not all of these costs, by any means, are inherent in the automation; many have resulted from humans' deficient mental models of that automation. Other problems result from cumbersome interfaces between the humans and their automated tools. Whatever the reasons for these problems, they tend to make the human–machine system less effective, less reliable, or less safe. As noted in the Preface, this document is not a study only of humans who use automation, nor of the automation itself, but of the *system* in which both attempt to work cooperatively to accomplish social objectives.

In chapter 9, I summarize and generalize some of the problems introduced in part II to remind the reader of what they are, where they are seen, and why they occur. Chapter 10 discusses in more detail a central question in human–machine system design and operation: the respective roles of the human and machine, and how responsibility and authority are apportioned in this system. Chapter 11 discusses an important issue with regard to aviation system architecture: whether the future aviation system should be more tightly coupled, or whether it should remain integrated but uncoupled, as at present. The points made in these chapters are the basis for the human-centered automation concepts presented in chapter 3, and for the requirements and guidelines in part IV.

CHAPTER 9

BENEFITS AND COSTS
OF AVIATION AUTOMATION

INTRODUCTION

The NASA Aviation Safety/Automation research initiative (NASA, 1990), the work of Wiener and Curry that preceded it (Wiener & Curry, 1980; Curry, 1985; Wiener, 1985a, 1989, 1993); studies by Rouse and colleagues (1980, 1988; Rouse & Rouse, 1983; Rouse, Geddes, & Curry, 1987), research by Sarter and Woods (1991; 1992a, 1992b; 1994), Sheridan's (1984; 1987; 1988) studies of supervisory control, and contributions by Rasmussen (1988), Reason (1990), and many others form the theoretical and empirical foundations for these comments on humans and automation. There is now a substantial body of data concerning human cognitive function in complex, dynamic environments. I hope that this chapter demonstrates to designers and operators working in the aviation domain that there is considerable knowledge that can help them to do their respective jobs more effectively.

As noted before, it is necessary to examine the unwanted behavior both of automation and people in an automated system, because it is only through such study that we can minimize the costs, while increasing the already considerable benefits, of this technology. It is important that we not lose sight of the benefits (see next section), for aviation cannot advance without automation if we are to meet future challenges, which will tax our ingenuity to the utmost. We must not throw the baby out with the bath water.

But it is equally important that we not ignore the potential costs of yet more sophisticated automation, for if it is not designed and used properly it can make the future aviation system less flexible, less effective, and less able to meet those challenges. In recent years, it has become evident that our operators do not always understand or properly manage the automation they now have at their disposal. It is essential that we make every effort to understand why this is true, if we are to design future automation so that it will be more effective and error tolerant than what we now have.

BENEFITS OF AVIATION AUTOMATION

I have referred in several places to the benefits derived from aviation automation to date. Let me summarize explicitly what these benefits are, to keep this discussion of problems in context. In a landmark paper, Wiener and Curry (1980) discussed system goals. Paraphrased, they are:

- Safety
- Reliability
- Economy
- Comfort

I briefly cite demonstrated benefits with respect to each of these system goals. This list is not inclusive, but it will provide some insights into the extent to which we rely upon automation to accomplish our objectives.

Safety has always been proclaimed by the aviation industry as its primary objective. An examination of air carrier accidents by Lautmann and colleagues (Lautmann & Gallimore, 1987) suggests that newer, more highly automated aircraft have had substantially fewer accidents than earlier aircraft (Fig. 9.1). In their first decade of operation, the widely used Boeing 757/767 models were involved in only one fatal accident (Thailand, 1991). (Two recent accidents have marred this record, however: A B757 suffered a controlled flight into terrain accident near Cali, Colombia on December 20, 1995; this accident involved both human error and

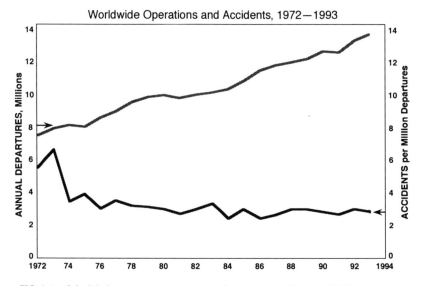

FIG. 9.1.   Scheduled air transport operations and accident rates (Boeing, 1994).

human–machine interaction problems. Another 757 crashed into the sea after takeoff from Puerto Plata, Dominican Republic, on February 6, 1996. The mishap involved faulty airspeed indications presented to the pilot flying; these incorrect indications may also have affected autopilot operation and may have been due to a plugged pitot tube. The first officer's airspeed system was probably reading correctly. Both accidents are under investigation.) Other new types have been involved in more accidents, but the record is still generally good. (For a balanced discussion of this question, see Automated cockpits, 1995a, 1995b)

Reliability has been improved; autoland-capable automation, head-up displays, and other innovations have increased the number of flights able to operate at destinations obscured by very low visibility. Newer systems (GNSS, enhanced vision) have the potential to improve approach and landing safety worldwide. Improvements in ATC also have the potential to increase reliability, as well as efficiency, in the future system.

Economy has been improved by flight management systems that can take costs into account in constructing flight plans (although the benefits possible from such computations have been diluted by the inability of the present ATC system to permit aircraft to operate routinely on most cost-efficient profiles). Despite this limitation, significant economies are being achieved in the United States by coordination of nonpreferred and direct routes between air carrier systems operations centers and the FAA's System Command Center.

Comfort has been improved by gust alleviation algorithms in the newest aircraft, as well as by the ability of newer aircraft to fly at higher altitudes, above most weather. Greater flexibility enabled by ATC automation will permit pilots to utilize a wider range of options to achieve more comfortable flight paths.

In what respects are we still deficient with respect to these system goals? Most of our accidents can be traced to the human operators of the system, and increasing numbers can be traced to the interactions of humans with automated systems. More can be done to make aircraft automation more human-centered, but perhaps even more important, advanced automation can be used to make the system as a whole more resistant to and tolerant of human errors, be they in the implementation or the operation of these systems.

COSTS OF AVIATION AUTOMATION

The 1989 ATA Human Factors Task Force report stated,

> During the 1970s and early 1980s, the concept of automating as much as possible was considered appropriate. The expected benefits were a reduction in pilot workload and increased safety…. Although many of these benefits have been

realized, serious questions have arisen and incidents/accidents have occurred which question the underlying assumption that the maximum available automation is always appropriate, or that we understand how to design automated systems so that they are fully compatible with the capabilities and limitations of the humans in the system. (pp. 4–5)

Let us examine this statement, which was largely responsible for the inquiry described in this book and its predecessors (Billings, 1991, 1996).

At the time the ATA report was prepared, the outlines of the A320 and B-747-400 automation suites were just becoming visible to the knowledgeable observers on the Task Force. The MD-11 was at an early stage of development and its cockpit design was not yet firm. It is clear that in the A320 and MD-11, the concept of automating as much as possible, with the intent of reducing flight crew workload and minimizing human errors, was in fact considered appropriate, though the two design teams took different approaches. The 747-400 was more conservative in its automation philosophy and more evolutionary than revolutionary in its application.

It is clear, with the hindsight afforded by 5 years of operational experience, that at least some pilots have found certain of the automation features in this new generation of aircraft difficult to understand and to manage. The difficulties that have been experienced appear to me to have been due in large part to five factors. Four are design factors: complexity, brittleness, opacity, and literalism. A fifth related factor is training, which in turn is related to understanding. Each is considered in more detail here. A discussion of other relevant factors follows.

Complexity

As indicated in chapters 3 and 4, today's aircraft automation suites are very capable, increasingly flexible, and very complex. Tactical control automation (enabled through a mode control panel, as in Fig. 5.11) is tightly coupled to strategic flight management automation (the FMS, with its CDU interface) in ways that are not always obvious. The FMS itself is capable of autonomous operation through several phases of flight. Both parts of the system are "mode-rich" (Sarter & Woods, 1994); default and reversion options vary among modes.

When these interactions cause unwanted behavior (from the pilot's viewpoint), the pilot may not have a mental model that allows him or her to correct the situation short of reverting to a lower level of management (see chapter 10) or turning the automation off, which is not always desirable and may not be possible in some circumstances. "Turning it off" (Curry, 1985), for instance, may disable certain protective features such as FMS knowledge of altitude restrictions during a descent into a terminal area, or the automation's intent to level the aircraft at a given altitude during a climb. Pilots of recent, very powerful aircraft have become concerned about the rate at which the airplane was approaching a level-off altitude and have reverted to autopilot vertical speed mode to slow the climb as they approached the new altitude, unaware that this reversion also canceled the altitude capture mode. The result has often been a deviation above assigned altitude.

Another aspect of automation complexity is the great flexibility found in the modern flight management and autoflight system. Modern systems may have several modes for each of several control elements (Fig. 9.2). These modes interact in ways not always obvious to pilots. Operators must learn about, remember, and be able to access information concerning each mode in order to use it effectively; this imposes a considerable cognitive burden, makes it less likely that the operator will have an appropriate mental model of the automation, and increases the likelihood that modes may be used improperly. In addition, the capability of the modern FMS means that the system may direct the airplane through successive modes of operation autonomously, in ways that may leave the pilots uncertain of exactly why the automation is behaving in a certain manner at a particular point in time.

Sarter and Woods (1995) and Sarter (1994) discussed mode errors and mode awareness. Figure 9.2 is adapted from their paper. It illustrates the mode flexibility (and complexity) in a modern transport aircraft. Compare this with the relatively small number of flight modes in the Lockheed L-1011 automation shown in Fig. 5.12.

AIRCRAFT FLIGHT MODES: A320

Autothrust Modes	Vertical Modes	Lateral Modes
TOGA	SRS	RWY
FLX 42	CLB	NAV
MCT	DES	HDG/TRK
CLB	OPEN CLB	LOC*
IDLE	OPEN DES	LOC/APP NAV
THR	EXPEDITE	LAND
SPD/MACH	ALT	ROLLOUT
ALPHA FLOOR	V/S-FPA	
TOGA LK	G/S-FINAL	
	FLARE	

FIG. 9.2.   FMS and autoflight modes in the Airbus A320. From "FMS and Autoflight Modes in the Airbus A320," by N. B. Sarter and D. D. Woods, 1995, *Human Factors* 37(1), p. 13. Copyright 1995 Human Factors and Ergonomics Society. All rights reserved.

Each of the modes listed represents a different set of operating instructions for the automation. The mode in use (or armed, ready for use) is displayed in an alphanumeric legend on a flight-mode annunciator panel, usually located at the top of the primary flight display. In their conclusions, the authors of this very useful paper stated,

> As technology allows for the proliferation of more automated modes of operation...human supervisory control faces new challenges. The flexibility achieved through these mode-rich systems has a price: it increases the need for mode awareness—human supervisory controllers tracking what their machine counterparts are doing, what they will do next, and why they are doing it.... While we understand a great deal about mode problems, the research to examine specific classes of countermeasures in more detail and to determine what is required to use them effectively, singly or in combination, is just beginning. (Sarter & Woods, 1994)

Hollnagel (1993) suggested that increasing system complexity leads to increasing task complexity. This leads to an increasing opportunity for malfunctions and errors, which leads to an increasing number of unwanted consequences, which in turn leads to solutions that ultimately increase system complexity still further. He noted that this is sometimes humorously referred to as the "law of unintended consequences." The "law" states that the effort to fix things sometimes worsens the damage. Although we are perhaps not there yet in this domain, the quantum increase in complexity of aircraft automation has unquestionably created new opportunities for human errors, both slips or mistakes by the operator and those that result from deficient or "buggy" knowledge of the system being utilized.

I believe that automation complexity has been at least part of the problem in several incidents and accidents involving this new generation of aircraft (in Appendix 1, see Mulhouse-Habsheim, 1988; Bangalore, 1990; Strasbourg, 1992; Manchester, 1994; Paris, 1994; Toulouse, 1994). This is not to say that the automation has not functioned as it was intended to function; it has usually done exactly what its designers and programmers told it to do. The problem has been rather that the human operators have not understood its intended functioning and consequently have used it either beyond its capabilities or without regard to its constraints or rules. In another recent example of this problem, an A300-600 crashed at Nagoya, Japan (1994), after the pilot flying inadvertently engaged an autopilot mode (TOGA), then provided opposing inputs to the airplane's autoflight systems, which were counteracted by the autopilot when it was engaged to stabilize the flight path (Mecham, 1994).

The likelihood that all of the subtleties of such complex systems will be fully comprehended by pilots, even after considerable line experience with the systems, is not high (Sarter, 1994; Sarter & Woods, 1992b; Wiener, 1989); the likelihood that they will be understood after a few weeks of training is very small indeed. Uchtdorf and Heldt (1989), studying pilot understanding of the A310, indicated that a year or so of line experience may be required before pilots feel fully comfortable with the automation features—and this does not guarantee that they understand the entire system, only that they feel comfortable enough with its modes to operate it effectively.

Brittleness

As software becomes more and more complex, it becomes more and more difficult to verify that it will always function as desired throughout the full operating range of the aircraft in which it will be placed. The reason for this is that there is an almost infinite variety of circumstances that can affect its operation, only a subset of which can be evaluated prior to certification even if they are known to the evaluators. Even then, there will be conditions not thought of by the designers, that will inevitably arise at some point in the course of the airplane's operation. Brittleness is an attribute of a system that works well under normal or usual conditions but that does not have desired behavior at or close to some margin of its operating envelope.

An example might be a pitch control system that was selected, then reverted or defaulted to "vertical speed" mode while an airplane was climbing. The autoflight system would attempt to maintain constant vertical speed by increasing pitch angle at the expense of airspeed, which would gradually decay to unsafe levels. One of several examples was an Aeromexico DC-10 whose autoflight system maintained a climb at constant vertical speed until the airplane stalled; the pilots were thought to have improperly programmed the autopilot for constant vertical speed instead of constant airspeed and subsequently failed to notice the decaying airspeed until too late to maintain control (Luxembourg, 1979). Another example would be a descent mode that involved idle power without safeguards to ensure that such a descent could not continue all the way to the ground (see Bangalore, 1990), or an autothrust system that permitted power to remain at idle after descending onto and capturing the glide slope followed by a decrease in descent rate and a consequent decrease in airspeed to unsafe levels.

An example of brittle automation was present in the TCAS software when it was first implemented in civil transports. Under certain circumstances, the TCAS logic was able to recognize a hazard but was unable to advise a safe maneuver to resolve the conflict. When this occurred, the system simply gave up and indicated to the pilot that there was a conflict but the system could not resolve it. FAA certification pilots raised serious objections to such a mode and the software was modified to exclude this problem, although at the expense of commanding much more drastic avoidance actions under such circumstances, which has caused greater altitude excursions. This problem has still not been fully resolved, although the TCAS automation is no longer able to "walk away" from a conflict that requires a resolution advisory.

Yet another example of brittleness at the margins of the operating envelope was seen, I believe, in the crash of an A320 at Mulhouse-Habsheim after an experienced pilot made a low pass over the airfield at minimum airspeed during an air show (1988). During this maneuver, he descended below 100 ft above ground level and was unable to obtain full engine power in time to avoid trees at the far end of the runway. The automation prevented the airplane from stalling, but when the pilot descended below 100 ft the automation disabled the angle-of-attack protection also built into the airplane's flight control system. This feature, which under any other circumstances would have applied full power and rotated the airplane into a climb, must be disabled to permit the machine to land.

Opacity

Three questions with which Wiener (1989) paraphrased the frequent responses of pilots to automation surprises—"What is it doing?", "Why is it doing that?", and "What's it going to do next?"—may be indicative of either or both of two problems. One is a deficient mental model of the automation—a lack of understanding of how and why it functions as it does. This can be due to automation complexity, or to inadequate training, or both.

Another problem, however, is not that the operators do not understand the behavior being observed, but rather that the automation does not help them by telling them what it's doing (and if necessary, why). Sarter and Woods (1994, p. 24) observed that "The interpretation of data on the automation as process is apparently a cognitively demanding one rather than a mentally economical one given the 'strong and silent' character of the machine agent."

This problem represents a failure in communication or coordination between the machine and human elements of the system. Regardless of the cause, the net effect is diminished awareness of the situation, a serious problem in a dynamic environment.

In earlier times, less capable automation simply controlled the airplane's attitude and path; pilots could usually understand exactly what it was doing by observing the same instruments they used when they were controlling the airplane manually. Today's automation may use a combination of several modes to accomplish the objectives it has been ordered to reach. The information about what it is doing is almost always available somewhere in some form, although not necessarily in terms that the pilot can easily decipher. Why it is behaving in that manner is often not available except in the requirements document that motivated it. What it is going to do next is often, although not always, unavailable on the instrument panel.

In short, as automation complexity increases, it becomes more difficult for the designer to provide obvious, unambiguous information about its processes to the monitoring pilot (even if the designer believes that the pilot needs this information and therefore tries to provide it). I call this *opacity*. Others have referred to it as a *lack of transparency*; the two terms are synonymous in this context. Norman (1989) argued that the problem is not automation complexity, but lack of feedback to its operators.

As noted in chapter 4, automation opacity may be deliberate: One sure way to keep the operator from intervening in a process is to deny him or her the information necessary to permit intervention in that process. Much more commonly, I think, it is the desire, and need, to avoid overburdening the operator with information that is not essential to the performance of his or her necessary functions (as those functions are understood by the designer). The capabilities of the computer and its screens have made it possible for designers to overwhelm pilots with information and data. Opacity at some level is required to avoid overwhelming the pilot with data. We know that the ability of pilots to assimilate information is context dependent, and that when we provide more data without adequate consideration of context we simply make it less certain that they will attend to that which they really need to know (Woods, 1993c).

The mode awareness problems cited by Sarter and Woods (1992a) are in part due to opacity, although most modes are announced on mode annunciator panels. In part, the problem is one of salience: Alphanumeric symbols must not only be attended to, but must be read, to convey information. Hutchins (1993) has attempted to ease this problem by using iconic representations, with some experimental success (see Automated cockpits, 1995b, for an illustration of this approach). But Woods (1996) wrote of "*apparent simplicity* (of the system as represented), *real complexity* (of the system's actual behavior)" as one of our more serious problems with advanced automation.

There have been some notable examples of the effects of opacity on advanced flight decks, although it must be noted that in most of the cases, the information could have been found had there been time to look for it. This tends to reinforce the notion that drowning the operator in information isn't a wise way to design a system. Perhaps the most notable recent example is an accident that occurred during an approach to Strasbourg (1992), when the flight crew inadvertently commanded the autopilot to descend at a 3300 ft/min vertical speed rather than at a 3.3° flight path angle.

The FCU display read "–33" instead of "–3.3," although smaller letters on the LCD display also read "HDG/VS" instead of "T/FPA" and the symbology on the primary flight display was different in the two modes. The fact remains that the pilots, already heavily loaded because of late ATC instructions and inexperience in the airplane, missed these discrepancies and descended into the ground several miles from their destination. Changes have been made in later cockpits of this type to display "–3300" versus "–3.3" in the hope of eliminating this possible source of confusion. Another example is the TOGA (takeoff/go-around) indication in the A300 at Nagoya (1994), which was initially missed by the pilot flying. (It is worth noting that in both these cases, the flight crew provided the autoflight system with an incorrect indication of intent; see chapter 3.)

Literalism

A fourth attribute of automation (and of computers in general) could be described as its literalism or "narrow-mindedness" (S. W. A. Dekker, personal communication, January 1994). Automation is able only to do exactly what it is programmed to do, as it did in the two cases just cited. Human problem solvers are creative in their reasoning and their search for solutions to a problem. They can and will draw knowledge or evidence from any available source (either in memory or external to themselves: reference books, manuals, contact with others by radio, etc.), as long as that knowledge is relevant to the problem to be solved. Automation, on the other hand, is constrained by its instructions and is insensitive to unanticipated changes in goals and world states that may fall well within its usual operating range but were unanticipated by the designers of its software. It is in this sense that computer literality contrasts with brittleness; the latter term refers to undesired automation behavior at the margins of the operating envelope.

As an example of this, some flight management systems with vertical navigation capability will calculate an optimal descent point, based on cost factors, that is closer to a destination airport than pilots may wish for a smooth, gradual descent. The pilots may be unaware of the logic that drives this decision and action, but they learn through experience that they can "trick" the automation by programming a higher tailwind than is actually present. This false information causes the automation to begin the airplane's descent at an earlier point in time, thus achieving the pilots' desired ends. Human operators have always shaped the tools at hand to assist in accomplishing their objectives, but this shaping also increases task demand and cognitive workload, and increases the opportunity for errors.

Training

I indicated earlier that a fifth relevant factor is training. Let me preface this discussion by saying that if we cannot *show* the pilot what he or she needs to know in a given situation, then the pilot needs to *know* what he or she needs to know. The only way this knowledge can be acquired is through education and training.

In the early 1960s, Trans World Airlines ordered its first DC-9 aircraft, also its first jets with a two-person crew complement. The airline decided to undertake a major revision of its training philosophy for the new airplane; its new, and highly successful, training program emphasized the specific behavioral objectives (SBOs) required of pilots, rather than the older (and until then universal) approach of "teaching the pilot how to build the airplane." (Previous training programs emphasized detailed knowledge of how airplane systems were constructed, how the various parts contributed to the whole, and based, on this knowledge, how to operate them.) The new approach provided significant economies in training time, which is expensive, and appeared to be fully as successful in teaching pilots how to operate the new airplanes without burdening them with more systems knowledge than they "needed to know." United Airlines later adopted a similar philosophy, with similar success, and a training revolution was underway.

There has been continual pressure to minimize training time for the last 30 years. Pilots are paid virtually the same amount for training as for line flying, and when they are in training they are not flying trips that produce revenue for their company. There is no question that the SBO concept has been effective and efficient. Until recently, there has been no reason to question the concept.

The complexity of advanced automation, however, gives rise to questions about this approach to training. As indicated earlier, pilots must have an adequate mental model of the behavior of the equipment they are flying. I believe that our experience to date with advanced automated aircraft suggests that the training we now provide does not always give them a sufficient basis for forming such models. One example of this, in the MD-11, was that takeoff speeds could be incorrectly calculated by the FMS if engine anti-ice significantly warmed certain sensors. An error message was generated, but this message was inhibited by flap extension. If flaps were lowered at the beginning of taxi, before airflow over the sensors had time to cool them, the erroneous speeds were locked in and takeoff speeds were incorrectly displayed on the speed tape of the PFD.

There is no question about the growing complexity and opacity of automated systems in these aircraft. I believe that questions must be raised about whether present training in *how to operate* these more complex and less transparent systems, as opposed to *how they operate*, is sufficient to provide pilots with the information they need when the systems reach their limits or behave unpredictably. If a pilot does not have an adequate internal model about how the computer works when it is functioning properly, it will be far more difficult for the pilot to detect a subtle failure. We cannot always predict failure modes in these complex digital systems, so we must provide pilots with adequate understanding of how and why aircraft automation functions as it does.

Comments about automation (McClumpha, James, Green, & Belyavin, 1991; Rudisill, 1994, 1995) make it plain that many pilots do not understand the reasons why aircraft and avionics manufacturers have built their automation as they have—and there are usually very good reasons, although they may not be known to the users of the automation. This, again, represents a failure of training to explain how the system operates and why, rather than simply how to operate the system.

OTHER OBSERVED PROBLEMS
WITH AVIATION AUTOMATION

Several other problems, some associated with or caused by those just enumerated, deserve mention here. Each has been associated with undesired outcomes in line operations; all can be mitigated to some extent by effective human–machine interface design.

Reliance on Automation

Several examples showed that pilots given highly reliable automated devices (and most are) will come, over time, to rely on the assistance they provide. They rely on the correct function of configuration warning systems, altitude alerters, and other information automation to which they have become accustomed. When GPWS was first introduced, the nuisance warnings to which it was prone caused pilots to distrust it; conformance with its warnings had to be mandated by company standard operating procedures. Later models have proved themselves more trustworthy, and they are relied on. Pilots have long been served reliably by autopilots and are sometimes less alert in monitoring their behavior than they should be, as evidenced by the failure to detect a few uncommanded roll inputs in early 747s (e.g., Nakina, Ontario, 1991). In some cases, pilots have continued to use automation even when they had every reason to mistrust it. This misplaced reliance has led to at least one accident, a runway overrun on landing at J. F. Kennedy Airport (New York, 1984). The NTSB discussed overreliance on automation at length in its report on this mishap (National Transportation Safety Board, 1984).

Air traffic controllers likewise rely on the data presented to them on their CRTs, even though much automation is required to present the synthetic images with which they work. They are surprised by occasional "tag swaps" and other misrepresentations of the data when they occur.

It does little good to remind human operators that automation is not always reliable or trustworthy when their own experience tells them it can be trusted to perform correctly over long periods of time. Many pilots have never seen these automation elements fail, just as many of them have never had to shut down a malfunctioning engine except in a simulator, and in any case, humans are not good monitors of infrequent events. The solutions to the "human failings" of trust, and of inattentiveness, must be found elsewhere. If we are to continue to provide operators with automation aids, we must make the system in which they are embedded more error tolerant so that such "failings" will not compromise safety of flight. In this area, there is much more we can do, even though much has been accomplished in the past.

Clumsy Automation

Wiener (1989) coined this descriptor to denote automation that lightens crew workload when it is already low, but requires more attention and interaction at times when workload is already high (see also Tenney, Rogers, & Pew, 1995). He and others have cited today's flight management systems as having this characteristic, as I noted in chapter 5. In the aviation context, it is in locations where traffic density is highest that ATC will most often have to change clearances to adjust to unexpected problems. It is also in these areas that aircraft are often climbing or descending and preparing to land.

These are the phases of flight that involve the highest likelihood of conflicts with other aircraft and that therefore demand that as much attention as possible be devoted to scanning for such traffic. Programming a flight management computer requires that the nonflying pilot's attention be inside the cockpit and focused on the CDU for some period of time. This is an attentional requirement that directly competes with outside surveillance and monitoring the activities of the flying pilot. Although efforts have been made in the newest FMSs to lighten this burden, reprogramming, often required to meet ATC requirements during transition to terminal areas, can still be cumbersome. Flights into Los Angeles, which may well be the world's most heavily traveled airspace, are often cited by pilots as perhaps the most taxing example of this problem.

Digital Versus Analog Control

I mentioned earlier the criticism by pilots of automation that makes it necessary for them to enter new navigation radio frequencies through alphanumeric keystrokes on the CDU rather than by turning rotary selectors as they did on older radio control units. Whether digital frequency entry actually takes longer has not been studied, to my knowledge, but I must confess that I share the bias of these pilots. At this time, communications frequencies are still accessed through the older types of control devices, most of which also show and make available both the old and new frequencies. This is a help to pilots if they are unable to establish radio contact on a new channel, but communication frequencies also may be accessed in future through the FMS.

In the autoflight control wheel steering (CWS) mode, pilots manipulate their control columns to instruct the automation what rates of change are desired for a maneuver. Once placed in a certain attitude, the autopilot will hold that attitude until other control instructions are received. This "rate command" function is all accomplished digitally in newer aircraft, but the pilot perceives a graded input that produces a continuous response. In contrast, the command mode of the autopilot is controlled by providing it with digital numeric targets representing airspeed, desired altitude and heading, and sometimes desired vertical rate. In today's aircraft, these digital values can be specified either through rotary switches on the mode control panel in a manner quite similar to the selection of new radio frequencies in older aircraft, or by digital numeric input to the FMS.

The control wheel steering mode can be a trap, as was evidenced in a DC-10 incident in which, after a close-in turn to final approach, the flying pilot, who was heavily loaded, forgot that he was in that mode, continued to command an increasing pitch *rate*, and incurred a tail strike during the subsequent landing (NASA ASRS, 1976). It is the normal mode of autopilot control in older Boeing 737-200 series aircraft, however; it permits quick tactical changes to flight path, and it therefore represents a potentially useful intermediate between fully manual and fully automatic flight. It is shown as assisted control in my control and management continuum (see chapter 10). In at least one new airplane, the MD-11, all longitudinal control is carried out through the CWS function of the autoflight system, and full-time CWS for lateral (roll) control is also available as a customer-specified option.

Fully Autonomous Automation

Some automation elements have been essentially autonomous for a long time. No pilot would think of hand-flying a jet throughout cruise, as one instance. Many airlines require the use of the autobraking function for all landings, and autospoilers are also used routinely. Several other automatic functions that are used at all times have been mentioned in chapters 5 and 6. Despite this, concern has been expressed in various quarters about more complex functions that are now essentially autonomous, several of which can be turned off only with difficulty or not at all.

Among these functions is the full-time envelope protection system in the A320, which in effect prevents pilots from exceeding certain flight control parameters. This could more accurately be called an *envelope limiting* system. Several current and planned aircraft have systems that fulfill similar functions, although in a somewhat different manner. The MD-11's automatic systems control computers, as noted in chapter 5, will reconfigure aircraft subsystems autonomously if they sense specific malfunctions in those systems. Systems such as these give rise to questions concerning pilot authority and responsibility (Billings, 1996; Tenney et al., 1995. These questions are discussed in more detail in chapter 10.

Skill Degradation

One potentially serious problem in human–machine systems with highly capable automation is a loss of certain skills by the human when the automation routinely performs tasks that require such skills. This effect has been observed in numerous contexts (e.g., Cooley, 1987; see also chapter 15, case 1). It may be due largely to lack of practice of the particular skill by the human operator, although in certain contexts, other factors may play a part.

Psychomotor skill decrements were observed by pilots transitioning from copilot positions in the DC-10, a fairly automated airplane, to command positions in less automated aircraft such as the 727. After some failures to complete this transition, air carrier training personnel suggested to pilots approaching transition that they should forego the use of the automation for a couple of months prior to transition, in order to obtain more practice in manual control. The pilots took this advice and were able thereafter to complete transition training without difficulty. Note, in this example, that the pilots coming to transition all had extensive flying experience in older, relatively unautomated, aircraft. Their problem was to reacquire skills that they had already possessed in adequate measure before their transition to the more automated DC-10.

The advent of the new generation of highly automated aircraft, and the replacement of the older machines by such airplanes, implies that at some point in the near future, pilots may begin their airline careers flying as first officers on advanced aircraft that incorporate envelope protection and a variety of other control automation. Such automation may include limits on rate of roll, bank angle, pitch rate as a function of speed, gust alleviation, and other functions.

Will pilots who have never had to acquire the finely tuned manual skills that older pilots take for granted be able to demonstrate such skills at an acceptable level if they must transition to another aircraft that lacks these advanced features? Similarly, will they have learned the cognitive skills necessary for unassisted navigation if the flight management software fails? Finally, and perhaps most important given the high reliability of today's aircraft, will they acquire the judgmental skills and experience that alone can enable them to make wise decisions in the face of uncertainty or serious mechanical or environmental problems? At this point, no one knows the answers to such questions, but we do know that it is these skills, collectively called airmanship, that provide the last line of defense against catastrophes in aviation operations.

Similar questions can be asked about some air carriers that effectively require their pilots of advanced aircraft to utilize the automation on a full-time basis. *Flight International*, in its Letters columns, carried a brisk debate on this topic early in 1993: "Excessive reliance on equipment to help pilots fly 'smarter and safer' has become institutionalized to the point of becoming dangerous" (Hopkins, 1993, p. 40). "…I remember being admonished by the chief pilot for daring to hand-fly a raw-data standard instrument departure, and, worse still, for practising enroute VOR tracking by hand flying for 10 min in the cruise" (Laming, 1993, p. 140).

Some operators suggest to their pilots that they should exercise as many options as possible, and that they should fly at each level of automation on a periodic basis, to remain familiar with the systems and to maintain proficiency. Delta Airlines has stated these goals formally in its statement of automation philosophy: "Pilots must be proficient in operating their airplanes at all levels of automation. They must be knowledgeable in the selection of the appropriate degree of automation and must have the skills needed to move from one level of automation to another" (Byrnes & Black, 1993, p. 443). Many airline pilots make it a point to fly at least part of each flight segment manually to maintain their skills, regardless of the policies and preferences of their carriers.

Recall that similar questions were raised with respect to the ability of air traffic controllers, trained only in a full radar environment, to transition to procedural control of air traffic in the event of a complete radar failure. The ability of the FAA System Command Center to offload controllers during such failures has lessened this concern to some extent, but it is still possible for controllers to be grossly overloaded by system contingencies such as occurred after ATC communications and data transfer were suddenly shut down by a massive failure of communications facilities in New York (Lee, 1992), or by a 1-hour total power failure during morning rush hour at Oakland Air Route Traffic Control Center on August 9, 1995.

Crew Coordination

Wiener (1993) discussed crew coordination and resource management in the context of automated aircraft. In his extensive cockpit observations in advanced aircraft (Wiener, 1989), he noted several crew coordination issues (pp. 177–178):

- "Compared to traditional models, it is physically difficult for one pilot to see what the other is doing [on the CDU].… Though some carriers have a procedure that requires the captain (or pilot flying) to approve any changes entered into the CDU before they are executed, this is seldom done; often he or she is working on the CDU on another page at the same time."

- "It is more difficult for the captain to monitor the work of the first officer and to understand what he is doing, and vice versa."
- "Automation tends to induce a breakdown of the traditional (and stated) roles and duties of the pilot-flying versus pilot-not-flying and a less clear demarcation of 'who does what' than in traditional cockpits. In aircraft in the past, the standardization of allocation of duties and functions has been one of the foundations of cockpit safety."
- "There is a tendency for the crew to 'help' each other with programming duties when workload increases. This may or may not be a good thing…but it clearly tends to dissolve the clear demarcation of duties."

Costley, Johnson, and Lawson (1989) found in flight observations in 737 and 757 aircraft that less communication occurred in more advanced cockpits. Wiener interpreted these findings in terms of extremely low workload during cruise in advanced automated aircraft, and expressed concern "because of the presumed vulnerability of crews to boredom and complacency" (Wiener et al., 1993; p. 26). Wiener's findings agree with others reported here: that our traditional models of the behavior of competent air transport pilots may be insufficient guides to behavior in automated aircraft, because the machines themselves are, in certain respects, qualitatively different from older aircraft. New cognitive models that emphasize the increased cognitive loading on pilots are needed to guide our designs and implementation in the future.

We may have been shielded to some extent from problems in this realm by the very high experience levels of many first officers, as well as captains, in today's system. Many former captains with extensive command experience are now flying as copilots after having been laid off by defunct or bankrupt carriers. This will lessen during coming years, however.

Monitoring Requirements

Pilots (and increasingly, air traffic controllers as well) must monitor flight progress closely, for others, human and machine, are monitoring as well, to an extent unprecedented in the history of the industry. One problem inherent in automation is that pilots cannot usually detect that it is not going to do what they expected it to do until after it has failed to do it. It is only after automation has "misbehaved" that operators can detect its "misbehavior" and correct it. This is another aspect of the opacity problem. Unfortunately, when this occurs in aviation, the airplane may already be in a position from which rapid reactions may be necessary to return it to nominal conditions.

During an idle power descent, an airplane may descend 50 ft during each second it takes the crew to recognize an anomaly, decide to take action, make a control input, and wait for an appropriate response. Aircraft are separated by only 1,000 ft vertically below 29,000 ft; deviations of 500 ft or more are not uncommon after an autopilot has failed to capture an altitude. Such a deviation can be easily observed by air traffic control personnel and, if there is a conflict, by ATC automated conflict alert software. If the deviation is reported, pilots may face disciplinary or enforcement action from FAA.

For these reasons as well as others, pilots must closely monitor the behavior of their automated systems, but if an anomaly occurs, they must sometimes take very prompt action. Present automation (except the ubiquitous altitude alerting system) provides no predictive or premonitory warning that a failure is likely to occur in the immediate future; such information would give pilots time to prevent, rather than correct, the problem. Fortunately or unfortunately, flight path automation is reliable enough so that pilots may be tempted to relax their guard on the (justified) assumption that it will almost always behave correctly. Moray, Lee and Hiskes (1994) even suggested that this is the logical and appropriate strategy for pilots to adopt, because it is rare for such malfunctions to occur; thus, pilots are better advised to spend more time monitoring aspects of their flight that involve more uncontrolled variability.

Without question, the most effective monitoring of pilots flying is by a nonflying pilot in the same cockpit. This redundancy is absolutely critical. The vast majority of errors in the cockpit are detected, announced, and corrected without adverse consequences, often before any sort of anomaly can occur. When this fails, air traffic controllers often detect and warn of small deviations, permitting the pilot to correct them at an early stage. All of this cross-monitoring assumes that the monitoring agents understand the intent of the monitored agents (see chapter 3). Newer automation can do more than it has thus far been called on to do to strengthen still further the redundancy and thus the error tolerance of the aviation system.

Automated System Navigation Problems

Although manufacturers of the latest flight management systems have gone to considerable effort to simplify the operation of these systems, they are still exceedingly complex, and all interaction with them must be through several displays brought up sequentially on a single small CDU screen containing a large amount of alphanumeric information. As more functions have been implemented, more and more screens have been designed, each requiring serial access by the operator (see Fig. 5.31). In today's system, a great deal of information must be accessed through a very small "keyhole." As a consequence, navigating among the many screens has become complicated. This requirement imposes yet another cognitive burden on operators, who must remember enough of the FMS architecture to recall how to get to specific information when it is needed.

One method that designers have utilized to lessen the memory burden is to increase the number of modes in the FMS itself. This simplifies the navigation problem within the FMS but increases the requirement to remember the various modes and what each is used for. As these remarkable devices become still more capable, this cognitive burden imposed by the need for mode awareness can be expected to increase, unless a different approach is taken to their design (Woods et al., 1994).

Data Overload

Automation and the glass cockpit have increased considerably the amount of information available to pilots. The information is of much higher quality than was available in the past, a blessing for it decreases ambiguity and uncertainty, but the quantity imposes much higher attentional demands than in the past. The flight navigation displays on today's panels integrate a great deal of data into a clear and intuitive representation of the aircraft's location, directional trend, and chosen course—but this screen may also contain data regarding severe weather, wind shears, waypoints, airfields, obstructions, and other traffic, almost none of which was explicit in earlier aircraft. Depending on the circumstances of the flight, any part or all of this information may be relevant. Much of it, fortunately, can be turned off when it is not needed. Nonetheless, pilots must now manage a potential glut of information, where in the past, they simply had to wonder about it.

Pilots have often demonstrated that they want access to *all* information that may be relevant to their decision processes in flight, and that they are willing to accept a higher workload to deal with it. Unfortunately, as Fadden noted, if they have too much data, it become less certain that they will be able to attend to and integrate the appropriate data in time to address the problem that is most important. Particularly when virtually all information is visual in form, this is a serious potential problem for designers. Some have suggested adaptive displays that can be automatically decluttered as the pilot becomes more heavily loaded, but this poses other problems relating to operator authority (see chapters 10 and 12).

COMMENT

I have tried here to summarize some attributes of contemporary aircraft automation that appear to have been associated with problems in pilot cognitive behavior. Few of these problems represent failures of the automation as such; most represent either conceptual failures at the design or operator level, or problems in the implementation of these concepts. As machines grow more complex and difficult to understand, operators are more likely to err in their operation, so the net effect of these problems is often seen as human error at the sharp end. As Reason (1990) and Woods et al. (1994) pointed out so clearly, to say this and stop is simply to insure that the latent organizational and other factors that lie behind human error will go unnoticed, and that attempts to insulate the system against such errors will not get at the systemic and conceptual problems that cause most of them.

It is for this reason that I have tried, in this chapter, to generalize from the particular problems cited in earlier chapters to the conceptual issues that appear to me to underlie many or most of those problems. These issues, I believe, are the "latent factors" that we must attack if we are to make aviation automation more human-centered.

I have said little here about problems associated with ATC automation, simply because at this time there is relatively little automation to help the controller perform the primary task of directing air traffic (although much of the data management in ATC is automated). Controllers still work in a largely linear system whose peculiarities and nonlinearities they understand. Perrow (1984) cited the ATC system, in which "interactive complexity and tight coupling have been reduced by better organization and 'technological fixes' " (p. 5).

> The goal of preventing mid-air collisions conflicts with the production demands placed upon the airways system ... The problem... for ATC has been to keep collision risks low while increasing the occasions for collisions. This they have done with remarkable success. The density increases steadily, but the number of mid-air collisions has been reduced to near-zero (especially those where both airplanes are controlled by ATC). (p. 158)

As I said in the preface, it is necessary that we look not only at the human or at the machines, but at the system, if we are to correct system faults or to design and implement more effective systems in the future. If we do not take this approach, our present systems, as tightly integrated as they are, will simply acquire more layers of "Band-Aids" as we attempt to solve specific problems one by one, without considering the effects of those solutions on the system as a whole, or on the competing demands upon both pilots and controllers. I am frankly worried that this may be what we are doing in our present attempts to improve TCAS, a very tightly coupled system, by adding more and more software to lessen nuisance warnings while trying to extend the basic usefulness of the device by placing new requirements on it.

Chapter 10

Human and Machine Roles:
Responsibility and Authority

INTRODUCTION

Much industrial automation has been implemented on the implicit assumption that machines could be substituted for humans in the workplace (chapter 4). The Fitts (1951) list of functions that are best performed by humans and those best performed by machines exemplifies this concept. Jordan (1963) proposed that humans and machines should be considered as complementary, rather than competitive. The design and operation of the modern transport airplane exemplifies the concept of complementarity, yet in certain respects its automation very much exemplifies the principle of the interchangeability of parts. There are good reasons for this, but we must question whether we should still be designing and operating machines in that manner and whether a different approach could solve some of the problems we now observe in the aviation system.

Today's aircraft automation controls an airplane more or less as the pilot does. It navigates as the pilot does, or would if pilots could carry out in real time the complex calculations now performed by the computer. It operates the systems as the pilots do, or would do if they do not forget or overlook any of the procedural steps. In the near future, it will communicate with ATC computers, accept and execute ATC clearances, and report its location when not under radar coverage, just as pilots do now. Some have noted that automation usually performs all of these functions correctly, that it does not become tired or distracted or bored or irritable, that it often speaks more clearly and succinctly than pilots do, that its data stream will be easily comprehended by ATC computers in any nation, and that it does all these things without complaints. They have concluded that automation is as capable as the human for these functions, and some air carriers have mandated that it be used whenever possible. Are these "parts" interchangeable? That is the subject of this chapter.

THE PILOT AS CONTROLLER AND MANAGER

It should be clear from chapters 5 and 6 that pilots may play any of a variety of roles in the control and management of a highly automated airplane. These roles range from direct manual control of flight path and aircraft systems to a largely autonomous operation in which the pilot's active role is minimal. This range of allocation of functions between human and machine can be expressed as a control–management continuum, as shown in Fig. 10.1.

None of today's aircraft can be operated entirely at either end of this spectrum of control and management. Indeed, a complex airplane operated even by *direct manual control* may incorporate several kinds of control automation such as yaw dampers, a Mach trim compensator, automated configuration warning systems, and so forth. Conversely, even remotely piloted vehicles are not fully autonomous; the locus of control of these aircraft has simply been moved to another location.

Most transport flying today is *assisted* to a greater or lesser extent, by hydraulic amplification of control inputs and often by computer-implemented flight control laws. Flight directors, stability augmentation systems, enhanced displays, and, in newer aircraft, various degrees of envelope protection assist the pilot in his or her manual control tasks. To some extent, pilots can specify the degree of assistance desired, but much of it operates full-time and some of it is not intended to be

	MANAGEMENT MODE	AUTOMATION FUNCTIONS	HUMAN FUNCTIONS	
VERY HIGH	AUTONOMOUS OPERATION	Fully autonomous operation; Pilot not usually informed; System may or may not be capable of being disabled.	Pilot generally has no role in operation Monitoring is limited to fault detection Goals are self-defined; pilot normally has no reason to intervene.	VERY LOW
	MANAGEMENT BY EXCEPTION	Essentially autonomous operation; Automatic reconfiguration. System informs pilot and monitors responses.	Pilot informed of system intent; Must consent to critical decisions; May intervene by reverting to lower level of management.	
LEVEL OF AUTOMATION	MANAGEMENT BY CONSENT	Full automatic control of aircraft and flight. Intent, diagnostic and prompting functions provided.	Pilot must consent to state changes, checklist execution, anomaly resolution; Manual execution of critical actions.	LEVEL OF INVOLVEMENT
	MANAGEMENT BY DELEGATION	Autopilot & autothrottle control of flight path. Automatic communications and nav following.	Pilot commands hdg, alt, speed; Manual or coupled navigation; Commands system operations, checklists, communications.	
	SHARED CONTROL	Enhanced control and guidance; Smart advisory systems; Potential flight path and other predictor displays.	Pilot in control through CWS or envelope-protected system; May utilize advisory systems; System management is manual.	
	ASSISTED MANUAL CONTROL	Flight director, FMS, nav modules; Data link with manual messages; Monitoring of flight path control and aircraft systems.	Direct authority over all systems; Manual control, aided by F/D and enhanced navigation displays; FMS is available; trend info on request.	
VERY LOW	DIRECT MANUAL CONTROL	Normal warnings and alerts; Voice communication with ATC; Routine ACARS communications performed automatically.	Direct authority over all systems; Manual control utilizing raw data; Unaided decision-making; Manual communications.	VERY HIGH

FIG. 10.1.   A continuum of aircraft control and management for pilots.

bypassed. The pilot remains in the control loop, although it is an intermediate rather than the inner loop.

Whether pilots of limited experience should be required to have and demonstrate direct manual control ability in today's airplanes, which incorporate highly redundant automated control assistance, is a reasonable question but beyond the scope of this document. Airbus has rendered this issue moot to some extent by providing *shared control* as the A320's basic control mode. Pilots' control inputs are considerably modified and shaped by the flight control computers; envelope limits prevent them from exceeding predetermined parameters. In this airplane, pilots are provided with considerable assistance even during control failure modes; true manual flight capability is limited to rudder control and horizontal stabilizer trim and is designed only to maintain controlled flight while the automated systems are restored to operation. Under normal circumstances, the aircraft automation is responsible for much of the inner-loop control, although control laws are tailored to respond in ways that seem natural to the pilot. In the MD-11, a combination of longitudinal stability augmentation and control wheel steering is in operation at all times.

When an autopilot is used to perform the flight path (or power) control tasks, the pilot becomes a manager rather than a controller (this is also true to some extent of the shared control option). The pilot may elect to have the autopilot perform only the most basic functions: pitch, roll, and yaw control (this most basic autoflight level is no longer available in all systems); he or she may command the automation to maintain or alter heading, altitude, or speed, or may direct the autopilot to capture and follow navigation paths, either horizontal or vertical. This is *management by delegation*, although at differing levels of management, from fairly immediate to fairly remote. In all cases, however, the aircraft is carrying out a set of tactical directions supplied by the pilot. It will not deviate from these directions unless it is incapable of executing them.

As always, there are exceptions to the generalizations. Several aircraft will not initiate a programmed descent from cruise altitude without an enabling action by the pilot. Other modern flight management systems require that the pilots provide certain inputs before they will accept certain conditional instructions. *Management by consent* describes a mode of operation in which automation, once provided with goals to be achieved, operates autonomously, but requires consent from its supervisor before instituting successive phases of flight, or certain critical procedures. The consent principle has important theoretical advantages, in that it keeps pilots involved and aware of system intent, and provides them the opportunity to intervene if they believe the intended action is inappropriate at that point in time.

This management mode may become more important as intelligent decision-aiding or decision-making systems come into use (see chapter 12). A protracted period of close monitoring of these systems will be necessary; requiring consent is one way to monitor and moderate the potential influence of these systems. Although management by consent is an attractive option worthy of further exploration, it must be *informed* consent. More fundamental human factors research is needed to identify how to implement it without the consent becoming perfunctory.

Management by exception refers to a management/control mode in which the automation possesses the capability to perform all actions required for mission completion and performs them unless the pilot takes exception. Today's very capable flight management systems will conduct an operation in accordance with prepro-grammed instructions unless a change in goals is provided to the flight management system and is enabled by the pilots. Such revisions occur relatively frequently when air traffic control requires changes in the previously cleared flight path, most often during descent into a terminal area. Some FMS lateral and waypoint management tasks now operate by exception.

The desire to lighten the pilot's workload and decrease the required bandwidth of pilot actions led to much of the control automation now installed in transport aircraft. The more capable control and management automation now in service has certainly achieved this objective. It also has the capacity, however, to decrease markedly the pilot's involvement with the flying task and even with the mission. Today's aircraft can be operated for long periods of time with very little pilot activity. Flight path control, navigation, and in some aircraft subsystems management are almost entirely automatic. The capable, alert pilot will remain conversant with flight progress despite the low level of required activity, but even capable, motivated pilots get tired, lose their concentration and become diverted, or worry about personal problems unrelated to the flight. A critical task for designers is to find ways to maintain and enhance pilot involvement during operation at higher levels of automation.

This is less simple than it sounds, for pilots will both resent and find ways to bypass tasks that are imposed merely for the purpose of ascertaining that they are still "present in the cockpit." Tasks to maintain involvement must be flight-relevant and, equally important, must be *perceived by pilots* to be relevant. Designing pilot involvement into highly automated systems will not be easy but must be accomplished to minimize boredom and complacency, particularly in very-long-range aircraft that spend many hours in overwater cruise. The progress of avionics, satellite navigation and communications, and data link will very likely have an opposite effect unless this uniquely human factor receives more consideration than it has to date.

Fully autonomous operation denotes operation in accordance with instructions provided by system designers; no attention or management is required of the pilots. Until recently, relatively few complex systems operated fully autonomously. With the introduction of the A320 and MD-11, however, major systems operate in this way.

A fundamental question is how wide a range of control and management options should be provided. This may well vary across functions; indeed, pilots often prefer to operate using a mix of levels, for example, controlling thrust manually while managing the autopilot and using the flight director to monitor navigation. Pilot cognitive styles vary; their skill levels also vary somewhat as a function of the amount of recent flying they have done, how tired they are, and so on. These factors lead me to argue that a reasonable range of control/management options must be provided, but widening that range is expensive in terms of training time and time required to maintain familiarity with a broader spectrum of automation capabilities, as well as in terms of equipment costs.

One possible way to keep pilots involved in the operation of an aircraft is to limit their ability to withdraw from it by invoking very high levels of management. Another, perhaps preferable way is to structure those higher levels of management so that they still require planning, decision making, and procedural tasks. The use of a management by consent approach, rather than management by exception, could be structured to insure that pilots must enable each successive flight phase or aircraft change of status, as an instance. It has been suggested by one air carrier that long-haul pilots should be given the tools with which to become involved in flight planning for maximum economy on an ongoing basis; this is another approach to maintaining higher levels of involvement, but it is presently being implemented as a dispatcher/AOC function.

THE ROLE OF THE AIR TRAFFIC CONTROLLER

When a more highly automated ATC system is implemented, its computers will be able to search for traffic conflicts and to provide at least decision support in resolving them. This is the foundation of the FAA's automated en route air traffic control system (formerly referred to as AERA), and it is a key feature of the free flight proposal (chapter 8). Direct ATC computer-to-flight management computer data transfers, and direct negotiations between these computers, will likewise be a part of such a system, which opens the possibility of direct control of air traffic by ATC automation without involvement of either controllers or pilots.

I have discussed a control–management continuum in terms of pilot roles in an automated system. A similar construct can be proposed for air traffic controllers and their automation (Fig. 10.2), although it should be kept in mind that air traffic controllers actually *direct* and *coordinate* the movements of aircraft; only pilots control them. In this respect, the controller's task is fundamentally different from that of the pilot.

MANAGEMENT MODE	AUTOMATION FUNCTIONS	HUMAN FUNCTIONS
AUTONOMOUS OPERATION	Fully autonomous operation; Controller not usually informed. System may or may not be capable of being bypassed.	Controller has no active role in operation Monitoring is limited to fault detection. Goals are self-defined; controller normally has no reason to intervene.
MANAGEMENT BY EXCEPTION	Essentially autonomous operation. Automatic decision selection. System informs controller and monitors responses.	Controller is informed of system intent; May intervene by reverting to lower level.
MANAGEMENT BY CONSENT	Decisions are made by automation. Controller must assent to decisions before implementation.	Controller must consent to decisions. Controller may select alternative decision options.
MANAGEMENT BY DELEGATION	Automation takes action only as directed by controller. Level of assistance is selectable.	Controller specifies strategy and may specify level of computer authority.
ASSISTED CONTROL	Control automation is not available. Processed radar imagery is available. Backup computer data is available.	Direct authority over all decisions; Voice control and coordination.
UNASSISTED CONTROL	Complete computer failure; No assistance is available.	Procedural control of all traffic. Unaided decision-making; Voice communications.

(Left axis: LEVEL OF AUTOMATION — VERY HIGH to VERY LOW. Right axis: LEVEL OF INVOLVEMENT — VERY LOW to VERY HIGH.)

FIG. 10.2.   A continuum of system control and management for air traffic controllers.

As in the case of pilots, a broad range of roles is theoretically possible, ranging from unassisted procedural control without visualization aids such as radar, all the way to autonomous machine control of air traffic. Indeed, the former option will probably continue in some parts of the world, even while other areas adopt advanced automation. The important point is that the role of the controller can vary greatly, from absolute direct authority over the entire operation to a relatively passive oversight function in which air traffic control tactics are purely the computer's task.

Whether such a broad range of roles is desirable is another matter entirely. The first principles of human-centered automation indicate that involvement is necessary if the human operator is to remain in command of the operation. I question the controller's ability to remain actively involved for very long if he or she has no active role in the conduct of an almost entirely automated process. On the other hand, some range of options should be permitted, to account for differences in cognitive style, variations in workload, and a wide range of controller experience levels.

The potential for increased opacity of new air traffic management automation is great, as indicated in chapters 8 and 13. Controllers cannot remain in command of air traffic unless they are both informed and involved, not only when automation fails but when it is performing normally. This again argues against placing the human operator in a role at the high extreme of this control–management continuum.

HUMAN AND MACHINE ROLES

Present aircraft automation does not plan flights, although it is able to execute them and to assist in replanning (e.g., after an engine failure). It cannot configure an airplane for flight or start the engines. It knows with great precision where runways are, but not how to get to them from a gate, nor from a runway turnoff to a gate after landing. Automation does not, at this time, accomplish the checklists required before and during flight. Flight control automation is locked out during the takeoff sequence, although thrust is under automatic control from early in the process in some aircraft. Automation controls neither the landing gear nor the flaps during takeoff and approach. From shortly after takeoff until the airplane touches down at a destination, however, automation is fully capable of executing virtually all the required tasks in a flight.

There is no reason, of course, why automation could not perform taxi maneuvers, although implementing this function would be extremely costly. There is absolutely no reason why landing gear and flap actuation could not be automatic. The few aspects of subsystem management that are still manual in some of the newer aircraft (e.g., the MD-11) could certainly be automated as well. Why, then, have they not been? The answer does not lie in the inadequacies of technology, but in the intricate domains of sociology, psychology, and politics.

Pilots are perceived to be essential because passengers are not willing to fly in an autonomous, unmanned airplane—although millions entrust themselves every day to the Bay Area Rapid Transit, the Washington Metro, and other mass transit systems in which the locus of control has shifted from the on-board operator station to a central control room. The trains on these systems do carry a human operator, but under normal circumstances, the operator does not operate the vehicles and is proscribed from doing so. Airport "people-movers," some of which travel over several miles of dedicated track or roadway, do not have on-board operators; the voice announcements are recorded or synthetic. Note that these systems are not fully autonomous; humans control them, as they always did, but the control is supervisory and remote (Sheridan, 1984).

The flight environment, however, is far more complex and variable than that of a modern light-rail system, and many of the variables are not under the control of system managers. Pilots are essential because they are trained to compensate for unexpected variability. Automation does fail, and unlike surface vehicles, airplanes cannot simply come to a stop while the automation is fixed. Once in flight, they must be guided to a landing. In other words, pilots and air traffic controllers are essential because they are able to make good decisions and take appropriate actions in difficult situations. We have not yet devised a computer that can cope with the variability inherent in the flight and air traffic environment.

The human role, then, is to do what the automation cannot do: to plan, to oversee, to reflect and make intelligent decisions in the face of uncertainty, and to make passengers (and air carrier management, and the FAA) feel comfortable about air transportation.

RESPONSIBILITY AND AUTHORITY

If a controller fails to maintain separation because of a tag swap or a radar outage, is the computer "grounded"? No; the controller remains responsible for traffic separation regardless of the circumstances. There may be mitigating circumstances, but this responsibility cannot be delegated.

If an automated airplane gets lost and lands at the wrong airport, or runs out of fuel and crash lands, or violates regulations for whatever reason, is the flight management computer held to account? Not to my knowledge. The pilot, not the autopilot, is in command of the flight and is responsible for its safe conduct.

Does the pilot have the authority required to fulfill this responsibility? What responsibility, and how much authority, does the pilot have in today's system and today's airplanes? It is a maxim of military command that authority can be delegated by a commander. Responsibility for the outcome cannot be delegated to others. It remains with the commander.

These precepts are extremely important in aviation. Although aviation involves a widely distributed system in which no individual can get the job done alone, the roles of all the humans in the system come together in the process of flight. In that process, the pilot and dispatcher share responsibility for the plan that guides the flight. The pilot is solely responsible for its safe execution, and the air traffic controller is solely responsible for keeping the flight safely separated from other air traffic.

Part 91.3 of the Federal Aviation Regulations (Federal Aviation Administration) describes the responsibility and authority of the pilot in command. It is brief and succinct:

(a) The pilot in command of an aircraft is directly responsible for, and is the final authority as to, the operation of that aircraft.

(b) In an in-flight emergency requiring immediate action, the pilot in command may deviate from *any rule* [italics added] of this part to the extent required to meet that emergency.

(c) Each pilot in command who deviates from a rule under paragraph (b) of this section shall, upon the request of the Administrator, send a written report of that deviation to the Administrator.

This regulation confers upon the pilot essentially unlimited authority to depart from the accepted rules for the conduct of flights if that pilot believes that an emergency exists. Under emergency authority, the pilot is permitted to request whatever assistance is necessary, to declare for his or her flight absolute priority for any maneuver, flight path, or action, and to take whatever steps are necessary, in the pilot's view, to protect the passengers. The pilot's decisions may be questioned afterward, but the authority remains and is recognized without question at the time.

It is a matter of record that pilots have sometimes not used their emergency authority when hindsight says they should have done so. Some situations, like the undeclared fuel emergency that led to the loss of Avianca flight 107 (Cove Neck, New York, 1990), seem obvious to anyone, although the NTSB raised the question of whether the pilot's very limited English competency may have permitted him to think that he had made such a declaration when the proper enabling words ("Mayday" or "Emergency") were not used. In other cases, pilots have been inhibited by fear of the paperwork and questions that inevitably follow such a declaration (although onerous questions after a safe landing are a great deal easier to walk away from than an aircraft accident).

Pilots, then, have as much authority as they need to permit them to fulfill their responsibility for flight safety—or do they? Does a pilot whose control authority is limited by software encoded in the flight control computer have full authority to do whatever is necessary to avoid an imminent collision, or ground contact? United States transport aviation involving jet aircraft was scarcely 4 months old in 1959 when a Boeing 707 entered a vertical dive over the North Atlantic Ocean. The pilots recovered from the dive and landed the airplane safely at Gander, Newfoundland. Postflight inspections revealed severe structural damage of the wing and horizontal stabilizer, but all the passengers survived and the airplane flew again after major repairs (NTSB, 1960). Would this have been possible if flight control software had limited the forces that could be applied to levels within the permitted flight envelope of the airplane?

LIMITATIONS ON PILOT AUTHORITY

In the A320/330/340 series aircraft, the flight control system incorporates envelope limitation. Certain parameters (bank angle, pitch, or angle of attack) cannot be exceeded by the pilot except by turning off portions of the flight control computer systems or flying outside their cutoff values, as was done during the low-altitude flyover prior to the Mulhouse-Habsheim accident (1988). Predetermined thrust parameters also cannot be exceeded.

Systems designed for autonomous operation pose serious philosophical questions with respect to pilot authority as well as pilot involvement. These questions arose first in the design of fighter aircraft and were discussed succinctly in an unsigned editorial in *Flight International* ("Hard limits, soft options," 1990). The American F-16 fighter's fly-by-wire control system incorporates "hard" limits that "preserve the aircraft's flying qualities right to the limit of its closely defined envelope" but do not permit the pilot to maneuver beyond those limits. The *Flight* editorial pointed out that

> There is, however, another approach available: to develop a "softer" fly-by-wire system which allows the aircraft to go to higher limits than before but with a progressive degradation of flying qualities as those higher limits are approached. It is this latter philosophy which was adopted by the Soviets with fighters like the MiG-29 and Sukhoi Su-27. It is not, as Mikoyan's chief test pilot ... admits, "necessarily a philosophy which an air force will prefer." [He] says, however: "Although this ... approach requires greater efforts ... it guarantees a significant increase in the overall quality of the aircraft-pilot combination. This method also allows a pilot to use his intellect and initiative to their fullest extent." (p. 3)

The softer approach was taken in the MD-11 (Hopkins, 1990) and Boeing 777, which permit pilots to override automatic protection mechanisms by application of additional control forces. The flying qualities are degraded under these circumstances, but the pilots retain control authority. The MD-11 incorporates angle of attach protection, as do the A320/330/340, but the MD-11's limits can be overridden by the pilot, as can the limits of its autothrust system. The 777 incorporates hard limits on engine power, for reasons that are not clear to me. In the MD-11, as noted earlier, many aircraft systems operate autonomously; subsystem reconfiguration after failures is also autonomous if the ASCs are enabled (the normal condition). Any of these systems can also be operated manually, but the protections provided by the ASC computers are not available during manual operation.

Although civil aircraft do not face the threat posed to a fighter under attack if its maneuverability is limited, their pilots do on occasion have to take violent evasive or corrective action, and they may on rare occasions need control or power authority up to (or even beyond) normal structural and engine limits (e.g., Pacific Ocean, 1985) to cope with very serious problems. The issue is whether the pilot, who is ultimately responsible for safe mission completion, should be permitted to operate to or even beyond airplane limits when he or she determines that a dire emergency requires such operation. The counterargument, raised elsewhere in this book, is of course that some pilot, sometime, for some reason, will exceed these limits unnecessarily, raising an equal safety hazard. This issue will not be simply resolved, and the rarity of such emergencies makes it difficult to obtain empirical support for one or the other philosophy. Nonetheless, the issue is a fundamental one.

COMMENT

The increasing capabilities of advanced automation pose a severe temptation to new aircraft design teams. They can decide that the safety of the airplane makes it important that they limit the authority of the pilots, and they can implement that limitation very easily in airplane software. They could match the software limits to the structural parameters of the airplane insofar as possible, although this is an approach that has not yet been implemented. Whether they have considered all of the circumstances that may confront a pilot in line operations is a question that may only be answered when totally unforeseen circumstances arise, perhaps years after the airplane has left the factory. It is in such circumstances that the "parts" of the human–machine system will be found to be not interchangeable but complementary.

Given that pilots bear the ultimate responsibility for the outcome, it would seem that their authority to do whatever is necessary to insure that the outcome is favorable should be foreclosed only with extreme reluctance. The concept of "soft limits" on control authority may represent one useful and constructive approach to this dilemma. What is important is to realize how easily the pilot's authority can be compromised, given the technologies that are now available. It may take only a line or two of software and may or may not be known or obvious to the pilot. (See Mårtensson, 1995, for a trenchant example of a critical automated function of which the pilots were unaware.)

The same dilemma will face us in the near future with respect to air traffic controllers, as the tools they use are automated in the AAS. This question has not received the attention it deserves, and the rarity of situations that define the issue makes it very difficult to provide good data in support of any extreme position. It is necessary that we realize, however, that issues involving such rare events must sometimes be handled on the basis of the best available a priori reasoning. The views of pilots and controllers on this issue are clear: If they have the responsibility, they want knowledge and the authority necessary to remain in command.

Chapter 11

Integration and Coupling in the Future Aviation System

INTRODUCTION

The technical challenge of integrating advanced automation in aircraft pales in the face of the challenge posed by the need for a highly integrated air traffic management system. Simply developing a set of agreed-on standards for such a system has already taken 5 years, and the task is far from finished. FAA, ICAO and other organizations must produce standards and requirements for data link technologies, the aeronautical telecommunications network, automatic dependent surveillance, future ATC procedures, satellite surveillance, navigation and communications, ground communication links, integration of satellite and radar surveillance, the necessary airborne equipment, and assessment of the problems posed by a mix of airborne capabilities (Fitzsimons, 1993). "Harmonizing" all of the pieces needed for a truly integrated aviation system will be a staggering task.

Despite, or perhaps because of, the magnitude of this task, many system planners have proposed that the ATC and aircraft computers in the aviation system can function more effectively if they can exchange data directly and can negotiate clearance revisions automatically. This concept would tightly couple the various system elements.

Perrow (1984) discussed automation complexity and coupling at length. He characterized tightly-coupled systems as having more time-dependent processes, requiring invariant sequences, "unifinality" (little flexibility regarding ways to reach a goal state), little slack, and limited to those buffers and redundancy built in to the system. He pointed out the many nonlinearities in such systems, and the difficulty in modeling them. He also noted that more ATC automation "will lead to much tighter coupling—that is, less resources for recovery from incidents" (p. 161). In accordance with Perrow's cautions, I examine issues related to coupling and complexity, as well as integration, in this system.

The UK National Air Traffic Service (NATS) has supported studies to ensure that a variety of technologies can "play together" in a future environment. In October 1991 Eurocontrol and the UK CAA demonstrated the automatic delivery of clearance data, weather interrogation by pilots, and the transmission of ATC instructions and pilot acknowledgements using a BAC 1-11 airplane belonging to the Royal Aircraft Establishment.

> Downlinked autopilot settings were automatically checked against the controllers' original instructions, enhancing safety, while the downlinking of other avionics data (such as true airspeed, heading and vertical rates) reduced voice traffic and the controller's workload. The Volmet [weather] messages were printed in the cockpit, reducing the pilots' workload, and the downlinking of pilot acknowledgements gave the controllers assurance that the message had reached the correct recipient and was unlikely to be misinterpreted. (Fitzsimons, 1993, p. 23)

In 1991, I proposed that ATC clearances transmitted to aircraft by datalink be downlinked to ATC computers as they were executed, to provide confirmation of FMS and presumably pilot intentions and to provide positive confirmation that the aircraft would proceed in accordance with ATC intentions (Billings, 1991).

"Studies suggest that aircraft-derived data could provide additional inputs to ground-based trackers, reducing position uncertainty and enabling improved conflict alert algorithms to reduce the number of nuisance alerts while giving earlier warning of potential conflicts" (Fitzsimons, 1993, p. 23). Although limited, the UK experiments represent an encouraging start on the task of integrating the ground and airborne components of modern aviation systems. Since 1991, a number of other demonstrations have been conducted to examine various elements of an integrated system. In this chapter, I examine the implications of creating such a system for the humans who must operate within it.

ELEMENTS OF AN INTEGRATED AVIATION SYSTEM

A very large number of functional capabilities must be in place in a future aviation system if it is to accomplish the tasks assigned to it. Briefly, these functions are to facilitate the movement and tracking of large numbers of variably equipped aircraft on or over any part of the earth's surface, to assist them in landing and taking off from airports, and to provide all assistance necessary during contingency operations. These tasks must be accomplished in all extremes of weather, across national boundaries, with limited resources. The aviation system is information bound, and the complexity of the system results largely from the complexity of moving all necessary information in real time to all system participants who have a need for it.

Avionics data have been downlinked and processed automatically during the UK NATS mode S trials. At certain airports, predeparture clearance delivery is now routinely accomplished by ACARS. Two carriers successfully tested automatic dependent surveillance over the Pacific, transmitting data through satellites to air traffic control facilities on land. Other elements of the system have also been tested in simulation; some have had flight trials. Large-scale global positioning system testing has been performed, and A330 and A340 aircraft have been certificated for satellite navigation by the JAA in Europe.

There appear to be no insurmountable technological barriers to the implementation of the technologies required for a more highly integrated system. The barriers that remain are in the areas of standards, procedures, software, and harmonization across nations. The knotty issue of how ATC will cope with a broad mix of aircraft capabilities is more difficult in a constrained economic climate. ICAO's Required Navigation Performance concept may help to some extent, although retrofit of advanced equipment in a large number of regional and commuter aircraft may not be economically possible in the near term.

The software issue is critical; the elements of this system must be able to communicate, and the design and verification of software to make this happen throughout the system will be immensely difficult tasks. The AAS system will incorporate more than 2 million lines of code; a system for the ground support of free flight is likely to be substantially more complex. A long period of debugging will be required. Some verification work may not be able to be performed until the system is on line with live traffic, for the new system may be difficult to integrate with the present one. The overall system will be extremely complex, distributed across a great many nodes. Integration of such a system is far different from integration of the many control and display modules in even as complicated a process control system as a nuclear power plant.

COUPLING AND COMPLEXITY

In our present aviation system, the various automation elements are not necessarily coupled except by their information content. That is to say that the various elements operate *independently*. The coupling among them (more properly, the integration) is procedural: It is agreed among the various system participants that on receipt of a given instruction or request, a system component will take certain actions. The results of those actions may be visible in many parts of the system, but they are not predetermined. Although the various system components may be very complex in and of themselves, they are not physically or virtually linked at this time.

As noted, most officials in the air traffic system and an increasing number in the air carrier technical community envision direct communications between ATC computers (and perhaps, in the future, AOC computers as well) and aircraft flight management computers, although it is generally accepted at this time that when clearance modifications are uplinked to an aircraft, they will be subject to consent by the pilots. Direct negotiation of such clearance modifications between computers is also envisioned by FAA, however, and forms a part of the free flight concept (IATA, 1994). Such a process could confront both controllers and pilots with decisions arrived at by processes that were opaque to them.

It is also planned to require acceptance of datalinked messages within a certain short time interval (40 sec has been mentioned), although presumably execution of an uplinked clearance could be delayed for some further period of time to permit more review by the pilots. Nonetheless, the clearance execution process can be time-critical under some conditions.

These proposals present potentially serious problems for human operators. It is not always easy to understand a complicated clearance, particularly if it involves waypoints or instructions that depart from expectations. The process may require, for instance, that the pilots consult navigation charts, their dispatchers, or the FMS map display, even though the FMC may contain sufficient data to comply with the clearance. A new clearance may not comport with the pilot's view of the environment; it may require the expenditure of extra fuel or may take the airplane too close to the limits of an operating envelope. These factors will sometimes require deliberation and decision making by the flight crew, which will take time.

Executing an uplinked flight plan is simple, requiring only a single keystroke on the FMS CDU. If procedures for voice or data link negotiations with ATC to secure a revision are difficult or time-consuming, a flight crew already busy with another problem may not have the time and may accept an undesirable clearance rather than argue about it. A controller may also need time to understand a complex recommended clearance revision and may not have the time at that moment due to the pressure of other tasks. These are problems that occur now; they can be dealt with by methods similar to those used now, but in a future more automated system these problems can be dealt with *only if provision is made for them in its design.*

Figure 11.1 is a schematic representation of the present air traffic control and management process (solid arrows), in which pilots control aircraft by direction from air traffic controllers, with strategic oversight by the ATC System Command Center. Airline dispatchers and airline operations centers may coordinate aircraft movements with the SCC; revised flight plans are worked out with ATC traffic movements units at ATC facilities.

In a future automated air traffic management system such as the AERA system, ATC computers would negotiate necessary flight plan revisions directly with aircraft (FMS) computers (the vertically shaded arrows at left on the diagram). In a free flight system, in contrast, the locus of control of the system would normally reside

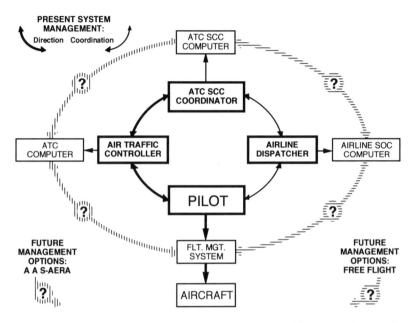

FIG. 11.1.   Current process of air traffic control (inner ellipse), and future air traffic control options (AERA and Free Flight; outer ellipse).

in AOCs (horizontally shaded arrows at the right). The AOCs would communicate with aircraft FMSs and with the SCC or air traffic management facilities, which would monitor the flow of traffic for near-term conflicts. The communications technology for this system now exists. Together with data link, new flight management systems for the Boeing 737-700/800 will enable AOCs "to control data-link information from the ground. Operations could request information from the aircraft computer without assistance [from] or knowledge of the pilots" (Nordwall, 1995, p. 48).

At this time, it is envisioned that flight paths and routings will be executed only with pilot and controller consent. Future technology will be able to perform these functions autonomously, however, and automatic execution of such clearances might assist ATC by insuring prompt responses to ATC commands. In this case, the ATC and airborne components of the system would be *coupled* as well as integrated.

In an automated ground-centered system, airplane flight paths would be managed by exception rather than consent (pilots would presumably still be able to countermand the actions of the FMS, although they might not necessarily be given advance notice of its intent). The pilot's role in such a hypothetical situation begins to resemble the monitoring role of the air traffic manager under the free flight concept.

THE AUTOMATED AIR TRAFFIC
MANAGEMENT SYSTEM CONCEPT

As noted in chapter 8, NASA and FAA are presently defining the elements of a new automated air transportation technology (AATT) research and development program devoted to advanced air traffic management. The objective of the program is to develop advanced "conflict-free, knowledge-based automated systems for real-time adaptive scheduling and sequencing, for global flow control of large numbers and varieties of aircraft, and for terminal area and ground operations that are compatible with 'free flight' enroute operations" (J. V. Lebacqz, personal communication, October 1994). This system will involve much tighter coupling, not merely integration, of the ground and airborne elements of the aviation system by virtue of the tighter linking of air and ground computers. These concepts run a very real (and very high) risk of infringing significantly on the authority of both air traffic controllers and pilots, despite their proponents' claims that the new automation will be human-centered.

Issues Raised by Tightly Coupled Systems Concepts

In a much more tightly coupled hypothetical system involving intercomputer negotiation and automatic execution of clearances, it would unquestionably be more difficult for pilots to understand how a clearance was arrived at and why it was given, because they would not have access to the ATC computer's reasoning. Similarly, it would be much more difficult for responsible controllers to understand the rules by which the clearance was derived, because they would not have access to the FMS data. This is the complexity-coupling problem discussed by Perrow (1984). It would certainly result in more surprises for the human operators, and would seriously diminish their ability to develop mental models of the ATC automation.

Although cruise flight is a comparatively low workload period for pilots of advanced aircraft, it is quite likely that the cognitive burdens, and workload, now placed on en route controllers will be transferred to pilots, not mitigated, if a free flight concept comes to fruition. This has happened before, when profile descents were imposed in busy terminal areas. Controllers found their workloads lightened by the new procedures, but pilots found their task demands to be considerably increased.

At this time, pilots do not have in their cockpits the information necessary to permit them to accomplish "air traffic control" other than short-range conflict avoidance using TCAS, which provides less than a fully adequate representation even of immediate potential threats (Fig. 6.7). Despite their limitations, which are considerable, TCAS displays are now being used on a test basis for in-trail climb separation over the Pacific Ocean. Other uses, to include lateral separation during closely spaced (1700 ft) parallel approaches to landing, are being actively considered (FAA, 1994), and displays for this function are in development. Note that none of this new functionality has been integrated into the cockpit task flow, nor have the displays and tasking been looked at in the larger context of cockpit and flight management, as so often happens when new functions are considered for retrofit on present flight decks. (See also discussion of FMS in chapter 5.)

COMMENT

Removing pilots or controllers from the command loop, under constrained conditions, would be a comparatively small step from a technical viewpoint. It would represent, however, a *qualitative* change in the rules by which the aviation system has been governed throughout its history. It would diminish the authority of the human operators appreciably, and it would change the dominant mode of system management as much as would the free flight concept. It would, however, be technologically feasible and implementable, and it could result in decreased workload for either pilots or controllers or possibly both—for which reasons, it will probably be seriously considered at some point in the future. This is the reason I have chosen to raise the specter here.

The differences between *integration of independent systems* and *coupling of interdependent systems* need to be clearly understood. The disadvantage of an uncoupled system is that its elements may, or may not, always behave predictably when particular instructions are issued. A pilot may turn too slowly after receiving a controller's request for an immediate maneuver, as an instance (and this is probably more likely when a data linked instruction is received than when a controller issues an urgent voice instruction). The most significant advantage of an integrated but uncoupled system is that operators are much more likely to understand it, and therefore less likely to be surprised by its behavior.

Given a system as complex as the future aviation system will be, however, attempts to couple its ground and airborne elements will inevitably make it more difficult for operators to predict its behavior, particularly under other than nominal conditions. I believe that this would be quite likely to result in less safe rather than more safe operations.

Part IV

Issues for Future Aviation Automation

The last part of this book deals with some issues facing system designers and operators, primarily in the aviation system but also in other domains. Chapter 12 contains a brief overview and discussion of newer computational concepts and techniques, including artificial intelligence (AI) and expert systems (ES), which have been proposed for use in future aviation system automation. In chapters 13 and 14, I attempt to summarize some "lessons learned" from the studies presented in chapters 1 through 11 and to suggest some requirements and guidance for future aviation system automation. Chapter 15 discusses other domains in which humans and complex computational machinery must solve difficult problems in real time, and considers whether the lessons learned in the aviation domain could be helpful in these domains. Chapter 16 contains some general comments and a brief conclusion.

CHAPTER 12

ADVANCED AND NOVEL AUTOMATION CONCEPTS FOR THE FUTURE SYSTEM

Charles E. Billings and Sidney W. A. Dekker
The Ohio State University

INTRODUCTION

Today's tightly coupled automation systems have become extremely complex and in many cases, relatively opaque to humans. At the same time, these systems have limits which may or may not be clear to their operators. An example of the problems that can be created is seen in this information extracted from a 1991 incident report:

> Flight XXX departed on schedule; heavy rain and gusty winds were experienced on takeoff and during the departure. The climbout was normal until approximately flight level 240 [24,000 feet] when numerous caution/warning messages began to appear, indicating a deteriorating mechanical condition. The first...was OVHT ENG 1 NAC, closely followed by BLEED DUCT LEAK L, ENG 1 OIL PRESSURE, FLAPS PRIMARY, FMC L, STARTER CUTOUT 1, and others. No. 1 generator tripped off line and the #1 engine amber "**REV**" [reverser unstowed] indication appeared. However, no yaw control problems were noted. The maximum and minimum speed references on the airspeed [tape] came together, followed by stick shaker activation.

> At approximately FL 260 [26,000 feet altitude], the cabin was climbing rapidly [pressurization had been lost] and could not be controlled. The Captain initiated an emergency descent and turned back to the departure airport. The crew began to perform emergency procedures and declared an emergency. During the descent, the stick shaker (indicative of an impending stall) activated several times but ceased below FL 200. Due to the abnormal flap indication and the #1 engine reverse, airspeed during the descent was limited to 260–270 knots.

The Captain called upon the two augmented crew pilots[1] to assist during the remainder of the flight. While maintaining control of the aircraft, he directed the first officer to handle ATC communications and to accomplish multiple abnormal procedures with the help of the additional first officer. The additional captain maintained communications with the lead flight attendant and company operations as the emergency progressed and later assisted in the passenger evacuation.

Fuel dumping began on descent below 10,000 ft. The fuel jettison procedure was complicated as the left dump nozzle appeared inoperative. The crew dumped 160,000 lb of fuel; this action took about 40 minutes. When the fuel dumping was completed, the captain requested vectors for a 20 mile final for runway XX.

The crew extended flaps early using alternate procedures due to an abnormal leading edge indication and the FLAPS PRIMARY message ... a final approach speed of $V_{ref}+20$[2] and 25° of trailing edge flaps was planned. They selected auto brakes number 4 [maximum setting]. The weather was still bad with strong, gusty winds and heavy rain causing moderate turbulence during the approach.

The ILS approach and landing were normal. At touchdown, maximum reverse was selected on #2 and #3 engines and about half reverse on #4 engine .... As the aircraft passed a taxiway turnoff, the tower advised that they saw fire on the left side of the aircraft ... this was the first time crew members were aware of any fire ... a runway turnoff was used, and the aircraft stopped on a taxiway ... [a difficult but successful evacuation followed].

[Comment by author of report] This incident is an example of an electronic system "nightmare." The crew received and had to sort out 42 EICAS [alerting] messages, 12 caution/warning indications, repeated stick shaker activation and abnormal speed reference information on the primary flight display. Many of these indications were conflicting, leading the crew to suspect number one engine problems when that engine was actually functioning normally. There was no indication of fire presented to the crew when a fire actually existed.[3]

Aviation automation to this time has been accomplished using conventional numerical computational methods and conventional software architectures. These have yielded remarkable capabilities, but numerical methods have inherent limits. It has been difficult to provide decision support using numerical techniques, and

---

[1] The planned flight, because of its length, was required to carry two additional flight crew members to permit the primary crew a rest period during flight.

[2] Reference speed, a fixed percentage above stall speed, plus 20 knots because of the surface winds and the airplane's condition, requiring partial flaps for landing.

[3] To preserve the anonymity of the source and the airline, the source of this quotation remains confidential. Identifying characteristics are also expunged.

many human factors researchers have argued that in cases like this, decision support technology is needed by pilots to avoid serious overload. Note, incidentally, in the foregoing occurrence, that four pilots were fully occupied in dealing with this emergency; one can only ponder how the outcome might have been affected had only the normal crew complement of two persons been in the cockpit.

Such concerns have motivated proposals to apply novel computational concepts and techniques to aircraft automation. These approaches, generally speaking, are designed to enable machines to carry out reasoning tasks we normally ascribe to human intelligence. During the past 30 years, newer classes of computational technology have been developed, using symbolic rather than numerical manipulation of the behavior of "objects." Their purpose is to free computation from the narrow, inflexible bounds of numerical and arithmetic deduction and permit a broader, inferential approach to computer reasoning.

Cognitive assistance (the ability to reason, plan, and allocate resources) has been accepted in several domains; these computational methods have been successful in a variety of applications (e.g., Proctor, 1995). They are often resource intensive; complex programs may run slowly because of the large knowledge bases that must be searched. They are imperfect and limited, but many believe them to be the wave of the future. Their advocates have suggested that they have clear advantages for certain aviation applications, and for that reason they are considered here.

DIAGNOSIS OF AIRCRAFT SYSTEM FAULTS

The management of disturbances, and the presentation to pilots of information concerning them, is a function that appears to be well suited to artificial intelligence (AI) approaches. It has been examined in depth by several researchers, stimulated in large part by leadership at NASA's Langley Research Center. Before considering this work, a word should be said about the constraints that a dynamic problem-solving environment (such as aviation) imposes on any diagnostic process.

The diagnosis of faults on a flight deck differs fundamentally from static systems in which a malfunctioning device can be taken off-line for troubleshooting. In a dynamic system, the process must go on while the fault is handled; an airplane cannot be "parked at a waypoint" while the trouble is dealt with. Fault scenarios are event driven; symptoms emerge over time in a fluid, sometimes cascading fashion (Woods, 1993a). In some cases, disturbance management requires that faults be ignored temporarily while the process is kept under control. In others, the true nature of the fault is not known and cannot be discerned until some outcome has ensued (as in the case just cited).

The challenges associated with the nature of dynamic faults are legion, as indicated by this incident report. I review various AI proposals set forth to address some of these challenges. It should be remarked that an evaluation much longer than the one that follows would not do justice to the complexity of the work that has been done in this area of AI. Further, it must always be kept in mind that in the cockpit or an ATC facility, the issue is not just a single human working with a computer, but rather multiple humans and often multiple tools working cooperatively to supervise and maintain the operation of a complex system. In such settings, each human must evaluate what others are doing, as well as what the total system is doing.

Rule-Based Diagnostic Systems

The first general AI proposal for aiding diagnosis in real-time systems was the use of rule-based expert systems (RBES). Machine expert systems are designed to support troubleshooting by human problem solvers (Clancey, 1983). Their strength is a large knowledge base, built up from domain information and experience provided to it by many domain experts. All the known faults in the domain, and all their associated symptoms and root causes, are enumerated and encoded in a knowledge database. The reasoning performed on the knowledge base during diagnosis by an "inference engine" is typically rule-based (e.g., "*if* oil temp 200° *and if* oil pressure 65 psi, *then execute* 'critical lubrication problem' rules"). This means, in simple terms, that rules guide the machine problem solver from symptom to symptom until a root cause for the observed fault has been found. The human may have to function as a data gatherer for the machine, and is the critic of its results.

The locus of control in this type of diagnostic reasoning resides with the machine, not with the human. Such a constellation has been called the paradigm of the intelligent system as prosthesis (Roth, Bennett, & Woods, 1987), where the RBES functions as a replacement or remedy for a presumed deficiency in the human reasoner. Experience with such systems in context has indicated that, as might be expected, the human and the expert system typically proceed in parallel to try to diagnose and solve the problem using whatever data is available (see chapter 15, case 4). Intelligent agents do not typically work as team players with humans.

The degradation of joint human–machine performance in such a system has been well documented (e.g., Roth et al., 1987). However, this is not the only reason why an ES as aid in the diagnosis of in-flight faults is ineffective. The time needed to accumulate experience and gather knowledge on all of the subsystems that make up a commercial aircraft is prohibitive. Pre-enumeration of all possible faults and all of their symptoms is simply not possible for any but the most simple or longest serving airframes still flying. ESs cannot deal with novel faults at all. The models that motivate the machine's decisions are implicit rather than explicit, which renders the machine's results both brittle and difficult for the human to understand.

In dynamic settings, the computer's progression through many low-level symptoms, and the conversation-style interface with most ESs, is unsuitable for time-pressured situations in which symptoms can emerge in a cascading and seemingly unconnected fashion. Although various expert systems have been and are being developed for aerospace applications (see, e.g., Pilot's Associate later in this chapter), none is in use today (see Malin et al., 1991 for an extensive evaluation of fault management systems in primarily space applications).

Model-Based Diagnostic Systems

Contrasting sharply with rule-based diagnosis is model-based diagnosis. This AI approach has also been called *reasoning from first principles*, or *deep reasoning*, as it relies on only a limited number of basic assumptions or principles about causality in the underlying system. Central to model-based diagnosis is the ability to view malfunctioning as anything other than what the system is supposed to do (Davis & Hamscher, 1988). The behavior of the system is observed with appropriate sensors on the one hand, while it is predicted on the basis of a model of the system on the other. Discrepancies between observations and predictions are called symptoms. The fundamental assumption is that if the model of the system is indeed correct, then all symptoms arise from actual malfunctions in the system.

Model-based diagnosis is much more robust than rule-based reasoning. Among the aviation studies done in this domain (see e.g., Malin et al., 1991; Ovenden, 1991; Rogers, 1990), Kathy Abbott (1990) studied and described a model-based diagnostician for aircraft systems: DRAPhyS (Diagnostic Reasoning About Physical Systems). DRAPhyS is part of a larger fault management research program supported by NASA Langley Research Center. Some of the modules developed under the NASA Faultfinder program have been taken up by others and restructured and enhanced (e.g., a Boeing project on the Flight Deck Engine Advisor using elaborations of DRAPhyS and its monitoring cousin MONITAUR). The goal is to develop a system that advises the crew of inconsistencies, adverse performance trends, or nonnormal situations before the conditions become critical and then assists the crew in system diagnosis while recommending applicable procedures in response to the situation (Shontz, Records, & Antonelli, 1992). DRAPhyS is discussed next in more detail in order to contrast the model-based approach (including its promise and problems) with rule-based expert systems.

DRAPhyS generates candidate hypotheses about the root causes of faults in an incremental, constructive approach, following the cascading emergence of symptoms. In that respect, DRAPhyS has the capability of degrading gracefully, just as human problem solvers would. If it decides it can no longer generate useful hypotheses at a more detailed level of system description, it confines its troubleshooting to a higher level of system description.

DRAPhyS knows that not all faults should be approached using the same underlying model as its criterion of right behavior. Faults can propagate through a system functionally (due to functional connections) as well as physically (due to physical proximity of affected components, e.g., a fractured turbofan blade severing a hydraulic line), and DRAPhyS has different underlying models to aid in the successful diagnosis of both classes of problems.

More exotic symptom scenarios are presented by faults that propagate and interact physically as well as functionally. DRAPhyS is able to utilize these classes of models in such a way that (hybrid) interactions between the various types of progressions (i.e., functional and physical) can be captured and reasoned upon. Another proposal for how to deal with this (Bylander, 1988) goes back to the use of knowledge bases: Although model-based diagnosis is suitable to determine which hypotheses explain which symptoms, many model-based systems cannot reason with uncertainty. That is, they cannot order or rank their hypotheses according to their plausibility relative to each other. The interaction with a knowledge base may be able to suggest which of several hypotheses is more likely than others relative to what is known about the domain.

A problem with model-based diagnosis is the grain of analysis of the reasoning. Information about an underlying fault may very well reside in the rate at which a symptom changes its behavior. In DRAPhyS, there is no difference between a slowly decreasing and a rapidly oscillating fan speed; both are called "abnormal." Yet diagnosis of an underlying fault can be different on the basis of the behavior of the symptom at a finer grain of analysis. The tradeoff here, of course, is the increasing complexity of the model with the incorporation of more detailed system behavior.

Ultimately, the need for a finer grain of reasoning depends on the context in which diagnosis takes place. In cases where full, consistent engine performance is absolutely critical (such as takeoff), the difference between rapidly fluctuating and steadily decreasing turbine speed does not matter. These issues, together with intermittent faults and faulty sensors, are further challenges to and future research targets for model-based diagnosis methods.

Finally, AI systems of these types need a monitoring front end that can decide which of the system's findings are to be pursued further, and which are trivial or redundant. The introduction of faults into a complex, tightly coupled system such as an aircraft can lead to symptoms in many parts of the system, and thus to an explosion of hypotheses regarding the root causes of the disturbance. Such a front end is extremely sensitive to how the system's hypotheses are represented. For example, under acute time pressure, pilots have been found to attend to the first line of computer output and begin looking immediately for a prescribed procedure with which to solve the problem represented. If the AI system offers too many hypotheses or presents no procedural solution, it cannot work cooperatively with the humans to solve the problem.

THE ELECTRONIC CREW MEMBER

In the early 1970s, investigators became interested in the interaction process between humans and AI systems. Rouse (1988, p. 432) described the criteria for what are now called *adaptive aiding systems*:

> The level of aiding, as well as the ways in which human and aid interact, should change as task demands vary. More specifically, the level of aiding should increase as task demands become such that human performance will unacceptably degrade without aiding. Further, the ways in which human and aid interact should become increasingly streamlined as task demands increase. Finally, it is quite likely that variations in level of aiding and modes of interaction will have to be initiated by the aid rather than by the human whose excess task demands have created a situation requiring aiding. The term *adaptive aiding* is used to denote aiding concepts that meet [these] requirements.

It is implied here that the pilot who needs such assistance will usually be too busy to ask for it, a premise that needs careful examination.

Following development of the concept and modeling studies of human performance (Rouse, 1980), several empirical studies were performed to evaluate and expand the concept and its potential applications. These led to the elaboration of a comprehensive "framework for adaptive aiding" (Rouse & Rouse, 1983). This work, in turn, was embodied in the Pilot's Associate program, carried out by the Lockheed-Georgia Company under the sponsorship of the Defense Advanced Research Projects Agency.

In this application, adaptive aiding

> Is an element of an overall intelligent interface, which includes AI modules for display management, error monitoring, and adaptive aiding .... One particularly interesting aspect of this effort is the nature of the expertise embedded in the many expert systems that make up the Pilot's Associate. There are suites of expert systems for mission planning, tactics planning, situation assessment, and systems status monitoring that include expertise on aircraft, flying, military doctrine, and so on. In contrast, the primary expertise within the six expert systems that make up the pilot–vehicle interface is expertise on human information processing and performance, with special emphasis on how situational characteristics and information presentation affect the formulation of intentions and subsequent plans. Thus, to an extent, the pilot–vehicle interface is a highly specialized human factors expert. (p. 433)

Rouse (1988, p. 441) concluded that,

> In retrospect, the notion of adaptive aiding is much more evolutionary than revolutionary. User-initiated adaptation has long been the norm in aerospace system (e.g., autopilots). There are also many everyday examples of humans adapting their automobiles and appliances. Thus the primary innovation of adaptive aiding is not adaptation *per se* but the possibility of aid-initiated adaptation.

(There is a fundamental difference, however, between *user*-initiated adaptation and *machine*-initiated adaptation. The user almost always has more knowledge of the world state and its implications than the machine.)

Building on its Pilot Associate program, Lockheed has continued its interest in this class of computer aids. Work is in progress on new "associate" technologies for dispatchers, air traffic controllers, and others. The Air Force's Armstrong Laboratory has continued to study adaptive aiding systems for pilots, and there is a "surface movements advisor" element in NASA's Terminal Area Productivity research program.

ISSUES RAISED
BY ADVANCED COMPUTATIONAL CONCEPTS

Human and Machine Roles

Let us first return to the paradigm of the human or machine as prostheses of one another. In chapter 3, and repeated in chapter 4 and implicitly elsewhere in this document, the prosthesis paradigm is contrasted with what could be called the paradigm of "the cognitive instrument" (Roth et al., 1987). In the cognitive instrument paradigm, automation is not in place to supplant human functions. Rather, automation consists of tools to assist human beings in their problem-solving tasks. Machines should be considered as complementary, instead of competitive (Jordan, 1963). We should ask ourselves again: Is the effort of AI in diagnosis directed toward supplanting the human diagnostician? Or is it aimed at aiding the human problem solver?

In most relevant AI research, emphasis is placed on how to conduct automated diagnosis, and less attention is usually paid to how the information from such automated diagnostic processes could benefit flight crew in various contexts. Such issues as the flight crew information requirements for fault management on the commercial flight deck are addressed within the NASA Faultfinder program, however (Abbott & Rogers, 1992; Rogers, 1990). The study of information presentation in this program is focused on understanding the cognitive activities associated with fault management, so that needed support of human information processing and decision making can be offered.

Note that such issues are embedded in the question in chapter 3 about whether crews in newer aircraft are sufficiently involved, or "drawn in," to their operations. Following the cognitive instrument paradigm, the aim of automated fault diagnosis should *not* be to interpose more automated processes between pilot and aircraft. Diagnostic systems should bring pilots closer to what is going on within a subsystem, rather than distancing them from the process.

Autonomous Intelligence

Whether such systems perform their assigned functions autonomously or are able to work as "team players" is often less related to their inherent capabilities than to the design of the interface between the systems and the humans responsible for management of the overall process. A conversational representation of AI behavior is a grossly inadequate communications tool for a pilot or controller under time pressure. As noted earlier, pilots cannot sort out the multiple symptoms in tightly coupled systems and are unlikely to have time to decide which of 10 or more possible faults is the culprit in a particular anomaly. Here, as elsewhere, "representations are never neutral" (Woods, in press); if they do not help solve the problem, they will be perceived as part of the problem.

Adaptability versus Adaptation

Adaptability (the ability to adapt autonomously given certain input conditions) is a characteristic of some of these computational concepts. An example was the mode annunciator panel decluttering in the A330 accident at Toulouse in 1994. It is this characteristic that gives rise to certain concerns about their use in a high-risk, dynamic environment such as aviation.

Many machine systems are designed to adapt autonomously. In the A330, autospoiler extension occurs slowly on landing until reverse thrust is selected, and rapidly thereafter. In many aircraft, trailing-edge wing flaps will not extend (or will retract) above a certain airspeed to avoid excessive airloads on the surfaces. Warning systems are inhibited in most newer aircraft during takeoff; some function only during cruise flight. The brightness of newer cockpit displays is controlled as a function of ambient light in the cockpit. These systems, however, adapt in known ways to known stimuli; they remain predictable, and if they do not behave in the expected way, the pilot is alerted to the presence of a malfunction and can compensate for it.

I have suggested throughout that the machine component of a human–machine system must be predictable, so that the human can understand and form a clear mental model of the machine's present and expected behavior. There is a good deal of difference between a machine system that can be adapted or modified by its operator and a system that can adapt autonomously in perhaps unpredictable ways. In the former case, the human operator is at the locus of control; in the latter case, the machine is at the locus of control. With regard to maintaining command of the process, the difference is crucial. Machines that behave in unexpected ways produce surprises for their operators. In the systems under discussion here, surprises can also occur because it is not possible to fully characterize the ways in which complex AI systems may behave when confronted with novel circumstances.

Roth et al. (1987) discussed this in the context of intelligent decision systems:

> Psychologists are fond of discovering biases in human decision making. One judgmental bias is the overconfidence bias where people at all levels of expertise overestimate how much they know. However, we sometimes forget that these biases can apply to the designers of machines as well as to the users of machines. This means that the designer of an intelligent decision support system is likely to overestimate his/her ability to capture all relevant aspects of the actual problem solving situation in the behavior of the machine expert. (p. 502)

Aid-initiated adaptation was a factor in the Charlotte wind shear accident (1994); it also posed problems in the Tarom Airlines A310 incident at Orly Airport (Paris, 1994).

COMMENT

These factors have led me over the past decade to a position of extreme conservatism with regard to the potential of AI systems as autonomous agents, and particularly self-adapting systems, for flight-critical applications. I recognize that these newer computational architectures have considerable promise for defined tasks that can be bounded (such as some of the diagnostic tasks discussed earlier). I also realize that object-oriented programming may significantly decrease the enormous software development cost involved in the programming of some of today's very complex, integrated systems. To the extent that these software technologies can ease the large and growing development burden without making verification of software even more difficult than it is today, they should be adopted.

I believe, however, that,

> In high-risk, dynamic environments … technology-centered automation has tended to decrease human involvement in system tasks, and has thus impaired

human situation awareness; both are unwanted consequences of today's system designs, but both are dangerous in high-risk systems. [At its present state of development,] adaptive ("self-adapting") automation represents a potentially serious threat ... to the authority that the human pilot must have to fulfill his or her responsibility for flight safety. (Billings & Woods, 1994)

In civil aviation, at least, it is unlikely that artificial intelligence concepts will find their way into flight-critical automation systems until they have been thoroughly proven in less critical applications. In one such application, the Boeing Company is now implementing a knowledge-based engineering (KBE) program in which a computer-aided system automatically designs aircraft components using various Boeing design rules, manufacturing criteria, and stress analyses. The software is said to have the potential to dramatically reduce design costs; component design may take 30 to 60 min instead of weeks (Proctor, 1995). Another application is the use of AI to assist airline systems operation centers and dispatchers in resolving flight replanning problems (Smith, McCoy, Layton, & Bihari, 1993; Layton, Smith, & McCoy, 1994). A third is the use of AI to create more adaptive and individualized computer-assisted training modules. These and other applications will give airlines and manufacturers the opportunity to evaluate these technologies and to gain confidence regarding their usefulness and limitations.

Chapter 13

Requirements for Aviation Automation

INTRODUCTION

Since 1991, when my NASA Technical Memorandum on human-centered automation was published, I have become steadily more convinced that specific design guidelines can only be proposed in the context of a particular system being designed to meet specific requirements, subject to specific constraints. That being the case, can a book such as this offer any useful guidance to those who must design, build, and operate future aviation automation? I am not certain of the answer to this question, but I am indebted to those colleagues who have tried to guide me in proposing such guidance. The entire document to this point is in reality an attempt to motivate lessons that can be learned from our experience in this domain.

I suggested in chapter 3 some "common factors" I believe are found in automation-associated incidents and mishaps, namely, loss of situation or state awareness, associated with:

- Complexity
- Coupling
- Autonomy
- Inadequate feedback.

In chapter 9, I elaborated on these and on other automation attributes that appear to have been associated with problems in the operation of highly automated systems, namely, automation characteristics of

- Brittleness
- Opacity
- Literalism

- Clumsiness
- Monitoring requirements
- Data overload.

To generalize still further, the data reported here indicate that a fundamental problem in this human–machine system is that *human operators sometimes do not understand what their automated tools are doing, and why.* For that reason, they have difficulty in using automation effectively to attain their objectives. There are many reasons for inadequate understanding; some are related to design complexity and coupling, some to inadequate training, and some to characteristics of humans, including their tendency to rely uncritically on these normally reliable tools. We must also consider that virtually all airline flying is done well within the operating envelope of the aircraft. It is extremely rare for air carrier pilots even to approach the limits of the operating envelope, except in training in simulators. For this reason also, pilots can perhaps be forgiven for a lack of familiarity with how their equipment operates under these conditions.

I said in chapter 3 that I believe a philosophy of human-centered automation can help to lessen these aviation system problems. The requirements set forth here are aimed at the conceptual and implementation issues underlying the (largely cognitive) problems presented in these pages. They are necessarily presented sequentially; they are like FMS screens that can only be accessed one at a time. They are not independent, however, and many or most of them have implications for at least several others. *These requirements and guidelines must be considered as a whole, not only as "stand-alone" statements.*

In a landmark paper in 1980, Earl Wiener and Renwick Curry discussed "Flight-Deck Automation: Promises and Problems" (see appendix B). Their contribution has been the stimulus for a great deal of research during the 15 years since it was published. After presenting candidate guidelines for control and information automation, the authors concluded that

> The rapid pace of automation is outstripping one's ability to comprehend all the implications for crew performance. It is unrealistic to call for a halt to cockpit automation until the manifestations are completely understood. We do, however, call for those designing, analyzing, and installing automatic systems in the cockpit to do so carefully; *to recognize the behavioral effects of automation* [italics added]; to avail themselves of present and future guidelines; and to be watchful for symptoms that might appear in training and operational settings.

Their statement is true today and their call is as appropriate as when it was written. This section is devoted to expanding on their guidelines with the benefit of an additional 15 years of experience and hindsight.

REQUIREMENTS
FOR HUMAN-CENTERED AIRCRAFT AUTOMATION

There are innumerable guides for aerospace system designers. All present more or less specific prescriptive guidance, often context-free, which may or may not meet the particular requirements of a design engineer working under specific constraints on a specific system. Designers frequently complain that most do not provide the guidance required in the design process, nor sufficiently firm reasons for taking a certain path in preference to others that may be easier or less expensive in a given project. Further, although designers are unquestionably at the "sharp end" when an airplane is conceived, they give way to the operators who purchase what is built, and to the ultimate users: the pilots or controllers who must make the system work. This book is as much for them, and they also need guidance to make the best possible use of the complex machinery that is provided to them.

Let me reiterate that I do not believe that specific "how-to" guidance is appropriate or particularly useful except in the context of a particular system, within which there may be several, perhaps equally effective, ways to implement a particular function. In this section, I have tried to provide guidance with respect to "what to do" (or not to do!) rather than "how to do it," for I believe that our knowledge of cognition and behavior is sufficient to provide some general outlines of what needs to be present in a human-centered aviation system. It is probably more appropriate to call these "requirements for human-centered automation," or guidelines for the development of requirements.

HUMAN-CENTERED AUTOMATION
FOR AIR TRAFFIC CONTROL

I hope it is obvious from chapter 3 that I believe in the importance of human-centered automation for air traffic controllers as well as for pilots. In fact, I believe this approach to automation may be more important in the ATC domain because of several factors. The opportunities for proper design are very great, because major elements of the automation system for ATC are not yet in place. The need is equally great, because ATC will be the instrumentality through which the required gains in air traffic capacity will have to be realized. Pilots can help, but they can only control one airplane at a time. Controllers will bear the burden of the additional traffic. They will need effective and highly efficient tools to help them manage the additional loads.

The first principles of human-centered automation have been discussed at some length in chapter 3. I am convinced that human air traffic controllers must remain in firm command of the future ATC system if it is to meet the challenges that will be imposed on it. They must retain authority commensurate with their great responsibilities. If future ATC automation is *not* human-centered, the entire aviation system will lose this focus and the flexibility that goes with it. That flexibility, both on the ground and in the air, is what has enabled the current system to cope with traffic demands to date. It will be even more necessary in the more crowded (and probably more tightly coupled) system of the future.

How can flexibility be maintained in the face of a need to move more aircraft, spaced more closely, with less room for error? I believe this can be accomplished by providing human controllers with decision and monitoring aids that will enhance their considerable cognitive capabilities while maintaining surveillance of traffic to insure that it is moving in accordance with their requirements. Controllers do not, in general, need to be told how to move airplanes; that is what they do best. What they do need, at a minimum, is confirmation that their plans are appropriate and assistance in keeping track of whether airplanes are moving in accordance with those plans. These are types of assistance that computers are quite capable of providing even without major advances in computational capability.

The NASA/FAA CTAS program (Erzberger & Nedell, 1988, 1989; Harwood & Sanford, 1993; Tobias & Scoggins, 1986) has demonstrated that properly designed decision aids, which take account of aircraft dynamic capabilities, can help arrival controllers to decrease traffic dispersions considerably, thereby increasing terminal area throughput. CTAS has also demonstrated that properly designed management and spacing aids can assist materially in the controller's planning processes, while leaving the human able to exercise his or her expertise and judgment as the traffic situation unfolds.

As noted in chapter 8, many system designers believe that only a truly radical reshaping of the traffic management architecture and infrastructure will be able to accommodate traffic expansion beyond perhaps 2010 (Whicher, as cited by Cooper, 1994a; RTCA, 1995). I am not convinced that the future ATC system must be so radically restructured to meet the demands that will be placed on it. On the contrary, I believe that the new system should perhaps rely more on controller expertise than is envisioned in present proposals for its architecture. I am more optimistic that a human-centered ATC system, in which human experts are complemented by the proper intelligent tools, can get the job done well beyond that time. What are the proper tools, and what must they be able to do to help the human operator do his or her job? That is the question that must be examined.

ISSUES SPECIFIC TO AIR TRAFFIC CONTROL

It is necessary to remember the important distinctions between pilot and air traffic controller tasks in the aviation system. The pilot receives essentially instantaneous feedback from an airplane and its displays once the pilot (or his or her automation) makes a control input. The "controller," on the other hand, directs traffic by giving instructions to an intermediary (the pilot); the controller must then wait an indeterminate period of time to observe whether the airplane appears to be executing the requested action. The difference in required lead time may be considerable and it can have major consequences for controller planning, as can the fact that controllers must usually manipulate several aircraft rather than only one. In these respects, the controller's tasks are conceptually more difficult than those of the pilot.

Also more difficult for the controller is the fact that he or she must ordinarily work entirely through representations of the monitored system rather than being able to observe its behavior directly. (Tower controllers in VMC are an exception.) Unlike controllers, pilots receive not only visual, but also tactile, proprioceptive, and auditory feedback from their airplane and environment. Woods and Holloway (in Woods, 1994a) illustrated the problem in this manner (Fig. 13.1). The controller is handicapped by having to view the monitored system entirely through a representation rather than being able to view its behavior directly:

> The viewport size is very small relative to the large size of the artificial data space
> . . . that could potentially be examined. This property is often referred to as the

FIG. 13.1. The keyhole property. From *Global Perspectives on the Ecology of Human–Machine Systems* (p. 167), by D. D. Woods, in J. Flach, P. Hancock, J. Caird, and K. Vicente (Eds.), 1995, Hillsdale, NJ: Lawrence Erlbaum Associates. Copyright 1995 by Lawrence Erlbaum Associates. Adapted with permission.

keyhole effect (Woods, 1984). Given this property, shifting one's "gaze" within the virtual perceptual field is carried out by selecting another part of the artificial data space and moving it into the limited viewport. (Woods, 1995, p. 167)

The controller sees only a plan view of the traffic space; the third dimension must be provided by alphanumeric data and by symbols on another screen. To assess traffic not yet visible, a second screen or strip bay that portrays flight data must be examined.

(The "keyhole" problem afflicts pilots as well; their view of the traffic situation is incomplete, and TCAS as implemented today is not an efficient means of providing them with information concerning other traffic not yet in conflict. The "party line" afforded by broadcast voice communications is of help, but some of the information conveyed to other aircraft may be ambiguous.)

Controllers become adept at constructing a mental model of the traffic under their direction from this imperfect view, but under the stress of heavy traffic or distractions, they are in danger of losing their internal model, a serious problem because "once the picture has been lost, the controller can seldom recall it in its entirety again but has to rebuild it painstakingly aircraft by aircraft" (Hopkin, 1994a, p. 173).

For all these reasons, the controller requires different tools to perform different tasks from those of the pilot. It is easy to argue that controllers need more help from automation than do pilots, but this neglects the obvious fact that they are now performing very well indeed without many of the aids that pilots take for granted. *Any attempt to produce guidance for ATC automation must take account of what controllers now do so effectively without such aiding, and why they are able to perform successfully in a largely "manual," or unaided, work environment.*

PRINCIPLES OF HUMAN-CENTERED AUTOMATION— GENERAL REQUIREMENTS

I return to the first principles of human-centered automation set forth in chapter 3, and repeat them here as general guidance, with some further discussion of each of them. These principles deal at a fundamental level with the relationship between human operators and the machines that assist them in carrying out their mission.

The Human Operator Must Be in Command

Fully autonomous transport aircraft are probably technically feasible, but are not politically possible at this time, in my view, because social factors would prevent them from being accepted by those who wished to utilize the services they offered. On the other hand, we accept and utilize the products made available by unmanned satellites without question, and their reliability at this time is of a high order. Were fully autonomous vehicles to become the dominant mode of transportation, this document would not be necessary, although a different document devoted to the human factors of ground control systems might be useful. I assume, for purposes of argument, that human "commanders" will continue to be responsible for the safety of air transport, and these guidelines are based on that premise. To the extent that it is true, I believe that the human operator must be given authority commensurate with that responsibility.

There are at least three ways that command authority can be compromised. A pilot in command can effectively relinquish that role, either to other humans or to the automation, by indecisiveness when a decision is required. This is fortunately rare, although it was observed both in simulation (Ruffell Smith, 1979) and in flight (Portland, Oregon, 1978) when the captain delayed making the decision to land until his fuel supply was insufficient to permit a controlled landing. Operators can provide command and crew or cockpit resource management (CRM) training to reduce the likelihood of such behavior, but it can still occur, especially if the first officer is a strong-willed, dominant person or the automation is highly authoritarian and the captain is relatively passive.

A second way in which command authority can be degraded is by overly restrictive operator policies and procedures that "hamstring" the commander's authority, or by operator failure to back its commanders when disputes arise with company, government, or other ground support personnel. The Air Ontario F-28 accident (Dryden, Ontario, 1989) grew out of a situation in which the captain was overruled and required by his company to off-load fuel to permit a full passenger load in an airplane whose auxiliary power unit (APU) was inoperative. This combination placed him in a classic "double bind" when he landed at Dryden to refuel.

A third way in which command authority can be eroded is by a system's designers or manufacturers. "Hard" airplane or engine operating limitations encoded in automation software can preclude a pilot from making full-capability maneuvers if they are required in an emergency. Inadequate feedback on cockpit displays can deny a pilot the information he or she needs to recognize, evaluate, and respond to a developing aircraft or automation problem, as may have occurred prior to the A330 accident at Toulouse, France (1994), when the pilots' mode annunciator panels "decluttered" after the airplane exceeded 25° of pitch during a test flight. In ATC, system architecture can deny operators the information they need to support effective decision making (as was the case in early numerical control machines; see chapter 4).

It is a fundamental tenet of human-centered automation that aircraft and ATC automation exists to *assist* pilots and controllers in carrying out their responsibilities as stated earlier. The reasoning is simple. Apart from the statutory responsibility of the human operators of the system, automation cannot deal with uncertainty and does not have comprehensive knowledge of world states. The human's responsibilities include detecting shortcomings in the automation's behavior, correcting it when necessary, and continuing the operation safely to a conclusion or until the automated systems can resume their normal functions.

To Command Effectively, the Human Operator Must Be Involved

To exercise effective command of a vehicle or operation, the commander must be involved in the operation. *Involved* is "to be drawn in;" the commander must have an active role, whether that role is to control the aircraft (or traffic) directly, or to manage the human or machine resources through which control is being exercised. The pilot's (or controller's) involvement, however, must be consistent with his or her command responsibilities; the priorities of the piloting or airplane management tasks remain inflexible. The pilot or controller must not be helped to become preoccupied by a welter of detail.

As we have implemented more capable and independent automation, particularly in long-haul aircraft, we have not made it appreciably harder for an alert, competent pilot to maintain situation awareness. What we have done, however, is to make it easier for a tired, bored, complacent or distracted pilot to distance him or herself from the situation. This is not new; Korean Air Lines flight 007 (Sakhalin Island, 1983) was probably flying in "heading" rather than "INS" mode for some considerable time before the first of its two incursions into Soviet airspace. What is important is that none of three crewmembers detected the mode error. They were not adequately involved. The same thing is likely to happen in a highly autonomous air traffic management system, if the automation is capable and reliable.

Ways must be found to keep pilots and controllers involved in their operations by requiring of them meaningful (not "make-work") tasks. Ideally, these tasks should have perceptual, cognitive, and psychomotor components so that pilots must perceive or detect, think about, and respond actively to some stimulus, and controllers must remain drawn in to the decision-making process. This may require that designers "un-automate" some tasks or functions that are or could be performed by the automation. Such a step always involves the risk that the tasks may be missed or performed wrong, but if we know enough about the task to have automated it, we also know enough to implement an error-detection module that will alert the pilot that a task was not performed or was performed incorrectly.

Modern aircraft automation is extremely capable; it has made it possible for the aircraft commander to delegate nearly all tactical control of an operation to the machine. Human-centered aircraft automation must be designed, and operated, in such a way that it does not permit the human operator to become too remote from operational details. *We know how to automate, and we know ways of keeping pilots involved. The goal here must be to do both simultaneously, a less easy task but an essential one.* In air traffic control, the future system can and must involve the working controller, rather than turning him or her into a monitor simply looking for rare events.

To Remain Involved, the Human Operator Must Be Appropriately Informed

Without appropriate information concerning the conduct of an operation, involvement becomes less immediate and decisions, if they are made, become unpredictable. The level of detail provided to the pilot or controller may vary, but certain information elements cannot be absent if the operator is to remain involved and, more important, is to remain able to resume direct control of the aircraft or operation in the event of automation failures.

Both the content of the information made available and the ways in which it is presented must reinforce the essential priorities of the operator's task. In particular, state and situation awareness must be supported and reinforced at all times. A quantity of data that could overwhelm an operator if presented poorly (see London, 1972) can be easily assimilated if displayed in a representation that requires less cognitive effort to understand. The navigation display is a good example of this.

In highly automated aircraft, and in the future AAS, an essential information element is information concerning the activity and present capability of the automation. Just as the pilot must be alert for performance decrements or incapacity in other human crew members, he or she must be alert for such decrements in automated systems that are assisting in the conduct of the operation. This leads to the next requirement.

The Human Operator Must Be Informed About Automated Systems Behavior

The essence of maintaining command of automated systems is the selection and use of appropriate means to accomplish an objective. Pilots must be able, from information about the state of aircraft subsystems, to determine that total system capability is, and will continue to be, appropriate to the flight situation and their selected strategies for its conduct. (For a case in which this information was available but was not used, see New York, 1984.)

In most aircraft systems to date, the human operator is informed only if there is a discrepancy between or among the units responsible for a particular function, or a failure of those units sufficient to disrupt or disable the performance of the function. In those cases, the operator is usually instructed to take over control of that function. To be able to do so without delay, it is necessary that the human operator have access to historical information concerning the operations to date if these are not evident from the behavior of the airplane or system controlled.

It is therefore necessary that the pilot and controllers be aware both of the function (or malfunction) of their automated system, and of the results of its processes, if they are to understand why complex automated systems are doing what they are doing. Wiener and Curry (1980) argued for displays of trend information to provide pilots with advance information concerning potential failures. They noted that the provision of such information would also increase pilots' trust of their automation. Fuel usage greater than nominal (as determined by the FMS knowledge of flight plan) might be a candidate; engine parameters that might later require shutdown may be another.

The ability to monitor the automation (and thus the need for automation predictability; see next guideline) is perhaps even more important in ATC than in flight, because the controller has less comprehensive feedback than the pilot with respect to the behavior of the controlled system. As Norman (1989, p. 1) observed:

> The problem is that operations under normal operating conditions are performed appropriately, but there is inadequate feedback and interaction with the humans who must control the overall conduct of the task. When the situations exceed the capabilities of the automatic equipment, then the inadequate feedback leads to difficulties for the human controllers.

Automated Systems Must Be Predictable

To know what automation to use (or not to use), the pilot as manager must be able to predict how the airplane will be affected by that automation, not only at the time of selection but throughout the flight. It is important that not only the nominal behavior, but also the full range of allowable behaviors, be understood; all unpredicted system behavior must be treated as possibly anomalous behavior. This was less difficult when automation only performed continuous flight control tasks; it becomes far more difficult when automation performs many discrete tasks. Its inability to perform those tasks may become evident only after it has failed to do so, as when it fails to level off at a selected altitude, and pilots are less likely to detect a failure to perform than aberrant performance.

Controllers must likewise understand the full range of permitted behaviors of the automated systems that will assist them. Their understanding may be made more difficult by automation that looks ahead farther than they can, because it may suggest or take actions for reasons that are beyond their area of surveillance. System designers must take account of this and must provide controllers with means to ascertain why a certain action was recommended or taken.

If operators are to monitor automation to protect against the likelihood of failures, they must be able to recognize such failures, either by means of specific warnings or by observation of aberrant behavior by the automated systems. Both are probably desirable for critical systems, to improve detection probability. To recognize aberrant behavior, the operator must know exactly what to expect of the automation when it is performing correctly. This requires that the normal behavior of automated systems be predictable and that the operator have access to feedback about their operation. It also argues strongly for *simplicity* in the design and behavior of such systems. (Tenney et al., 1995)

Automated Systems Must Also Monitor the Human Operators

Because human operators are prone to make errors, it is necessary that error detection, diagnosis, management (Wiener, 1993), and correction be integral parts of automated systems. Much effort has gone into making critical elements of the aviation system redundant. Pilots monitor the behavior of air traffic controllers, who in turn monitor the performance of pilots, as an important instance.

Automated devices already perform a variety of monitoring tasks in aircraft and ATC, as indicated throughout this book. It is indisputable, however, that failures of an automated warning system have enabled serious mishaps when the automation did not warn that it was disabled and pilots, relying on it because of its effective functioning over a long period, failed to notice the conditions it was designed to detect (Detroit, 1987; Dallas-Fort Worth, 1988). It is necessary to recognize the human tendency to rely on normally reliable assistance. We should consider whether safety-critical automated alerting systems should be duplicated in today's operating environment.

Data now resident in flight management and other aircraft computers can be used to monitor pilots more comprehensively and effectively, if specific attention is given to the design of the monitoring function. I mentioned the substantial number of nonobvious navigation data entry errors, some of which have had serious effects only long after they were committed. Research should be conducted using the growing body of accident and incident data to determine other areas in which errors are common or have particularly hazardous implications, and ways should be devised to detect such errors and alert pilots to their presence. Both Langley and Ames Research Centers have experimented with procedures monitors; some new electronic checklists alert pilots to items not performed. Verification of execution of an inappropriate clearance would be of help to pilots by making controllers aware of a discrepancy between their expectations and what the airplane is going to do; it might also be of help to the controllers by activating potential conflict detection modules within ATC automation.

The most difficult task, of course, is to monitor pilot or controller decision making. When a human operator consciously decides to do nothing, his or her decision cannot be differentiated by any algorithm from a failure to do something. Further, advanced automation has made the need for decisions and actions infrequent during cruising flight (too infrequent, perhaps; see involvement, earlier discussion). The advent of extremely long haul aircraft has emphasized the problem of monitoring human alertness and functionality.

There is no way to make the system totally error-proof, and each additional piece of hardware or software has a potential decremental effect on system reliability, but as Wiener (1993) put it, multiple "lines of defense" against errors are essential if we are to make the system relatively foolproof.

Each Agent in an Intelligent Human–Machine System Must Have Knowledge of the Intent of the Other Agents

Cross-monitoring (of machines by humans, of humans by machines, and of humans by humans) can only be effective if the agent doing the monitoring understands what the monitored agent is trying to accomplish, and in some cases, why. The intentions of both the automated systems and the human operators must be known *and communicated;* this applies equally to the monitoring of automated systems by pilots, of aircraft by human controllers on the ground, and of air traffic control by human pilots in flight.

Under normal circumstances, pilots communicate their intent to ATC by filing a flight plan, and to their FMS by inserting it into the computer or calling it up from the navigation data base. ATC, in turn, communicates its intent to pilots by granting a clearance to proceed; data link in the near future will make this information directly available to the FMS as well. Proposals for free flight, however, suggest that this may change in the future system (chapter 8).

It is when circumstances become abnormal that communication of intent among the various human and machine agents may break down, as occurred in the Avianca accident at New York (Cove Neck, New York, 1990). The communication of intent makes it possible for all involved parties to work cooperatively to solve problems. Cooperation among intelligent agents is the cornerstone of human-centered automation. Many controller problems occur simply because pilots do not understand what the controller is trying to accomplish, and the converse is also true. Finally, neither automation nor ATC can monitor pilot performance effectively unless they understand the pilot's intent, and this is most important when the operation departs from normality (e.g., during an airplane response to a TCAS resolution advisory).

In at least two recent accidents, the automation did not warn in unmistakable terms that it was behaving in a manner contrary to pilot intentions. It could not do so, because the pilots had inadvertently signaled contrary intentions to the automation (Strasbourg, 1992, when the pilots failed to switch from heading and vertical speed mode to track/flight path angle mode; Nagoya, 1994, in which the go-around mode was activated). We must ease the task of communicating intent to the machine components of the system, but we must also find better ways to protect the human–machine system against *mis*-communication of intent, which appears to have occurred in both mishaps.

To these "first principles," I add two others of a general nature that have emerged from this review of aviation automation.

Functions Should Be Automated Only If There Is a Good Reason for Doing So

In the past, to quote a Douglas (1990, p. 3) briefing, the dominant design philosophy has been, "If it is technically and economically feasible to automate a function, automate it." Similarly, "Airbus Industrie officials believe that if the technology exists to automate a function that would prevent a pilot from inadvertently exceeding safety limits, this should be done" (Hughes & Dornheim in Automated cockpits, 1995a, p. 54). The potential effects of this philosophy were warned against by the ATA Human Factors Task Force report (ATA, 1989; see preface) and are illustrated throughout this document. There are, however, tasks that pilots cannot accomplish by themselves (usually because of their complexity or because there is not time to do them), and other tasks that we know they do poorly, such as monotonous repetitive work or monitoring for rare events. Better criteria are needed to motivate the automation of functions on a human-centered flight deck. Among criteria that might be applied are the following:

- If the time within which action is required following a signal or stimulus is less than will normally be required for detection, diagnosis, and decision to act (less than perhaps 3–5 sec, depending on the context), the task should be considered for automation.
- If a task is very complex, requiring many rote steps, or if the task is very difficult to perform correctly, the task should be redesigned or considered for automation.
- If a complex task, improperly performed, will lead to a high probability of an adverse outcome, or if an adverse outcome will threaten the safety of the

mission, that task should be redesigned or, if this is not possible, considered for automation.

- If a task is boring, repetitive, or distracting, especially if it must be performed frequently, that task should be considered for automation.

Having said this, however, I quote from Wiener and Curry (1980, p. 2), "Any task can be automated. The question is whether it should be." Each of the following questions should be asked, and answered, prior to the implementation of any new element of automation in the aviation system:

- Why is this function being automated?
- Will automating the new function improve system capabilities or flight crew awareness?
- Would *not* doing so improve the pilot's or controller's involvement, information, or ability to remain in command?

Automation Should Be Designed to Be Simple to Train, to Learn, and to Operate

I believe that aircraft automation to date has not always been designed to be operated under difficult conditions in an unfavorable environment by tired and distracted operators of only average ability. Yet these are precisely the conditions where its assistance may be most needed. Simplicity, transparency, and intuitiveness should be among the cornerstones of automation design.

Training must be considered during the design of all cockpit or ATC systems and should reflect that design in practice. Particular care should be given to documenting automated systems in such a way that operators will be able to understand clearly how they operate and how they can best be exploited, as well as how to operate them. This will be particularly important in the future automated ATC system because controllers, unlike many pilots, have not had experience with advanced automation and will have to learn to operate within, and develop confidence in, a new and very different system. The propensity of designers to rely on keyboard entry (with its attendant visual and psychomotor workload) may be a problem in the advanced automation system; controllers rely to a great extent on their vision for real-time information transfer. Their present workstation management interfaces do not require a great deal of visual attention; it is important that their future system controls also be easy to operate.

These "first principles" are not absolutes; they are but one approach, intended to promote a more cooperative relationship between operators and automation that allows the humans in command of the system to utilize automated assistance to its fullest potential. It is vital that humans understand and be able to communicate with these tools; it is equally vital that the tools "understand" what the humans want and communicate with them as they are performing their tasks.

SPECIFIC REQUIREMENTS

Some more specific requirements for human-centered automation follow from the principles just given. These are the most important.

Automated Systems Must Be Comprehensible

As automation becomes more complex, it is likely to become more tightly coupled, with more potential interactions among modes. Pilots must be helped to understand the implications of those interactions, and especially to understand interactions that can be potentially hazardous at a critical point in flight (see Bangalore, 1990, regarding the interaction between flight director and open descent modes). Automated systems need to be as error resistant as possible in this respect, for the likelihood that pilots will remember all such potential interactions is low, especially if they are not encountered frequently. The memory burden imposed by complex automation is considerable; infrequently used knowledge may not be immediately available when it is needed. "Prompting" or brief explanations should be considered with regard to such knowledge items.

Designers, and operators, of the future ATC system should take a lesson from our experience with aircraft automation. Mode proliferation has led to complexity and opacity. Every effort should be made to limit the number of modes in which ATC automation can operate. Less automation complexity and coupling will speed the transition to the new system, decrease training requirements, and increase operator acceptance. More important, it will decrease the likelihood of human errors in its use, both initially and through its lifetime.

The ultimate solution to this problem lies in keeping the operation of the system, and of its automation, simple and predictable. If it is simple enough, it may not need to be automated at all. If it is predictable and reasonably intuitive, it may not need to be particularly simple, for operators will understand and remember it. Complexity is an enemy of comprehensibility.

Automation Must Insure That Operators Are Not Removed from the Command Role

Increasing automation of aircraft and of the ATC system, and increasing integration and coupling of the ground and airborne elements of that system, have the potential to bypass the humans who operate and manage the system. One way to guard against this is to design future flight management systems so that the pilot is shown the consequences of any clearance before accepting it; another is to continue to insure that the pilot must actively consent to any requested modification of a flight plan before it is executed. A third, more difficult way is to make it possible for pilots to negotiate easily with ATC on specific elements of a clearance, such as altitude changes, rather than having to accept or reject an entire clearance or modification. All three, and possibly other ways as well, may be required to keep pilots firmly in command of their operations in a future, more automated system.

I have indicated my concern with regard to the level of authority that controllers may have in the future ATM system. Although it may not always be appropriate for a controller to have to work one-on-one with each airplane under his or her control, it is vital that the controller remain in command of air traffic and able to modify its behavior as required to perform the mission.

The temptation to build more autonomous automation is pervasive, particularly when an express purpose of that automation is to "improve human productivity" (which in the past has usually meant to accomplish the same or greater throughput with less human involvement). But the losses in system productivity that will result if controllers are unable to remain "in the picture" are likely to negate any gains achieved by more autonomous machine systems.

The changes proposed for the future aviation system will require more than simply software changes. They will require detailed negotiations between the operating community and air traffic management system designers. In view of the rapidity with which the enabling technology is being pursued, the long-term goals and objectives of system designers and planners with respect to future human and machine roles in the system need to be known with precision. They have not been set forth with sufficient clarity thus far, and the potential consequences of fundamental changes in the locus of command of the system are so major as to require informed consensus before proceeding farther with system redesign.

A Primary Objective of Automation Is to Maintain and Enhance Situation Awareness. All Automation Elements and Displays Must Contribute to this Objective

The minimum elements of information required by pilots at all times are a knowledge of the airplane's position, velocity, attitude, error rate, status, threats, the status of the aircraft control automation and other aids, what must be done next, and when it must occur (Fig. 13.2). These are the elements of situation awareness. Many other information elements will be required in some form at specific times, however. The question is not whether these are needed, but in what

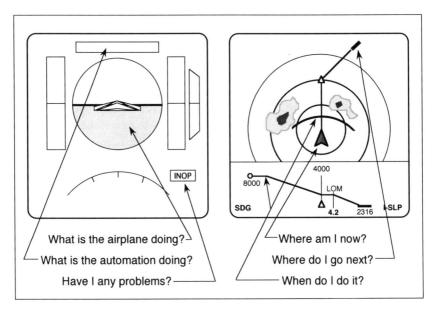

FIG. 13.2. Information elements required for maintenance of pilot situation awareness.

form and representation they will best reinforce the pilot's awareness of his or her situation and state.

The radar controller's video display units or radar scope are the sole means by which he or she can maintain cognizance of the system being controlled or monitored. Much progress has been made over the years in reducing ambiguity, clarifying presentations, providing aiding features, and improving the quality of the input data that are processed for use on this display. Ordinarily, these representations of system activity are sufficient to keep controllers "in the picture" by providing them with information they use to update their mental models of the traffic under their control.

Efforts have been made over the years to improve these representations. In particular, attempts have been made to provide three-dimensional representations of air traffic, thus far without notable success. A planform display remains the standard in air traffic management, as it does in most military command and control systems. The fact remains, however, that a good deal of cognitive activity is required to construct and maintain a three-dimensional mental model from the available imagery. Not all controllers accepted for training ever develop the ability, and this is one reason for high attrition rates in controller training.

The advent of advanced display media incorporating color and improved imagery offers the opportunity to examine again various aspects of ATC displays, with the aim of decreasing the mental effort involved in their interpretation. Among the features that have been proposed is more effective highlighting of significant events and new alerting techniques. Care must be taken, however, to insure that controllers can transition easily from the old to newer displays. Equally important, attempts should be made to assist the controller unfortunate enough to lose his or her mental model of traffic to regain it more easily and quickly, by systematic analyses of the cognitive processes now used by expert controllers and the provision of visual aids keyed to flight progress data.

Automation Must Never Be Permitted to Perform, or Fail, Silently

"Fail-passive" control automation represents a potential hazard in that its failure may not and usually does not change aircraft state if the airplane is in a stable condition at the time. Such failures must be announced unambiguously to insure that the pilots immediately resume active control of the machine. Automation must never encourage a situation in which "no one is in charge," for pilots must always "aviate" even if they have delegated control to the autopilot. The Everglades accident following an undetected autopilot disconnect was a good example of what can happen if this tenet is violated (Miami, 1972).

A particular case of this is uncoupled sidestick controllers, both of which can be operated simultaneously (the inputs are summed) without tactile or other feedback to either pilot. Consideration should be given to indicating to both pilots any situation in which commands are being input from both controllers.

Much of the activity of future ATC automation will be transparent to human operators. In particular, its ongoing or periodic monitoring of trajectories and its continuous searching for potential conflicts will not (and should not) be visible. The operator, however, must be informed that these activities are ongoing, for the absence of such information can mean either that the machine has not located any potential events of interest, or that it is not performing correctly. Ways must be found to keep controllers informed of these processes, and of their failure if the automation becomes degraded in any respect.

Management Automation Should Make Human–Machine Systems Easier to Manage

A major problem with flight management systems is that they are often cumbersome to operate. Under some circumstances, it is easier to operate without them than to use them, with the predictable result that they are apt to be bypassed under these circumstances. This is a pity, for the error resistance that they bring to flight path management is also bypassed.

Similarly, the controller must attend to substantial amounts of alphanumeric information, a major source of distraction from attending to traffic representations. If a controller does not have a data assistant, attention must be divided between the primary display and the flight data display (or flight data strips).

One partial solution to this problem for pilots is to improve the interfaces between system and pilot so that they can be manipulated more easily. This will not be a trivial task, for it may require establishing a different level of interface between the pilot and the system, one that involves a higher level interaction than the present point-by-point description of desired ends. On the other hand, data link may enable a higher level interaction and may even require it for effective interaction with ATC controllers, most of which may be through the FMS. One hopes that controller interaction with the advanced automation system, and particularly data entry, will be made less cumbersome than it sometimes is in today's aircraft. The lessons learned from clumsy avionics automation (see chapter 9) should be applied here.

Within the constraints of present-generation flight management systems, continued efforts to improve the ease of system programming and operation in high workload segments of flight would be most helpful to pilots, and would improve system safety. Much progress has been made in easing the task of modifying approach tasks to accommodate revised ATC instructions. The problem of tuning navigation and communications radio aids through the CDU, a digital interface, has been mentioned by pilots; providing alternate interfaces (similar to those available in older aircraft) through which these and other cumbersome tasks could be accomplished more readily is worthy of consideration.

The proliferation of modes in newer flight management systems has imposed an increasing cognitive burden on pilots. Both the modes themselves and their inadequate feedback have induced erroneous actions in flight crews not entirely familiar with them. Operators should determine whether all modes are required and should consider simplifying training and workload by eliminating those that are not entirely necessary.

Designers Must Assume That Human Operators Will Rely on Reliable Automation, Because They Will

Once pilots have flown an automated airplane long enough to become comfortable with it, they will come to "know" which control, information, and management elements can be trusted. Thereafter, many (although not all) pilots will become increasingly reliant on the continued functionality of those elements and therefore less liable to be suspicious of them if they become unreliable. It can be predicted with high confidence that the same phenomenon will be observed in controllers working in the future automated ATC system, if they find over a period of time that the automation provides generally dependable conflict resolution options. The problem may be greater in ATC, because the effects of clearance modifications proposed by the computers may not be visible to the controller who must approve them.

If information is derived or processed, the designer must insure that data from which it is derived are also either visible or accessible for verification. If it is not critical information for a particular flight phase or control decisions, make it available only on request, but insure that it remains accessible, as has been done with raw navigation data (Fig. 6.8).

Future automated decision support systems may pose a serious problem in this regard if pilots or controllers come over time to rely on the quality of the machine decisions. A poor machine decision may be much more difficult to detect than an aberrant subsystem operation. It may also be much more difficult to determine why a machine decision is correct, because of the complexity and opacity of the processes that motivate it.

It is not enough simply to warn pilots or controllers in training that their automation is not infallible. They will rely on their own experience to assign subjective probabilities of failure. If a machine has "always worked well" in their experience, they will assume that it will continue to work well, and they will usually be correct—but not always, which motivates the need for error (and failure) tolerant automation.

Requirements for Aircraft Control Automation

Several lessons learned relate specifically to requirements for control automation in aircraft. Among them are the following.

Control Automation Should Be Limited in Its Authority. It Must Not Be Permitted to Become "Insubordinate"

Control automation should not be allowed to endanger an aircraft or to make a difficult situation worse. It should not be able to assume a state that could cause an overspeed, a stall, or contact with the ground without explicit instructions from the pilot, and possibly not then. If either the pilot or the automation approaches safe operating limits, the automation should alert the pilot, giving him or her time to recognize the problem and take corrective action.

The pilot should not be permitted to select a potentially unsafe automatic operating mode without being challenged; automation either should foreclose the use of such modes or should alert the pilot that they may be hazardous. Many useful modes are "open-ended;" in these cases, continued pilot involvement is especially important. Alternatively, the designer should consider whether there is really a need for such a mode, or whether another way to accomplish the same function would be a safer, more error-resistant approach.

The edges of the operating envelope are a particular problem. Some accidents involving automation have occurred at, or outside, the normal range of operating conditions. Because designers cannot guarantee that aircraft will never reach these conditions, automation must be designed so that it is tolerant of such conditions, or in any event, does not worsen them. The phenomenon of "brittleness" is difficult to predict, but very serious when it occurs, usually during an emergency when there may not be time to compensate for it.

Designers Should Not Foreclose Pilot Authority to Override Normal Aircraft Operating Limits When Required for Safe Mission Completion

The Airbus Industrie statement quoted earlier, "if the technology exists to automate a function that would prevent a pilot from inadvertently exceeding safety limits, this should be done," does not consider that pilots may find it necessary to *deliberately* exceed safe operating limits. Limitations on pilot authority, under rare circumstances not considered by the designer, may leave the pilot unable to fulfill his or her responsibility for safety of flight.

An ASRS incident report, one of many, underscores the need to preserve pilot capability to do what is necessary; an abrupt 50° banked turn was required for collision avoidance in a wide-body airplane (NASA ASRS, 1986). There have been several cases in which pilots have violated aircraft structural limits in an acute emergency; in nearly all of these, the aircraft have been recovered, although with damage (Atlantic Ocean, 1959; Luxembourg, 1979; Pacific Ocean, 1985). These maneuvers might not have been possible had hard envelope limits been incorporated.

I believe that the "soft envelope limits" approach represents one way to avoid limiting pilot authority while enhancing flight safety. Other automated modules that "lock out" flight-critical functions should also be capable of being overridden in an emergency. The implementation of "hard" limiting functions should be undertaken only after extensive consultation with both test and line pilots. Any limiting or other autonomous function (e.g., Mårtensson, 1995) *must* be made known to all pilots who will fly the airplane.

Automation Should Provide the Human Operator with an Appropriate Range of Control and Management Options

The control and management of an airplane must be safely accomplished by pilots whose abilities and experience vary, under circumstances that vary widely. To provide effective assistance to whomever is flying, under whatever conditions, a degree of flexibility is required in aircraft automation. The aircraft control–management continuum was discussed in chapter 10; problems at the extremes of this continuum were discussed (very high workload at the low end of the spectrum, decreased involvement at the high end of the spectrum).

The range of control and management options appropriate to a given airplane must be wide enough to encompass the full range of pilots who may operate it, under the full range of operating conditions for which it is certificated. It should not be wider than is needed to provide an appropriate range of workloads, however, to avoid unnecessary complexity.

Several studies of controller operational errors have indicated that larger numbers of errors tend to occur during periods of low or moderate, as opposed to high, controller workload (usually measured as number of aircraft being controlled) (Rodgers & Nye, 1993). If this is the case, then future automation that relieves the human controller of most routine workload may tend to increase the likelihood of human error, even though the automation may assume a portion of the tasks in which errors might be committed.

There is an urgent need for further studies of the relationship between levels of automation and the probability of errors in human tasks. Some laboratory work has been done, but before automation design is predicated on the results it is necessary that studies in more naturalistic settings be performed. It is possible that the CTAS evaluations planned at Denver and Dallas-Fort Worth in the immediate future may yield new insights into the relationships between workload and error in more automated environments, but it is also quite possible that CTAS, which maintains a high level of human involvement by design, may not be an appropriate analog of a future en route system in which the controller is less actively involved in routine operations. See also the following item.

Designers Should Keep Human Operators Involved in an Operation by Requiring of Them Meaningful and Relevant Tasks, Regardless of the Level of Management Being Utilized by Them

As suggested earlier, high levels of strategic management have the potential to decrease pilot involvement beyond desirable limits. Automation should be designed to minimize this detachment, so that pilots are ready to reenter the loop in the event of its failure. Keeping pilots meaningfully involved may require less automation rather than more, but it may be critical to their ability to remain in command of an operation.

One of the stated objectives of the AAS is to improve controller, and therefore system, productivity. Whicher (cited in Cooper, 1994a) suggested that

> To permit unrestricted ATC growth we should first determine how to eliminate one-to-one coupling between a proactive sector controller and every aircraft in flight—and so avoid him becoming reminiscent of the man with a red flag in front of early motor vehicles .... Pilots can and are willing to take direct responsibility for ... track-keeping functions, freeing controllers to concentrate on the key areas where human skills have most to offer—traffic management, system safety assurance and dealing with the exceptional occurrence. (p. 8)

I would argue that the degree to which the controller becomes involved with individual aircraft should be, within reasonable bounds, his or her choice. *The controller should have the freedom to select the level of control and management to be exercised under particular circumstances, just as pilots now may select the level of automation assistance they wish to invoke.* There should certainly be levels that provide considerable assistance, to permit the controller to focus on specific problems. There should also be levels that permit the controller to direct traffic, in order to retain the skills necessary for minimally aided control. At each level, the controller must have meaningful tasks to perform. As noted earlier, each level of management should be a cooperative endeavor between the human and the machine, requiring active participation by both components of the human–machine system.

I have suggested that requiring management by consent rather than management by exception may be one way to maintain involvement, although it has also been pointed out that we do not yet know how to keep consent from becoming perfunctory, and this must also be avoided. One way to assist may be to give more attention to workload management, as is suggested below.

Aircraft Control Automation Should Be Designed to Be of Most Help During Times of Highest Workload, and Somewhat Less Help During Times of Lowest Workload

Some field studies of aircraft automation have suggested that it may appreciably lighten workload at times when it is already low, yet impose additional workload during times when it is already high, during climbs and particularly descents (e.g., Wiener, 1989). Although much of the additional burden relates to problems in interacting with the flight management system itself, the end product of that interaction is the control and guidance of the airplane as it moves toward its destination.

Avionics manufacturers have made appreciable strides in easing this workload by providing lists of departure, arrival, and runway options at particular destinations. Air traffic control authorities, however, pressed to increase capacity at busy terminals, may develop and utilize procedures that differ from those anticipated by the aircraft designers. In particular, "sidestep" maneuvers to alternate parallel or converging runways are a problem in this regard, especially if clearances are altered late in a descent. Easing such problems may require a better understanding by ATC of what is, and is not, reasonable to ask of the pilots of a highly automated airplane. Given the congestion at our busiest terminals, however, ATC is likely to continue to seek more, rather than less, flexibility from pilots, and any short-term improvements will have to be in the cockpit (see also management automation guidelines discussed later).

Aircraft and ATC Automation Should Be Designed Both for Maximum Error Resistance and Maximum Error Tolerance

Both automated control systems and their associated displays should be designed to be as error resistant as is feasible by incorporating the simplest possible architecture, clear, intuitive displays, and unambiguous responses to commands. Designers should incorporate clear, unambiguous statements for the intended uses of each control mode in their software documentation. The designs should incorporate the highest reasonable degree of error tolerance as well. More consideration should be given to embedding monitoring and error-trapping software in the systems. Accident and incident data should be reviewed on an ongoing basis to identify likely human and machine deficiencies, and these deficiencies should receive special attention in this process.

The future AAS will be designed with improved error resistance, in that automated conflict prediction will be an essential element. Controllers will remain responsible for insuring that conflicts do not occur, but computers will augment their watch over traffic and will probably provide decision options to assist them in resolving conflicts when they are detected. These automated functions will thus increase the redundancy of the ATC system. The automated safety functions in use today—conflict alert, minimum safe altitude warnings, and others—will still be there performing their vital monitoring functions and acting to improve the error tolerance of the system. Can more than this be done? I believe it can.

As indicated previously, data link architecture should be designed to insure that ATC computers receive confirmation of flight path changes when they are executed on the FMS control-display unit or through mode control panel entries. These data are indicative of aircraft intent; they should be automatically compared with previously issued ATC instructions to insure conformity with planned trajectories. If there is a conflict, a controller would be notified so that he or she can determine where the difference lies and resolve the problem. In today's system, detection of an incorrect or undesired flight path can only occur after the airplane has already strayed appreciably from the desired path. Prospective monitoring of airplane intent would make it possible, in many instances, to detect and correct these problems before they happen. This functionality could prevent a substantial fraction of the altitude deviations that plague today's system. The UK CAA flight demonstration in early 1991 indicated the feasibility of such an approach, using currently available equipment.

Proposals for free flight do not make use of a "flight plan contract" that would facilitate prospective monitoring. Automatic downlinking of flight management system instructions could serve an important function in such a system by making immediately available to ATC the intent of aircraft in the system, which otherwise would not have knowledge of that intent until pilots communicated with the air traffic management system.

In view of the known problems in data entry, FMS software should accomplish as much error trapping as is possible. When data link is available, the data entry process may be simplified, but that does not necessarily imply that data entry errors will be eliminated. Many intermediate altitude restrictions will still have to be entered manually (usually into the MCP). This task is also known to be error prone; the downlinking of such data when they are executed would trap many such errors, if ATC software were provided to verify the correctness of the entries.

As noted earlier, CDUs refuse to accept incorrectly formatted entries, but they do not provide feedback as to why an entry was rejected. If the computer knows, why doesn't it tell the pilot? Some data entry errors are obvious, but others may be less obvious and pilots may be tired or distracted by other problems. In general, the less often a pilot is required to perform a particular programming task, the more likely it is that the details of accomplishing that task will be forgotten. Infrequently performed tasks, therefore, should be the ones on which pilots, and in the future controllers, receive the most help. Prompting could be very useful under these circumstances.

Human errors, many enabled by equipment design, bring more aircraft to grief than any other factor. Error-resistant systems can protect against many of these errors, but it is necessary to give pilots and controllers authority to act contrary to normal operating practices when necessary, and this requires that designs also incorporate error tolerance. Automation should be used wherever possible to monitor operator actions and warn of mistakes, slips, and lapses (Norman, 1981).

Requirements for Information Automation

Some of the points made thus far relate to information provided to operators as well as to the control of the airplane and its subsystems. It is not always possible to draw a clear distinction between control and information automation, for all automation involves the requirement to keep operators informed. The following are suggested guidelines specifically for information automation.

Emphasize Information in Accordance with Its Importance

The most important information should be most obvious and most centrally located. In aircraft, information relevant to aircraft control deviations, power loss, or impending collisions with obstacles is always more important than information concerning other facets of the operation. Changes in state or status (e.g., a mode change) are more important than information concerning static states. Symbolic information should be redundantly coded (shape, size, color, use of two or more sensory modalities) to insure that it is detected. Auditory (sounds) or tactile information displays can be used to reinforce, or in some cases to substitute for, visual information; this can be particularly useful during periods of high visual workload.

A strenuous and largely successful attempt has been made to decrease the large number of discrete auditory warnings that were present in older cockpits. The use of discrete voice warnings is increasing, however; GPWS, TCAS, and wind shear alerts all incorporate voice signals, and an increasing number of aircraft also incorporate synthetic voice altitude callouts on final approach. This may be less of a potential problem when data link replaces some of the voice communications now required, but there remains the potential for interference among voice messages, as well as the potential for overuse of voice signals leading to diminished attentiveness to voice emergency messages.

More information displayed on VDUs, especially if it is alphanumeric, will increase controller workload. Consideration should be given to the use of a limited number of auditory signals to denote information of particular importance. One promising way to direct attention to an event of interest, for instance, is to use synthesized directional auditory signals to indicate the approximate azimuthal location of the event (Begault & Wenzel, 1992). The technology has been evaluated in flight simulations as a way of drawing attention to potential conflicts detected by TCAS (Begault, 1993); it has proved quite effective in that application. This approach may likewise offer potential benefits in air traffic control.

The use of color to increase the salience of displayed signals can be effective in attracting attention, but redundant coding of such signals should be implemented wherever possible. Size, shape, and brightness cues in addition to color will make it more likely that important information will be attended to.

Alerting and Warning Systems Should Be as Simple and Foolproof as Possible

Warning systems for discrete failures do not present a particular problem as long as they are annunciated in such a way that the pilot can determine the root cause.

This has not always been the case. Whether reconfiguration of aircraft systems following such a failure should be autonomous remains an open question awaiting more experience with the MD-11 systems. The problem of quantitative warning system sensitivity and specificity has been discussed: False or nuisance warnings must be kept to minimum levels to avoid the unwanted behavioral effects of excessive alarms.

Warnings and alerts must be unambiguous. When common signals are used to denote more than one condition (e.g., the master caution and master warning signals), there must be a clear indication of the specific condition which is responsible for the alert. This is not a problem in most newer aircraft, although large numbers of discrete messages may occur during emergencies (see incident report at beginning of chapter 12).

Alerting and warning systems in current ATC suites are fairly simple, in part because the monochrome displays permit only the use of blinking symbols and auditory warning tones as information transfer devices. The use of advanced color displays will permit the use of colors, new icons, and other symbols as alerting devices. It will be important to keep alerts to a minimum, in order that their meanings remain simple and universally understood. Wherever possible, the exact nature of the alert, and the aircraft involved, should be specified in a way that immediately makes the nature of the problem obvious to the responsible controller.

Consideration should be given, as noted earlier, to providing alerting trend information to the appropriate controller before mandated boundaries are encroached on. There is a danger that this will lead to an increased number of nuisance alerts under some circumstances; that danger should be balanced against the problem of not warning until a violation has occurred. Certainly the prevailing practice of broadcasting audible conflict alerts is undesirable from a psychological viewpoint; it holds an "offending" controller up to ridicule, and it distracts others who may be busy solving their own problems.

Integration of Information Does Not Mean
Simply Adding More elements to a Single Display

Integration in psychology means "the organization of various traits into one harmonious personality." An integrated display combines disparate information elements into a single representation that renders unnecessary many cognitive steps the pilot would otherwise have to perform to develop a concept. It thus relieves the pilot of mental workload. Glass cockpit navigation displays are very effectively integrated. In contrast, electronic primary flight displays are not integrated, although they combine a great deal of information, previously shown on many instruments, on a single screen. The elements, however, are still discrete and the cognitive workload of inferring aircraft state is still required. The same is true of most power displays in today's cockpits.

Clutter in displays is undesirable, for pilots may fail to notice the most important information or may focus on less important data. Pilots are able to add or remove display elements from navigation displays. Fairly radical (pilot-selectable) decluttering of the PFD would still provide the pilot flying at cruise on autopilot with all information required to monitor the autopilot and return to the control loop rapidly if required.

Many subsystem displays can also be made more simple and intuitive. Again, the controlling variable should be what the pilot needs to know under normal and abnormal circumstances. As long as all information necessary to take over manual control of these systems is available when required, it is not necessary that other data be visible in circumstances in which they are not central to the pilot's tasks.

If ATC displays are to be redesigned in any major way, consideration should be given to a higher degree of integration of the existing displays if this can be accomplished without compromising the integrity of the critical information. Data from on-the-job training with regard to task elements that are difficult for trainees to assimilate would be helpful to the designers of these displays. In order to effect maximum transfer of training from the present to the new controller suites, it is quite possible that essential elements of the old displays should be retained unchanged or only minimally modified.

New features should be displayed analogically or by means of icons where appropriate. Wherever possible, displays should focus attention on changes in the data and on events of potential interest, leaving static or less interesting data in the background.

Requirements for Management Automation

The overarching issue of automation complexity must guide design approaches to management automation. These issues have been discussed at length with respect to aircraft automation, but an additional requirement must be mentioned with respect to air traffic management; it applies to aircraft in the system as well.

Future ATC Automation Must Insure That Traffic Control and Management Remain Within the Capabilities of the Human Operators Who Must Accomplish the Task If Automation Fails

Today's air traffic control system, particularly in crowded terminal areas, is operating to tighter tolerances than have ever before been permitted. The trend toward decreasing tolerances still further will continue, not only around airports but in continental and oceanic en route airspace as well. The FAA has specified extremely high reliability for critical elements of the future automated system, but it can be predicted with total confidence that functional failures will continue to occur, whether due to software bugs, communications system failures, environmental contingencies, human errors, or acts of God (see, for instance, Atlanta, 1980). In those cases, human controllers will be required to "make do" safely despite degraded machine capabilities, as they have always had to.

The problem with this is that new automation may well enable the system, when it is functioning normally, to operate at higher capacities and to tighter spatial and temporal tolerances than are possible when the human pilot and controller are operating without the automated tools. If higher throughput is possible with new technology, it will become, over time, the normal and expected throughput. Hollnagel's concept of risk homeostasis (1993, pp. 3–4) applies here: "that advances in technology lead to a reduction in perceived risk, hence to behavior that is closer to the limits of acceptable performance—thereby effectively reducing the margin for safety." Among the conditions that will arise is machine failure, and the reversion procedures must take account of human capabilities and limitations.

AAS software will be subjected to exhaustive testing before it is placed on line. Care must be taken to explore the margins of its operational envelopes, however, to insure (insofar as is possible) that it is not brittle—that there are not conditions under which it begins to behave in ways that makes the controller's task more difficult, or places the pilot in a situation from which safe escape is not assured.

There are very few flight maneuvers that require such precision that they have been entrusted only to automation. Pilots generally have not been asked to engage in operations unless they can demonstrate their ability to perform them without machine aid. The limited capacity of the airspace system, however, has motivated intensive efforts to increase system throughput by making better use of presently available runways and terminal airspace. As noted earlier, this includes studies of closely spaced parallel approaches, the use of more complex approach paths, closer spacing in the terminal area, and other initiatives. At least some of these maneuvers will require extreme precision in flight path control. It is likely that automation will be called on to perform them and possible that it will be required.

This will be a safe approach if, and only if, pilots are provided with the monitoring capability required to maintain full situation awareness throughout the performance of the maneuvers, *and with ways of escaping from the maneuvers safely and expeditiously in the event of a contingency either within the airplane or the system.* New monitoring automation and enhanced displays may well be necessary in the cockpit if pilots are to remain in command during such maneuvers, just as higher scan-rate radar and enhanced displays will be necessary for the controllers who will direct and monitor such operations.

COMMENT

These requirements have implications for controllers and pilots as well as for airspace planners and flight deck designers. The Wiener (1993) and Woods et al. (1994) discussions of error management should receive careful attention from ATC system designers as well as the operators of the system. The future AAS has been widely espoused as a system that will minimize human errors; if it is implemented, it is more likely that it will transform them as automation has done in aircraft, foreclosing some while enabling others. What is critical is that the future system also be effective at detecting and mitigating the effects of those errors that will still occur.

Workload removed from one element of the system is often reflected in additional workload elsewhere. This has been the case in the past, and it may occur yet again if pilots are given more responsibility for traffic separation during the en route phase of flight. This concept has been seriously considered by the FAA in its airspace redesign efforts and more recently by airlines in their free flight proposals. During the past year, as an example, tests have been conducted to evaluate the use of TCAS as a separation aid for aircraft climbing through an altitude occupied by another airplane on oceanic routes not under radar surveillance.

A recurrent theme running through these suggested guidelines (and through this entire document) is that "simpler is often better." The overriding human factors problems in today's advanced aircraft are the complexity of the tools provided to help pilots do their job, and deficient understanding of how the tools work. More efforts devoted to simplifying the design and operation of these essential tools, both in aircraft and ATC, will decrease training required, improve the error resistance and error tolerance of the systems, ameliorate the increasing cognitive burdens placed on operators, and ultimately increase system safety.

It is worth pointing out again that en route and approach–departure controllers, unlike pilots, cannot "see out the window"—that their only contact with the real world is through the representations provided by their traffic and other displays. Woods (in press) discussed the heavy obligation this places on the designer, who must create virtual representations that provide all needed information under all circumstances. Human operators can visualize the processes they are controlling only through such representations—the "keyholes" provided by the computer.

I believe that to keep controllers actively involved in their tasks, it is necessary as well as desirable to provide them with a moderate degree of management flexibility, by permitting them to take a more, or less, direct role in controlling traffic. They should be able to be supervisory controllers when they wish, or to be more active in the process. This alone will maintain their skills if there are circumstances under which they may have to revert to a direct controlling role. Given the limited reliability of automation to date, I think it very unlikely that the future system will be infallible; no other advanced automated system has ever been, regardless of its specifications.

CHAPTER 14

REQUIREMENTS FOR CERTIFICATION OF AVIATION AUTOMATION

INTRODUCTION

This chapter contains extracts from a chapter in NASA Technical Memorandum 110,381 (Billings, 1996).[1] The chapter is devoted to a brief consideration of aircraft certification from the human factors viewpoint. I have included it here because people responsible for requirements development, proof testing, and acceptance of highly automated systems in any real-time domain are confronted with problems similar to those that arise in the certification of aircraft, although the latter process is perhaps more severely constrained by regulations than most others likely to be encountered. Although the guidance provided here is context specific, it applies to other domains as well, with minor modifications.

In the United States, the FAA is responsible for certifying new aircraft and avionics equipment. The certification role is a difficult one. FAR Part 25, Airworthiness Standards: Transport Category Airplanes, governs the certification process. Section 25.1, Applicability, says only the following:

(a) This part prescribes airworthiness standards for the issue of type certificates, and changes to those certificates, for transport category airplanes.

(b) Each person who applies under Part 21, for such a certificate or change, must show compliance with the applicable requirements in this part.

A manufacturer may choose to satisfy the requirements set forth in Part 25 in any of a number of ways. If compliance can be demonstrated, the airplane must be approved, even if the certifying authorities are less than comfortable with the

---

[1] I acknowledge with gratitude the help and support provided by the late Berk Greene, and the guidance so kindly made available by Donald Armstrong and Guy Thiel, all FAA certification pilots, in its preparation.

approach that has been taken. Although common sense usually prevails, certification staff cannot demand more than the regulation requires. The other regulation bearing directly on the type certification process is Part 21; Section 21.21 describes the conditions under which an applicant is entitled to a type certificate:

> (2) For an aircraft, that no feature or characteristic makes it unsafe for the category in which certification is requested.

This requirement is powerful but little used, because it shifts the burden of proof from the manufacturer to certifying authorities to show how a design feature or characteristic is both unsafe and not otherwise addressed in the basic regulation. More often, new technology is handled by the development of special conditions: rulemaking for particular novel or unusual design features that were not envisioned when the appropriate sections of Part 25 were adopted. This time-consuming process, unique to the United States, makes establishment of the complete body of airworthiness requirements for a new airplane occur much later in the design process than the manufacturer would prefer—akin to starting the ball game without knowing where the goal posts are.

Further, certification is usually the final step in a new airplane's development process. Given the schedule slips that invariably occur in the course of a complex airplane's years-long development and the financial burden on the manufacturer if initial deliveries of a new airplane are delayed, the FAA certification staff is routinely under enormous pressure throughout the latter phases of the certification process, especially if all does not go as planned or if some areas require further study or flight test. The certification process itself is extremely expensive because of the substantial amount of flying required, and this is another factor that places pressure on FAA personnel.

Finally, today's airplanes are software intensive; the Boeing 777 incorporates some 5 million lines of source code in its various computers. Software verification is extremely difficult, and "bugs" are bound to occur as the airplane goes through its flight testing, including certification. These can also complicate and delay the certification process.

THE OVERARCHING ISSUES

Certification personnel are not evaluating simply aircraft components. They are given *an airplane*—a product that must operate as an internally consistent entity. The machine is very complex, yet all of its functions must operate together harmoniously. As certification pilots examine all of these many functions, they must always consider how an average pilot operating under difficult circumstances might misunderstand, misread, or misinterpret what he or she sees; how such a pilot might be led to inappropriate decisions by the information provided by the machine; how he or she might make errors of omission or commission in executing those decisions; how line pilots might find it difficult to recover from failures or their own errors; and how tolerant the airplane will be of such mismanagement when it is in line service. They must do all of this in a comparatively short time, always under pressure, and they must then accept the responsibility of approving the airplane. This is not a job for the faint-hearted.

THE CERTIFICATION PROCESS

Faced with this mandate and these constraints, certification authorities attempt to evaluate flight-deck workload in comparative terms, measuring the difficulty of the flight crew's tasks in each new airplane against workload in earlier, "benchmark" airplanes certified for and successfully operated by a crew of two persons. This comparative evaluation is carried out by FAA certification pilots and other highly experienced air carrier inspectors. It necessarily yields subjective estimates of workload. Aircraft are evaluated in operational scenarios that simulate air carrier operations as much as possible. A variety of malfunctions is simulated in the course of the workload certification flights, including the incapacitation of one crewmember. Among the simulated malfunctions are failures and degraded operation of many elements of the automation.

Part 25 says little about either the range of conditions encountered in line flying, or about the capabilities of the range of air carrier pilots who will operate the new airplane. Although an attempt is made in certification to examine the widest range of environmental conditions and malfunctions possible, only a very limited subset of these conditions can be evaluated. Likewise, only a very limited number of FAA pilots, all highly experienced, can take part in the certification process, which means that until the first airplane is delivered, it will have been flown extensively only by company and FAA pilots of above-average knowledge, experience, and ability.

Transport aircraft are among our nation's most important exports. The United States has led the world in the design and production of aircraft throughout most of the history of aviation. Those involved in the certification process must continually be aware that they are certifying machines that will be operated throughout the world. The issue of cultural differences, and their impact on flight crew operations, is another aspect of the problems faced by certification personnel, who know that the aircraft will be used in different ways by operators worldwide.

I pointed out previously that operations well within the envelope may not show evidence of brittle automation (see chapter 9), nor for that matter of organizational latent factors, which may come to light only when a line crew is fatigued or distracted by other operational anomalies. All of these factors are considered by certification pilots, themselves operating under a different sort of pressure, but it is not surprising that potential problems with new aircraft sometimes are not recognized until they are operating in line service.

REQUIREMENTS FOR HUMAN FACTORS CERTIFICATION OF AIRCRAFT

Recognizing that FAA (or any other organization) can impose only a minimum standard or requirements as codified in regulations and standards that do not always incorporate the lessons learned during a period of rapid technological advances, can any generic guidance be offered that can help certification authorities, or experts in other domains who must evaluate and approve new products? They, after all, *are* the experts, and it is presumptuous to assume that this very difficult and exacting job can be thoroughly understood by anyone who has not "been there." Nonetheless, a careful examination of the process and requirements suggests certain areas in which guidance can be offered, if only to provoke argument.

I incorporate here the thoughts set forth in chapter 3, A Concept of Human-Centered Automation, as the overarching philosophy that should be applied. I believe this is justified by a careful reading of the regulations, which state repeatedly that flight must be possible under a great variety of conditions without exceptional skill or strength, and that it must be possible for the pilot to remain in command of the airplane under all but extremely improbable failures. The pilot must be warned of potentially unsafe conditions; the airplane's design must minimize crew errors. Crew members must be able to perform their duties without unreasonable concentration or fatigue.

The regulations discuss workload and factors that may increase it, but they do not consider the possibility of workload being too low, a factor not thought about very much prior to the introduction of advanced automation. It has become clear in the last decade or so that either overload or underload can pose hazards; both are considered here.

Principles for Certification of Human-Centered Automation

With regard to the "first principles" of human-centered automation (chapter 3), the following general requirements are suggested.

Automation Should Not Be Able to Remove the Pilots From Effective Command of Their Aircraft

It has been indicated elsewhere that sophisticated automation can decrease pilot authority in ways that may not be immediately evident (e.g., Mårtensson, 1995). I believe that pilots must be made aware of any modes or features that may act in this way, and that provisions should be incorporated to permit them to "quickly and positively" override these functions when an emergency requires it (e.g., NTSB, 1994). The central question of operator authority arises in many other domains as well, and must be carefully considered in each.

Automation Should Not Remove the Pilots From Direct Involvement in the Operation

In 1965, when the present Part 25 was implemented, the term "automation complacency" had not yet been invented, and such automation as was available accomplished only tactical functions under direct instruction from pilots. The regulation did take note of the ease with which pilots could mistake and misuse navigation modes and required that these modes be annunciated (and errors in selection and recognition of these modes continue to occur today). New automation should keep pilots meaningfully involved in the operation, by whatever means. I also believe ways should be found to increase that involvement to minimize the likelihood that they will accept and tolerate inappropriate mode selections, especially in long-range aircraft in which workload is likely to be very low.

Automation Must Keep the Pilots Informed of Its Actions

Recognizing that automation now performs a great number of discrete functions as well as the continuous task of flight path control, it is increasingly necessary that the automation inform the human operators of what it is doing. This guideline is intended to suggest that pilots must remain involved with, in part by being informed of, the actions of the automation that conducts the flights that they manage, as well as of failures in that automation. In the Nagoya accident (1994), autotrim was being applied, but there was no audible or other signal to indicate this activity (NTSB, 1994). When the pilot regained control, the pitch trim was in an extreme nose-up condition.

Automation Failures or Malfunctions Must Be Clearly Enunciated to the Pilots

Because automation performs discrete as well as continuous functions, its failure to continue performing these functions may not be obvious. This requirement is intended to suggest that such failures must be positively announced to the crew.

Automation Must Behave Predictably Under All Circumstances

Much of this book deals with the importance of predictability in the behavior of automated systems, so that operators can form a "correct" mental model of their functions. Very complex or infrequently used functions are especially likely to be misunderstood by pilots. It is important that they be able to follow, whether by dedicated displays or by the behavior of the airplane and its systems, the behavior of the automation. Means must be made available by which they can accomplish this critical monitoring task without undue attention. Equally important, the automation must behave predictably, in a manner that facilitates monitoring by its human operators.

Automation Should Inform Pilots of its Intentions and Should Request Consent for Actions that May Critically Affect the Conduct of the Flight

An essential ingredient of what Endsley (1994) called "deep" situation awareness is an understanding of the near-term future situation. Just as humans must be given the means by which to indicate their intent to the automation assisting them, I believe it is essential that automation indicate its near-term intent to the humans on the team, especially when mode changes, major changes in state or changes that could compromise the ability of the airplane to complete its mission are contemplated. This occurs today under some circumstances (e.g., the "flare" annunciation during coupled approaches, the green arc on the navigation display), but it should be applied as a general rule in complex systems.

Requirements for the Certification of Control Automation

Automation Involving Modes of Control Known to be Potentially Hazardous Should Contain Safeguards to Guard Against Its Use Under Inappropriate Conditions

Experience has shown that under certain circumstances, the "open descent" mode of operation can present predictable hazards. If such a mode is provided, should it be allowed to operate without restrictions to prevent it being used below a safe altitude? Similarly, it is known that the "vertical speed" mode can result in a critical decrease in airspeed during climbs at higher altitudes. The possibility of misuse of this mode should be guarded against when the mode is implemented.

Other automated modes may have to be used under circumstances that could present potential hazards. They should be identified and appropriate cautionary information should be provided to pilots under those circumstances. Design features that experience has shown to be potentially hazardous should not be utilized without such safeguards against misuse or inadvertent use in the stress of line operations.

Automation Design Must Permit Its Use at Some Lower Level of Authority if Stability Augmentation Systems Have Failed

Fly-by-wire technology has made direct manual control of aircraft impossible under most circumstances. Even "manual" flying is accomplished through computer assistance in these aircraft, and the normal mode of operation is the fully assisted mode. In the A320/330/340 series, a "direct" mode of control is available if the normal mode fails. In this mode, many of the protections built into the flight control system are bypassed, and the airplane flies as though directly controlled by the pilot through the proportional sidestick controller. The automation has less authority, but the pilot's authority is unchanged.

An analogous reversion mode should be available in any aircraft having highly augmented controls, but *it should be capable of being used under normal conditions as well,* so that pilots may remain proficient in its use through regular practice. Ideally, it should act and "feel" as much like the normal control modes as possible, so that the pilot is able to accomplish the control task without significant diversion of attention from other tasks.

Some newer aircraft have deliberately been designed to be neutrally stable or marginally unstable in the pitch axis. Any tendency to instability is compensated for by automated stability augmentation systems. Such aircraft can be difficult to fly if the augmentation fails. (The Space Shuttle is an extreme example of this problem.)

This may also become a problem in aircraft such as a future high-speed civil transport (HSCT) during flight on the "back side of the power curve,"[2] which will be required during low-speed flying during the approach to landing. To quote from Part 25, "The trim, stability, and stall characteristics are not impaired below a level needed to permit continued safe flight and landing;" this should be true in any mode or phase of operation.

---

[2]The wing configuration of a supersonic transport may make it necessary to fly approaches to landing at an angle of attack higher than that at which the wing generates optimum lift. This is called "the back side of the power curve." The airplane remains under control, but the stall margin is much decreased under these circumstances and extremely precise control of pitch and thrust is required.

Requirements for the Certification of Information Automation

Primary Flight Displays Have Become Very Complex. Certification Authorities Must Decide How Much Information Is Too Much

Although navigation displays in newer aircraft have been integrated and, in the process, have become easier to read, primary flight displays have become more cluttered through the addition of a considerable amount of additional information. Mandatory TCAS and wind shear advisory systems have added still more information to this display. Certification personnel should give consideration to whether this much information on the primary flight display is likely to distract, rather than inform, pilots when they are heavily task-loaded.

The PFD is critical for situation awareness, but the amount of information presented on this screen may be reaching limits using the conventional format. An increasing amount of data on this display is presented as alphanumerics, which must be read serially to be comprehended. This is also true of the mode annunciation panel that appears on this screen. Certification authorities should consider whether pilot duties can be performed "without unreasonable concentration or fatigue" using these displays.

Woods (in press) has pointed out that how information is represented is absolutely critical. The combination of discrete and continuous data into a more integrated display (such as the symbolic combination of velocity and acceleration data on some tape airspeed displays) can enable more precise speed control, particularly during turbulence. The use of actual and potential flight path vector symbols is another example long advocated by Bray (see Lauber et al., 1982) and others as an information management and integration tool.

Do the Most Important Information Elements Stand Out in Complex Displays?

Part 25 emphasizes "the ... conspicuity of ... failure warning devices" as a workload factor. Although the number of discrete warning and alerting devices has been decreased in glass cockpit aircraft, the number of discrete messages that can occur is still large and these alerts are almost invariably alphanumeric: They must be read to be comprehended. All are usually in the same size print; more important items may be boxed and shown in a different color. On busy displays, however, this may not be sufficient to draw attention to the most important items in what can sometimes be a lengthy list (see incident report that begins chapter 12). Is the most critical information always obvious? Is the busy pilot's attention drawn to the items requiring action?

The Status of Flight-Critical Automation Should Be Obvious at All Times, Not Only When Some Element Has Failed

This is a corollary of the first principles. It is an appropriate guideline in view of Part 25, which states that "the degree and duration of concentrated mental ... effort involved in normal operation" is to be considered as a workload factor. Affirmative information concerning automation activity, modes, and especially mode changes is much more easily monitored than the lack of such information. It is possible that an automation synoptic could be of help in view of the complexity and depth of the automated systems in advanced airplanes.

Requirements for Certification of Management Automation

Management automation was in its infancy when Part 25 was rewritten in 1965. It is not surprising that it was not considered in the regulation. Nonetheless, the regulation cited "the degree of automation provided in the aircraft systems" as a workload factor, and coupled navigation systems were discussed.

Bearing in mind that these management aids do relatively little that pilots cannot do without them aside from overwater navigation (although at the cost of much higher workload), what requirements are appropriate in this area? Those that I offer here have more to do with present systems than with those that may be implemented in the future, for the reasons stated elsewhere in the document about the direct and indirect costs of moving to a radically redesigned flight management system in a future aircraft.

Flight Management Systems and Their Associated Control-Display Units Should Assist Pilots in Programming, Particularly for Seldom-Performed Functions

As pilots gain experience with the FMS, they become facile in performing those tasks that they are required to perform frequently. Errors in programming these functions are usually slips or lapses rather than mistakes. Where possible, the CDU should indicate the error when it rejects an entry.

Rarely performed programming tasks are less likely to be recalled when needed. Prompting should be available in the software to assist pilots to perform such tasks rapidly and correctly. Contemporary CDUs rarely provide such assistance, which means that pilots must sometimes spend much longer than should be necessary in performing even relatively simple, but unfamiliar, tasks using the CDU. This can be an important workload factor.

Reprogramming Tasks That Must Be Performed at Busy
Times in Flight Should be Simplified Wherever Possible
to Minimize the Amount of "Head-Down" Time During
Flight at Low Altitudes

In the newest flight management systems, avionics manufacturers have gone to considerable effort to simplify reprogramming in and approaching terminal areas. Certification staff should be on the alert for functions that are still cumbersome if they must be performed at the expense of other monitoring functions important to safe flight at low altitudes. Pilots still "turn it off" rather than permit themselves to be distracted during the busiest periods of a flight, and thereby deprive themselves of the protective features in the systems. This should not be necessary.

Flight Management Systems Should Incorporate the
Maximum Practicable Amount of Internal Error-Checking
to Improve the Error Resistance of the Entire System

As has been noted elsewhere, computers are tireless and patient monitors. More use can and should be made of automation to monitor human performance. The flight management system "knows" a great deal about the aircraft and about navigation. The "knowledge" should be used to the maximum extent possible to increase error resistance and error tolerance. I have mentioned that "reasonableness checks" could catch some programming errors; nonsequential altitudes programmed into climb and descent profiles could also be questioned. A study of machine monitoring of human entry procedures and common errors would be useful as a basis for incorporating more systematic error trapping within the FMS.

Management Automation Should Be Standardized
Across Fleets to the Extent Possible, to Minimize
the Likelihood of Errors by Pilots Transitioning from
Other Aircraft

Certification staff, who work with many aircraft, are well positioned to advocate increased standardization of automation, as they have done with respect to primary flight and other displays in the past, although it is air carriers that will be most effective in enforcing standardization across their fleets, by requiring it when they purchase new aircraft.

QUESTIONS FOR CERTIFICATION AUTHORITIES

To summarize these requirements briefly, I have restated them as questions to be considered during the certification process. I recognize the lack of specificity and the trade-offs that are always necessary during design, but it seems to me that these questions need to be near the forefront of the certification pilot's mind as he or she examines a particular automation suite in a new airplane.

- Is the pilot truly in command under all circumstances?
- Is the pilot actively involved at all times?
- Does the automation always keep pilots informed of its actions?
- Are failures or malfunctions clearly announced?
- Is the automation always understandable and predictable?
- Does automation search for pilot errors and warn pilots about them?
- Does automation inform pilots of its intentions? Is it easy for pilots to inform automation about their intentions?
- Are there potentially hazardous modes? If so, are there safeguards against inappropriate use of such modes?
- Are all backup control modes usable without undue effort? Do they provide adequate assistance to pilots under all conditions?
- Are flight and systems displays easy to understand, or cluttered?
- Is the most important information always obvious?
- Is the status of control automation, and its mode changes, always obvious? Is it obvious enough?
- Do the FMS and its CDU assist pilots in programming?
- Are tasks that must be performed at busy times simple to execute?
- Does the FMS incorporate checks to guard against input errors?
- Is this FMS unique? Will it require extensive training to be used effectively?

It is probably worth pointing out that these questions are as relevant for designers as for certification personnel, and that with little modification they apply to systems in other domains as well.

COMMENT

It is worth stating again that safety is relative rather than absolute. Accidents are usually a conjunction of many factors operating together. Most of the organizational and other latent factors (Reason's "resident pathogens") are beyond the control of the manufacturers and those who certify their airplanes. All that builders and certification authorities can do is to produce (and authorize the use of) an airplane that is as resistant to, and tolerant of, both human errors and machine failures as is feasible given the state of the art, and insure that all systems are adequately documented and explained. Nonetheless, attention to first principles can be of help in these processes.

I believe the central problems for the human operators who work with today's aircraft automation are the complexity and opacity of these tools. Put in terms of the "first principles," the human operator must be able to understand the automation, and must be informed about its activities. To simplify the tools will require time and a better understanding of cognitive engineering. In the meantime, more and better training in how and why they operate as they do offers the best likelihood of ameliorating many of the problems outlined in this book. The opacity issue is also difficult and will ultimately require definitive solutions, but this is an area in which certification experts can be of real help by demanding that new automation keep the pilot informed of its activities and intentions.

Sarter and Woods had little difficulty in demonstrating deficient mode awareness and understanding in pilots in simulation studies in the Boeing 737-300 and Airbus A320, simply by using probes that require more than superficial understanding of how the flight management system and mode control panel actually function (Sarter, 1994; Sarter & Woods, 1992a). Even pilots experienced in these aircraft can get into trouble during nonroutine operations because of shallow knowledge of these systems. Certification pilots can do much to improve these automation deficiencies by utilizing such probes in their certification scenarios, and almost as much by simply being aware of the sorts of problems that line pilots are likely to have in handling their automation during routine and off-nominal operations. Information resident in safety databases should be routinely consulted as an aid in detecting such problems.

Chapter 15

Automated Systems
in Other Domains

INTRODUCTION

Most of this book focuses on the effects of automation in a single domain, civil aviation. This domain is by no means the only one in which humans and machines must engage in cooperative work involving high risk in a dynamic, time-paced environment, and it is certainly not the only one in which automation has been invoked, either to assist or to supplant human labor. Other mass transportation modes, nuclear power plant operation, and many other industries involving dynamic process control are among the settings in which automation technology is in use. What has been the experience in such domains? Has it been similar to our experience in aeronautics? Have lessons been learned from which we can profit, and have we in aviation perhaps learned things that can be of use elsewhere?

Although detailed answers to these questions would require another book, we will see, on pages 289–291 that there are many similarities, as well as some differences, between the aviation domain and others. I believe that the lessons to be learned about the effects of automation on human operators are common across at least several domains, and for that reason, that some of what we have learned can profitably be applied elsewhere. As I did in parts II and III of the book, I have selected examples of automation in a variety of domains, followed by a discussion of some of the more general, domain-independent issues that are raised by them.

In chapter 9, I identified automation attributes that seem to cause problems for at least some human operators. The most important are complexity, autonomy, and opacity, although there are others that also make it harder for even expert operators to understand what automation is doing. These problems are not new. What has changed over time is the degree to which automation has become a part of our lives. Most of us accept it, even when we do not understand it. All of us have found ourselves adapting to it, tailoring our activities around its idiosyncrasies and sometimes even mastering it. Much of today's automation is more capable; some of it is more user-friendly. All of it has opened new opportunities for us—and imposed new demands on us.

CASE 1: AN EXAMPLE FROM MANUFACTURING

In 1975–1976, a team of investigators from the British Steel Corporation, Hoogovens Steel, and the Technical University of Delft evaluated the effects of automation on human operators at a highly automated hot steel strip mill in the Netherlands, with the intent of making recommendations concerning the application of automation to other, similar plants elsewhere. Their report (ECSC, 1976) described the aim of the study in this way:

Aim of the Study

In recent years there has been considerable discussion about the positive and negative effects of automation on the operator. Particular issues which have been mentioned are:

- problems of maintaining vigilance over a process when very few incidents requiring manual intervention arise;
- the need for highly skilled and educated operators who, because of their prolonged inactivity, suffer from boredom and lack of motivation leading to high absenteeism and sickness;
- lack of social contact due to the reduction in manning;
- and the problem of maintaining highly developed manual control skills which need only be used infrequently in critical and stressful emergency conditions. (p. 4)

Methods of Work

Among the issues studied by the team were:

- "allocation of function between man and machine;"
- "operator influence over the control of the process;"
- "communication between operators and computers and between operators themselves;"
- "operators' understanding of the systems they are involved with;"
- "skill requirements of automated control jobs" (p. 6)

The results of this careful study were discussed at a meeting in October 1976, whose proceedings are the basis for these comments.

Summary of Comments and Conclusions

The investigating team reached several conclusions in the course of its study. Some are quoted here, with comments from the Proceedings concerning them.

- "The established principle of designing automation as a tool of the operator is reinforced by the results of this study, and the recommendations ... are intended to extend the operator's discretion over these tools." (p. 14)
- "The findings reinforce the well-known principle that the operational requirements of the work, task and job design, the social and organizational context of the work, and the selection and training of the operator should all be considered as an integrated whole in the design phase."
- "The automatic systems should be designed so that it is possible for the operator to anticipate potential problems and take preventive action, rather than react to correct problems only after they have already arisen." (p. 15)

[Comments] The operators were held responsible by management for the output of the mill and were required to take action if the quantity or quality of the product was threatened .... The mill was usually operated under computer control (over 90% of the time). This resulted in a job in which the operator may be idle for considerable periods of time; yet he is required to remain vigilant ... firstly, to take immediate action if a ... breakdown is thought to be imminent, secondly to take over certain controls if the computer is operating unsatisfactorily.

However, intervention in the automation was rare, except when circumstances demanded the operator to act. Detailed questioning and observation revealed that there were difficulties in intervention, as follows. (p.112)

Firstly, the complexity of the whole installation made it hard for the operator to identify the precise cause of malfunctions. For example a cobble [fault] might have been caused by a poor set up of the mill, bad speed control or irregularities in the slab itself. This confusion meant that the rolling continued with the operator in a state of uncertainty not knowing whether the automation, the mill or the slab was to blame.

A second difficulty lay with the adaptive system[.] When there was a problem[,] for example poor temperature control, the operator waited to see if the adaptive mechanisms within the system were adjusting and improving before taking over control himself ... Furthermore, when manual adjustments were made, the computer system no longer received feedback information so that it was not updated, making the return to computer control more difficult.

"Finally, the rollermen were reluctant even to use the manual controls because they did not know the effects their adjustments might have on the delicate adaptive systems. (p. 113)

- Information display(s) should be designed to help the operator predict future performance and also to help him understand the decisions being taken by the automation. This is in contrast to the present use of displays to advise operators of process states." (p. 15)

[Comments] In looking for reasons for these difficulties, it was apparent that there was a lack of suitable information displayed to the operator. Three main types of information were lacking (p.113)

Firstly, *there was little information on the functioning of the automation to tell the operator what the automated systems were doing and why* [italics added]

The second category refers to the poor feedback of performance. The direct view of the process was invaluable .... However for other parameters the only performance feedback information ... was in the shape of analogue displays ... and on chart recorders.

A limitation of these facilities was that there was no comparison of these results with quality control requirements ...,.

Thirdly, it was not easy for the operator to assess trends in performance. This hindered the operator in detecting a degradation in the performance of the computer system due for example to poor adaptation. (pp. 114–115)

- Performance feedback to the operator should be given attention in the design stage particularly in terms of frequency, accuracy and timeliness. Care should be taken that the physical layout of the plant does not constrain the provision of good feedback."
- "The use of meters to indicate deviations rather than absolute values should be evaluated, with the objective of making information on operator displays more easily assimilated." (p. 15)
- "The visual and auditory information which the operators use in the present mill is of paramount importance to their task, and special attention should be given in design to providing the best possible view of this process."
- "Manual set up controls ... should be abandoned ... in a future mill. Automation is so reliable ... and the problems of controlling an accelerating mill so great, that the operators' skill in using these controls has been almost lost." (p. 16)

[Comment] "Operators rarely have to use the manual back-up facilities available, and have gradually lost much of their previous skill at manual control." (p. 116)

- "Some form of back up is required should the process computer be out of action or the set ups inaccurate. This would probably be a form of 'preset' control. It is envisioned that the back up control system would be used very infrequently due to the reliability of future process computers." (p. 16)
- "The operator should have a number of facilities for interacting with the computer system which are not provided at present. These would enable him to improve the performance of the automation, and extend the situations under which automatic control could be used by amending incorrect or suspect data used by the computer in order to produce a better set-up; by 'game playing' with the computer to examine the desirability of alternative courses of action; and by using, where necessary, the operators' corrections to the automatic control as well as direct feedback loops, to update the computer set ups." (p. 17)

[Comments] "Looking again at the reasons for poor interaction between the operator and the automated systems, the lack of appropriate controls became apparent. There were very few opportunities for the operators to interact with the automatic system in order to improve its performance .... There were no facilities to modify the computer set-up, other than manual operation." (p. 17)

- "Operators of future mills should be selected on the basis of mental skills and in particular the ability to think in terms of systems, and technical education will be a requirement." (p. 18)
- "Emphasis should be placed in training on the function and operation of the automation" (p. 19)

[Comments] "Finally in categorizing the reasons for the difficulties in intervention, we noted that there were considerable differences between the operators in their description of the functioning of certain parts of the automation. This revealed a lack of understanding of the operation of the mill and poor training must be partly responsible. However *certain aspects of the automation were so complicated that nobody, rollermen, management or technical staff, fully understood them*. (italics added; pp. 115–116)

[Bainbridge (1987, quoted by Reason, 1990, p. 182) said, "Perhaps the final irony is that it is the most successful automated systems, with rare need for manual intervention, which may need the greatest investment in operator training."]

To sum up .... The difficulties in operating with the computer automation systems on this mill are threefold. They are namely, poor performance, lack of controls and inadequate training. *In fact the automation has been designed to function largely independent of the operator; instead of being in control of the automation, the operator was in many cases compelled to adopt a rather passive role, until a problem demanded his immediate action.* (pp. 116–117)

We have tried to suggest an approach to systems design which will give improved technical performance ... *by using the skill and judgement of the operator, with automation as an aid, to improving the control of the process* [italics added]. (p. 128)

CASE 2: AN EXAMPLE FROM MEDICINE

The introduction of a computerized radiation therapy machine, the Therac-25, was followed in 1985–1987 by at least six accidents involving massive overdoses of radiation. Using publicly available documents, Leveson and Turner conducted an investigation of these accidents. Their report, cited by Leveson (1995, pp. 515–553), provided instructive insights into problems of safety-critical systems based on software. With respect to our immediate concerns about the effects of automation on human operators, a number of lessons also emerge from this story.

Background

The Therac-25 is a linear accelerator that generates high-energy (up to 25 MeV) electrons. The electron beam, properly modulated, is used to treat superficial tumors. Alternatively, the electron beam is directed at a metallic target in which high-energy x-ray photons are generated. The x-ray beam is collimated and can be used to treat tumors in deeper tissues.

A turntable contains devices that modulate the electron beam or convert it to x-rays. The machine also contains ion chambers that measure the amount of radiation delivered, although these sensors can become saturated (and thus useless) if too much radiation passes through them. In the Therac-25, the electron accelerator, turntable and other machine components were controlled by software resident in a PDP-11 computer. Quoting from the summary report:

> Several features of the Therac-25 are important in understanding the accidents. First, as with like the Therac-6 and the Therac-20 (predecessor machines), the Therac-25 is controlled by a PDP-11 computer. However, AECL (Atomic Energy of Canada, Limited) designed the Therac-25 to take advantage of computer control from the outset; they did not build on a stand-alone machine. The Therac-6 and Therac-20 had been designed around machines that already had histories of clinical use without computer control.

In addition, the Therac-25 software has more responsibility for maintaining safety than the software in the previous machines .... AECL took advantage of the computer's abilities to control and monitor the hardware and decided not to duplicate all the existing hardware safety mechanisms and interlocks. (p. 5)

Extracts From Reports of the Accidents

Accident 1 (pp. 516–517).

The operator activated the machine, but the Therac shut down after five seconds with an HTILT error message. The Therac-25's console display read NO DOSE and indicated a TREATMENT PAUSE.

Since the machine did not suspend and the control display indicated no dose was delivered to the patient, the operator went ahead with a second attempt at treatment by pressing the "P" key (the Proceed command), expecting the machine to deliver the proper dose this time. This was standard operating procedure, and ... Therac-25 operators had become accustomed to frequent malfunctions that had no untoward consequences for the patient. Again, the machine shut down in the same manner. The operator repeated this process four times after the original attempt—the display showing NO DOSE delivered to the patient each time. After the fifth pause, the machine went into treatment suspend; a hospital technician found nothing wrong with the machine. This was not unusual.

After the treatment, the patient complained of a burning sensation ... An AECL technician later estimated the patient had received between 13,000 and 17,000 rads (radiation absorbed dose) [60–80 times the intended dose].

The patient died 3 months later of her cancer, but had she lived, a total hip replacement would have been required as a result of the radiation overdose.

Accident 2 (pp. 528–529).

The operator had held this job for some time, and her typing efficiency had increased with experience.... She entered the patient's prescription data quickly, then noticed that she had typed "x" [for x-ray] when she had intended "e" [for electron mode] .... The mistake was easy to fix; she merely used the [cursor up] key to edit the mode entry.

Because the other parameters she had entered were correct, she hit the return key several times and left their values unchanged. She reached the bottom of the screen where it was indicated that the parameters had been verified and the terminal displayed beam ready, as expected. She hit the one command key, "B" for "beam on," to begin the treatment. After a moment, the machine shut down and the console displayed the message MALFUNCTION 54. The machine also displayed a treatment pause … this malfunction … was a "dose input 2" error. The Medical Center did not have any other information available … to explain the meaning of Malfunction 54. An AECL technician later testified that 'dose input 2' meant that a dose had been delivered that was "either too high or too low."

The machine showed a substantial underdose on its dose monitor display…. The operator was accustomed to the quirks of the machine, which would frequently stop or delay treatment…. She immediately took the normal action when the machine merely paused which was to hit the "P" key to proceed with the treatment. The machine promptly shut down with the same Malfunction 54 error and the same underdose shown by the dosimetry.

The only way that the operator could be alerted to patient difficulty was through audio and video monitors. On this day, the video display was unplugged and the audio monitor was broken.

After the first attempt to treat him, the patient said that he felt like he had received an electric shock or that someone had poured hot coffee on his back. He went to the treatment room door and pounded on it. He appeared visibly shaken and upset. The patient may have received 16,000 to 25,000 rad in less than one second over an area of about one square centimeter. He was eventually hospitalized for radiation-induced myelitis of the cervical spinal cord and died from complications of the overdose 5 months after the accident.

Accident 3 (p. 530).

[same technician as in the preceding case] As with her former patient, she entered the prescription data and then noticed an error in the mode … After she finished editing … she saw the BEAM READY message displayed and turned the beam on.

Within a few seconds the machine shut down, making a loud noise audible via the [now working] intercom. The display showed MALFUNCTION 54 again. The operator rushed into the treatment room, hearing her patient moaning for help…. She asked him what he felt, and he replied, "fire" on the side of his face…. He was very agitated …

The patient died from the overdose … three weeks after the accident. An autopsy showed an acute high-dose radiation injury to the right temporal lobe of the brain and the brainstem.

Accident 4 (pp. 541–542).

The console displayed BEAM READY, and the operator hit the "B" key to turn the beam on.

The beam came on but the console displayed NO DOSE or dose rate. After 5 or 6 seconds, the unit shut down with a pause and displayed a message … since the machine merely paused, he was able to push the "P" key to proceed with treatment.

The machine paused again, this time displaying FLATNESS on the reason line. The operator went into the room to speak with the patient who reported "feeling a burning sensation" in the chest. The console displayed only the total dose of the two [previous x-ray] film exposures and nothing more. In two treatment attempts, the patient would have received 8,000-10,000 rad instead of the 86 rad prescribed. The patient died 3 months later of complications related to the overdose.

Comments Concerning Causal Factors From the Leveson Report

Patient reactions were the only real indications of the seriousness of the problems with the Therac-25. There were no independent checks that the software was operating correctly…. Such verification cannot be assigned to operators without providing them with some means of detecting errors: the Therac-25 software "lied" to the operators, and the machine itself was not capable of detecting that a massive overdose had occurred. (p. 550)

One of the lessons to be learned from the Therac-25 experience is that focusing on particular software design errors is not the way to make a system safe. Virtually all complex software can be made to behave in an unexpected fashion under some conditions. There will always be another software bug. (p. 550)

Safe Versus Friendly User Interfaces

Making the machine as easy as possible to use may conflict with safety goals. Certainly, the user interface design left much to be desired, but eliminating multiple data entry and asuming that operators would check the values more carefully before presssing the return key was unrealistic. (p. 553)

CASE 3: AN EXAMPLE FROM MARITIME OPERATIONS

On May 2, 1992, the German container ship *Berlin Express*, proceeding through a dredged cut in the South Channel of Port Phillips Bay en route to Melbourne, Australia, made a rapid turn to starboard and ran aground. The ship, assisted by two tugs, was successfully refloated on the next high tide 10 hr later. There was no structural damage and no pollution resulted from the grounding. Information concerning this incident is taken from a government report (Inspector of Marine Accidents, 1993).

The ship is equipped with an automatic steering system. A required course is entered into the steering computer by depressing and rotating a control knob; one degree of course adjustment requires about 15° of knob rotation. The actual and commanded courses (or tracks) are shown on digital displays; rudder position is shown on an analog display. Computer steering can be overridden by operating an autotiller on the same console. The ship is highly automated and was designed to be operated by a total crew of just 14 people. The bridge is designed for continuous one-person operation, including port entry and berthing. Although the ship is also fitted with a manual steering position or helm, at least one of the ship's Masters has used the automatic steering from berth to berth, including transits of the Suez Canal, right from the maiden voyage (except during maneuvering alongside).

History of the Incident

During the transit of the South Channel, an Apprentice, monitored by the Watch Officer, was operating the autopilot in response to orders from a Maritime Pilot from Port Phillips. The Pilot ordered a 2° turn from 105° to 107° to maintain the ship's position in the channel; this course adjustment was read back and entered into the autopilot by the Apprentice.

He then watched the ship's bow swinging across the shore-line ahead.

Everyone on the bridge realised simultaneously that the ship was swinging too rapidly to starboard. The Apprentice put the auto-tiller to port five and then port 10, cancelling the over-ride alarm as he did so. Both the Master and the Pilot ordered "hard to port," the Master leaning across the Apprentice and performing the function himself, noting that the rudder angle indicator showed starboard 15.... The Master then ordered the Mechanic to the manual steering position, ordered him to put the wheel hard to port and switched over from autopilot to manual.

The swing to starboard slowed then stopped .... Fearing that the stern would swing into the bank ... both the Master and the Pilot ordered stop engines and midships the wheel, and then "full astern" .... When the ship had swung back to a heading of about 111°, it grounded on the starboard bank of the cut, heeling over to port as it did so ...

Thorough testing of the steering gear system and the autopilot system by manufacturers' servicing agents while the ship was in Swanson Dock indicated that both systems were functioning normally. (p. 11)

Comments by the Investigator

It is considered that in acting as they did, those on the bridge acted quickly and correctly, preventing damage to the rudder and propeller, but were unable to prevent the grounding ...

Examination of the course and rudder angle recorder shows that the rudder moved to 20 degrees starboard at 1100, the time of the rapid swing to starboard .... There are a number of possible causes. [causes follow] (p. 13)

Incorrect Manual Input.

The Apprentice was manning the command station .... He had performed this duty at the three previous ports and the Master considered him competent in using the system. Had the Apprentice applied an input error ... this would have required a rotation of the control knob of about 210 degrees instead of about 30 degrees. Both the Apprentice and the Watch Officer are adamant that this did not occur.

While human input error cannot be totally discounted, the statements of the ship's staff directly involved do not support this as being the cause. (pp. 13–14)

Momentary System Failure.

The Autopilot and steering control were checked and tested .... No faults were found in the system.

It is considered unlikely that a momentary systems failure, either of the autopilot or the steering gear, occurred.

Outside Electrical Interference.

The electrical control systems installed aboard Berlin Express are more sophisti-cated than those normally found on merchant ships, being more similar to those in aircraft. Aircraft control systems are liable to interference from electrical equipment ... that may be on board the aircraft .... No such equipment was known to be in use adjacent to or on the bridge of Berlin Express, and ... the only other "near source" of radiation was not transmitting by radio at the relevant time.

It is, therefore, considered that electrical radiation interference of the electronic control systems from an outside source is unlikely to have been a causal factor. (pp. 14–15)

Analysis of Course and Rudder Angle Recorder Chart.

If the Apprentice made the course adjustment to 107 degrees as the ship's head was swinging to port through a heading of 105 degrees, it is probable that the autopilot computer, analysing this movement away from the new command course, would direct more starboard rudder.

Whether this would have been as great as 20 degrees is difficult to determine...it is not possible to determine whether this in fact had started to happen when the Apprentice activated the auto-tiller.

Based on the above hypothesis, it is conceivable that no malfunction occurred ... that it was the response of the autopilot, under the programmed settings, to that particular set of circumstances. (p. 19)

Manual/Autopilot Steering.

Although the Master on board at the time was in the practice of using the autopilot during port operations, this practice was not followed by all the masters who had command of the ship.

Had the ship been steered in manual...it is unlikely the Mechanic would have applied more helm than starboard 10, certainly not more than starboard 15.

If, as surmised, the starboard 20 degrees rudder was a natural function of the autopilot under those particular circumstances, or even if it was the result of an input error, the incident would most probably not have occurred had the steering been conducted manually. (p. 21)

Advice from Manufacturer's Agent.

Advice received...is that operationally 20 degrees of starboard rudder could not have been applied by the autopilot. (p. 21)

CASE 4: AN EXAMPLE OF DIAGNOSTIC AUTOMATION

This example differs from the preceding ones in that the automated system was designed to help humans troubleshoot a device off-line, rather than during an ongoing process. It also differs in that the diagnostic system made use of artificial intelligence—a rule-based expert system—to diagnose malfunctions in a new (digital) generation of a complex electromechanical device. The machine expert was designed for use by technicians familiar with the previous (analog) generation of the device, but unfamiliar with the new-generation device and with digital technology in general. The human operator acted as data gatherer for the expert system, prompted by system queries and directions; the computer analyzed data, then offered solutions to the symptoms and problems presented to it.

Background

This machine system was in the late stages of refinement and qualification testing when it was studied by Roth, Bennett, and Woods. (The information presented here is paraphrased from their 1987 report.) The expert system's designers, and management, had designated the system as ready to be deployed in the field on a regular basis.

The investigators collected data, analyzed, and reported on the interactions of four technicians having varying amounts of experience with the device as they attempted to use the expert system to help them troubleshoot and fix actual device problems. Eighteen instances of joint problem solving were studied by means of direct and video observation and protocol analysis.

Summary of Findings

Two of the four practitioners took an active role in initiating observations and measurements and in forming judgments beyond those requested by the machine expert (they tended to treat the system as an instrument which they could manipulate to accomplish goals). The other two technicians were more passive; they followed expert system directions as literally as possible and tended to act only as data gatherers for the system.

> We were able to observe human–machine interaction when novel situations arose (problems outside the machine's competence), when adaptation to special conditions was required, in the face of underspecified instructions, and when recovery from human or machine errors was demanded. While design for the prevention of trouble can minimize their frequency of occurrence, this is not sufficient to

guarantee that all such situations have been eliminated.... *Unanticipated variability* will arise as it did in this application...

Contrary to the implicit assumptions in the prosthesis paradigm [see chapters 4 and 12, this volume], the results revealed that trouble-shooters actively and substantially contributed to the diagnostic process. The more the human functioned as a passive data-gatherer for the machine, the more joint system performance was degraded. Those who passively followed the directives of the machine expert dwelled on unproductive paths and reached dead-ends more often than participants who took a more active role.

In contrast, active human participation in the problem solving process led to more successful and rapid solutions.... However, the design of the machine expert not only failed to support an active human role, it actually retarded technicians from taking or carrying out an active role. [This is the peripheralization problem discussed in chapter 3 and in case 1 in this chapter.]

The variability in performance...due to level of experience and degree of active participation makes clear that technician skill substantively contributed to overall person–machine performance.... Good performance depended heavily on the ability of the technicians to recognize and deal with situations unanticipated by the machine expert designers that led to deviations from the canonical path [the solution path embodied in the expert system's logic].

In order to wield a tool the user must know its boundaries and instrumental characteristics...what it can do and what are its limits, side effects, preconditions and post-conditions. The opacity of the machine expert hindered the human from effectively utilizing its capabilities to meet domain goals [see also case 2]. The machine expert provided few cues as to its intentions in pursuing a line of diagnosis and few means for redirecting its resources. This inhibited the human problem solver in his role as a manager for a semi-autonomous resource—to monitor the performance of the [machine] agent in order to detect when it is off track and to redirect it (again, the same result has occurred with the introduction of other forms of automation when the human's supervisory role has been ignored, e.g. the Hoogovens Report) [see case 1].

Whenever the goal is to automate the problem-solving process by assigning the human a passive role, the problem of unanticipated variability arises. Automated systems inevitably solve the simple cases and straightforward portions of the problem, but fall down in the more complex cases.... Thus, attempts to take the user out of the problem solving loop can inadvertently increase the user's burden by asking him to handle the difficult cases without the benefit of experience and the practice of solving the simpler cases...the costs can be high in risky domains where performance on the rare but catastrophic incident is important...These lessons emphasize the need to conceive of cognitive tools as an instrument in the hands of the human problem-solver.

Note the irony that increased automation is often justified on grounds of human incompetence, yet in practice it is often that same person who must now help the machine to cope with disturbances beyond its design range. [Author's note: This irony is explored in chapter 4.]

The results of this investigation show that questions of tool use cannot be treated as a secondary design problem, to be handled as only an interface issue that is relevant late in the development of an automated problem-solver. Rather, the characteristics of the joint person-machine cognitive system...have a fundamental impact on the ultimate effectiveness of the new system in the actual work context and on the definition and architecture of the tools themselves... [italics added]. (p. 48)

COMMENT

David Woods (1996) described automation problems succinctly: "Automated systems that are strong, silent, clumsy, and difficult to direct are not team players." He went on,

Automated systems are

Strong when they act autonomously.

Silent when they provide poor feedback about their activities and intentions.

Clumsy when they interrupt their human partners during high workload, high criticality periods or add new mental burdens during these high-tempo periods.

Difficult to direct when it is difficult and costly for the human supervisor to instruct the automation about how to change as circumstances change. (pp. 5–6)

Woods cited experience from many domains in deriving this maxim; the four cases cited here are merely examples. They illustrate, however, some of the characteristics enumerated earlier for automated systems in aviation, and the effects of these attributes upon the humans who operate them are similar in each of these domains.

Autonomous systems, by definition, are not "team players." The Hoogovens automation was largely autonomous by design, as was the expert diagnostic system. The Therac-25 was designed to be controlled by human operators, but its design deficiencies foreclosed the ability of experienced radiation therapy technicians to work cooperatively with it, or even to understand what it was doing. The *Berlin Express* autopilot was not intended to be autonomous, and indeed, it had been a reliable assistant—until it produced one fundamental surprise for its operators. The confidence that the Master had in the device (confidence not shared, the report pointed out, by the Maritime Pilot) was warranted by its previous faultless performance; humans will come to rely on normally reliable tools.

Automated systems, then, can be autonomous either by design, or by misadventure. In either case, to the extent that they are autonomous, they implicitly diminish the authority of their human operators. They may do so by not informing their operators of what they are doing, as at Hoogovens, or how they are doing it, as with the Therac machine, or by acting too quickly (the Berlin Express surprise), or by incorporating opaque methods and logic (the Expert System). They may diminish human authority by peripheralizing the operator so that he or she loses a central role in the process (Hoogovens, the Expert System), or by failing to provide the information necessary to prompt needed human intervention (the Therac).

Human authority will be compromised if the operator can only intervene by taking all control away from the automation, thus negating its ability to modulate the process (Hoogovens), or does not understand either how the machine operates (all four cases), or how to operate the machine (Therac). A lack of understanding can come about because of inadequate documentation (Therac), or inadequate training (Hoogovens), or because of the machine's inherent complexity (Expert System).

In all of these cases, there was a common failure: failure to consider "that the operational requirements of the work, task and job design, the social and organizational context of the work, and the selection and training of the operator should all be considered as an integrated whole in the design phase" (ECSC, 1976, p. 15). The problems in these case studies "emphasize the need to conceive of cognitive tools as an instrument in the hands of the human problem-solver" (Roth et al., 1987, p. 47). To the extent that this is not done during design, the machine component of the human–machine system is likely to inhibit effective cooperation between the human and the machine in accomplishing the work assigned to the system.

In the Therac case, the automation was designed to ease the application of radiation therapy. It was poorly designed, but its serious "bugs" showed up infrequently and in any event, operators were not given access to information concerning its idiosyncrasies and had been told that radiation overdosage was "impossible" (and the designers must have believed this as well). It acted autonomously once ordered to perform, and its interface design and lack of feedback made it easy for operators to order it to begin therapy again even when that was an inappropriate action.

Neither the Therac automation nor the Berlin Express automation had knowledge of world states that might have modified their behaviors. That was left to the humans, although the Therac device's sensors were grossly inadequate and could not tell operators what had actually happened. The expert diagnostic system utilized the human simply as a data gatherer rather than as an active collaborator in a cooperative problem-solving process, yet required more than just data to facilitate effective problem solving. These are examples of brittleness and literalism (see chapter 9). Roth et al. (1987) wrote of "unanticipated variability;" this variability is characteristic both of complex machines, and of the environments in which they must function. Expert human practitioners cope well with such variability; even very smart machines do not.

The *Berlin Express* investigator commented that had the automation not been in use, the grounding probably would not have occurred. It may be true that this automation was not designed to be used in tightly constrained situations, although this apparently was not known to the ship's Master. Yet surely this is the sort of situation in which automation may be most useful: when human operator workload is highest. Regardless of the designer's intentions, it was foreseeable that if the automation worked well, it would be used in situations beyond those contemplated by the designer. Humans will tailor their tools, and their behavior, as seems appropriate to them in the context of their actual practice.

We can observe, in these examples from other domains, many of the problems we see in aviation: losses of state and situation awareness, associated with automation complexity, coupling, autonomy, and opacity. The solutions in these domains are likewise similar to those required in our aeronautical enterprise: *automation must be designed to be a team player*. Its design and implementation must recognize the primacy of the human operator, who remains responsible for the safety and efficacy of the product or process. It must enhance system performance by cooperating in the performance of the task, rather than by acting autonomously, or by impeding human operators, or by adding to their tasks. Overall system effectiveness depends on a synergistic relationship among team members.

To return to the first principles of human-centered automation set forth in chapter 3, if automation is to function cooperatively with humans in the performance of useful work, it must involve the human operator, not only when it fails but during normal operations. It must inform the human of what it is doing so that the human can monitor its performance, even while it monitors the human's performance. Most important, it must keep the human informed of its intent, and the human must be able to inform the machine of his or her intent. The essence of human-centered automation in any domain is effective communication and information transfer among involved agents, regardless of the work being performed by the human–machine system. I believe that human-centered automation design would have aided task accomplishment in each of the systems described here, and in many other systems now functioning in our increasingly technological society.

CHAPTER 16

SOME FINAL THOUGHTS
AND CONCLUSIONS

INTRODUCTION

In this chapter, I append a few topics that need to be mentioned but do not fit well elsewhere. The comments are personal and represent my concerns regarding some issues that face us now, or are likely to in the near future. They are followed by a brief conclusion.

IS COCKPIT COMMONALITY
AN OPPORTUNITY, OR AN ISSUE?

I mentioned, in my dedication, Dr. Ruffell Smith's long crusade to improve the standardization of aircraft cockpits. Efforts to achieve this goal continued throughout the two decades after the war, stimulated from time to time by accidents in which a lack of standards for displays and device placement was a factor. Although there is still some variability across types, critical elements were fairly well standardized—until the advent of advanced automation. Technology has developed so rapidly that functional and physical standards for automated devices, displays, and controls have not kept up. Is this a problem?

An unattributed Associated Press report dated August 22, 1994, discusses fleet cockpit commonality and its economic implications for air carriers. It describes an Air Canada decision to order 25 new Airbus aircraft rather than refurbish its DC-9s despite an increased cost of $20 million per aircraft, which Air Canada estimates will save $3.5M per year due to decreases in spares inventory and the ease with which pilots can be exchanged with its fleet of 34 A320s already in service. Julius Maldutis, of Salomon Brothers, is quoted as saying that "Increasingly, you'll see airlines being supplied by a single manufacturer...the battle between manufacturers is increasingly not for the next 20 airplane deals, but for the next 200 airplane deals to convert airlines entirely to your product." Of course, the report might also have mentioned that in the same week, Northwest Airlines, another A320 operator, announced its intention to refurbish, modernize, and add "hush kits" to its sizeable DC-9 fleet, from which it estimated it could get perhaps two decades of additional service at much lower cost!

The report continued,

> "Boeing has not been left out...by flying only Boeing 737s, Southwest [Airlines] has been able to keep costs low by stocking only one type of [everything]. Southwest pilots only need to know their ways around one cockpit layout [because Southwest has also limited the authority of the flight management systems installed in its - 300 aircraft and has specified electromechanical rather than CRT displays in its newer aircraft].

> When Boeing set out to update the 737, a major customer told Boeing to change whatever it wants, "but put a padlock on the cockpit door" to keep the designers out. The different models of the 737 have identical cockpits[;] pilots can [also] move between the longer-range 757 and 767 with only an additional hour of training.

Although these statements are not quite correct, the article makes an important point that has major implications both for operators and for the human factors community. I have tried to indicate in this document that current automation suites are not free of human–machine interface problems. Are we, for economic reasons, at the point where air carriers would rather live with "the devil they know" than move toward correction of some of the acknowledged human factors problems on their flight decks?

I believe that at this point in time the answer to this question is probably a qualified "Yes." United Airlines estimated many years ago that each pilot retirement forced the movement (either upgrading or transition to another airplane), training, and qualification of some 13 other pilots. The expense was enormous, even at a time when air carriers were making money. Air carriers have spent years trying to minimize training costs; commonality among cockpits will certainly be of assistance.

As indicated earlier, Southwest has chosen to limit sharply the utility of its newer aircraft automation by limiting their flight management systems' functionality. The systems have to be initialized to set their inertial reference systems, but they are not used for direct navigation. The carrier has also stayed with electromechanical instruments to insure commonality between its 737-200s and the -300 series aircraft, which has precluded it from installing the more integrated navigation displays normally available in the later 737s. (See Fig. 5.20 and accompanying text for Boeing's proposed solution of this problem.)

Although Airbus cockpits do have a very high degree of commonality, the computer architecture and FMS functionality across types are not really identical (although the differences are normally transparent to pilots). I am concerned, however, about the behavior at the margins of the different hardware and software in these types, and the potential for surprises under difficult circumstances. In the initial report on the investigation of the A330 accident at Toulouse (1994), it was stated by Airbus that the combination of problems that occurred during that flight was unlikely or impossible in other Airbus models, for a variety of aerodynamic and other reasons. Further, the software involved in implementation of the altitude acquisition mode differed across types.

I doubt that line pilots of these or other closely related families of aircraft are consistently made aware of such differences during transition training, and many may not be of concern to them. On the other hand, this mishap need not have happened, and I am impelled to wonder what other occult problems may be lurking at the margins of operating envelopes, waiting to snare pilots who have operated successfully in another related type and who may therefore have been led to believe that they can operate in the same ways in this aircraft.

THE LIABILITY ISSUE IN AVIATION OPERATIONS

In recent years, we have seen an increasing number of criminal prosecutions of flight crew, and even air carrier managers, after aircraft accidents and even incidents in which they were alleged to have been negligent in the performance of their express or implied duties. As an instance, a manager was charged after the A320 Strasbourg accident (1992) for his failure to require ground proximity warning systems in Air Inter aircraft.

This trend has begun to appear in the United States as well, with the successful prosecution of three pilots for having detectable levels of blood alcohol in their bodies while engaged in flight duties. In the United States, the Federal Aviation Regulations (with one exception, interference with a flight member in the perform-ance of duty) are not criminal law, and their violation is almost always a civil rather than a criminal matter. The pilots mentioned were tried under a law prohibiting operation of a commercial motor vehicle under the influence of alcohol. Obviously, such issues are a matter of serious concern to pilots and can affect their behavior and decision-making processes.

In nations that govern under the Napoleonic code, and even in some common-law jurisdictions, violation of Air Navigation Orders is potentially a criminal offense. "Two Korean pilots were jailed in Libya in 1990 after landing short at Tripoli, killing 72 passengers and at least five others. In 1983, a Swissair crew was convicted and fined in Greece after skidding off the end of a wet runway in Athens; 14 passengers died" (Wilkinson, 1994, p. 8).

In a celebrated case in the United Kingdom (London, 1989), the pilot in command of a Boeing 747 was convicted of negligent endangerment of his passengers after an unstable autocoupled approach at London's Heathrow Airport during which the aircraft came within 70 ft of the ground outside the airport boundary (London, 1989). The airplane was landed safely from a second approach. The pilot in command was demoted to first officer by his company. After revoking his pilot in command license, the UK Civil Aviation Authority brought criminal charges, one of which was sustained in a split jury decision. The pilot was fined; his appeal was rejected. He subsequently committed suicide (Wilkinson, 1994). The two pilots involved in a recent A300 accident in Korea are under criminal investigation concerning their conduct of the flight and landing.

Air traffic controllers have not been immune. A Yugoslav controller was jailed following a midair collision over that nation, and others have also been prosecuted, although I do not have details concerning specific cases.

Automation has become more complex, autonomous, and opaque. In the 1994 accident at Charleston, North Carolina, the wind shear advisory system failed to warn the pilots of a microburst after they began a missed approach, because the airplane's flaps were in transit when the wind shear was detected. Human (pilot) error was found by the NTSB to be the probable cause of the resulting accident, despite the Board's knowledge of and previous recommendations concerning the WSAS software inhibition during flap movement, of which the pilots were unaware. Pilot responsibility is absolute—but who was responsible for not notifying the pilots of the automation's inability to warn them under those circumstances? (See also further discussion.)

My concern is that holding pilots or air traffic controllers liable for negligence is likely to inhibit seriously our ability to investigate air accidents. Regardless of what may be said about the duty of a professional person to disclose information that may compromise him or her but may save others, the fact is that many otherwise honest and upright people find it difficult or impossible to do so. *When aviation professionals know that their statements following an accident may cause them lasting harm, they are unlikely in many cases to be forthright with accident investigators. Today's legal climate insists that blame be apportioned, but the only way we are likely to continue to be able to learn lessons from accidents is to insure that the principals in such accidents can talk freely about what happened and why.*

HOW DO YOU PUNISH A COMPUTER?

Who is liable for the behavior of a highly automated system? If automation continues to become more pervasive and authoritative, who will be responsible for its actions? At this time, we simply say that the pilot and controller remain responsible, but if a more autonomous air traffic control system is put in place, can this a priori assignment of responsibility continue? The NTSB clearly thought so in the Charlotte case, but automation's failure to warn played a major, if not crucial, role, as it has in other accidents (see Fig. 6.8).

The Eurocontrol Experimental Center is France is pursuing long-term research into future air traffic management systems. One approach being explored is "complete air-ground automation of the separation assurance function;" the other is aircraft autonomy in an "open sky," using "electronic visual flight rules" (Maignan, as cited in Cooper, 1994a, p. 9). Our present concepts of responsibility and authority are silent on the implications of such automation, but I cannot imagine how a controller could be held entirely responsible for a loss of separation in a fully autonomous air traffic management system, nor even in a system such as my scenario 3 in chapter 8.

Nonetheless, it is unlikely that those inclined toward the assignment of blame will take much pleasure in suspending or fining a computer after an aviation incident. Given that our tort system requires the apportionment of liability, how will this be done? Will we continue to maintain a narrow focus on the "sharp end" of the system, or will we broaden our perspective to encompass the larger system in which the incident occurred, recognizing that nearly all such incidents involve many manifest and latent factors? These implications of increased automation of aircraft and the air traffic management system may well be among the more significant reverberations (see chapter 4) of automation technology in our present litigious environment.

CONCLUSIONS

I believe the central technical (and even social and legal) problems for the human operators who work with today's automation are the complexity and opacity of these tools. Put in terms of the "first principles," the human operator must be able to understand his or her tools, including automated devices, and must be informed about their activities, to remain involved in the process. As indicated in chapters 9, 13, and 15, this necessity for understanding by human operators of the capabilities and limitations of their tools is the conceptual need that must be attacked if humans and machines are to be able to work more cooperatively, as a team, in pursuit of system goals, whatever the system in which they are employed.

This book is by no means an exhaustive treatise on automation. It suggests requirements for new automation designs, but it does not (and, I have argued, cannot) specify how to satisfy those requirements in a specific setting. What I have tried to do is to suggest characteristics of automation that cause problems for at least some of its operators, the types of problems that are associated with these attributes, and means of mitigating some of those problems without compromising the effectiveness of automated tools.

In a future system in which the human does not play such a central and critical role, these human–automation interactions might be less of a problem. On the other hand, any such system is likely to remain under the control of humans at some level, and the problems posed by clumsy, brittle, or uninformative automation will still need to be solved at that level. *This problem will not go away; it will become more pervasive and vexing as technology continues its inexorable advance.*

Although aviation has become a remarkably safe way to move people and goods, preventable accidents continue to occur. To an increasing extent, these accidents involve both human operators and their machines, because the humans and machines have become more interdependent. They represent *system* failures, and they will only be prevented by a systematic approach to all components of the aviation system. Automation is now a central element in that system. It has been extremely successful in improving the reliability and productivity of the system. Like all technology, its successes have brought with them new problems to solve.

Although I hope that this book may point toward needed research into the cognitive issues raised by advanced automation, I will be content if the document improves the quality and depth of the dialogue about these problems, and an accelerated search for their solutions. The dialogue must be intensified at all levels: between system architects and the manufacturers who must realize their designs, between manufacturers and the customers who purchase their products, between the customers and the human operators who manage and control the system, and between all of them and the government officials who must certify the system and maintain oversight of its safety and effectiveness. That was the primary purpose of the predecessor documents, and that remains the purpose of this treatise.

Solution of these human–machine problems is critical to the continued success of the aviation system, on which so many millions of our citizens rely for transportation. Those citizens deserve the highest attainable level of safety. We in human factors, together with our colleagues in the aviation community, owe it to them to prevent these preventable accidents. I believe we now know enough about cognitive engineering to accomplish this goal. To do so, we must apply what we have learned to the management and operation of the systems we now have and to the design of every new automated system we put in place.

Appendix A:
Aircraft Accidents and Incidents

This appendix contains brief descriptions of the relevant aspects of aircraft incidents and mishaps cited in the text. The occurrences are ordered chronologically. Each summary is followed by one or more references to publications or reports.

Table A.1

Location	Date	
Grand Canyon, Arizona	6/30/1956	299
Atlantic Ocean	2/3/1959	300
Boston, Massachusetts	10/4/1960	300
Ellicott City, Maryland	11/23/1962	301
London, England	6/18/1972	301
Miami, Florida	12/29/1972	301
Boston, Massachusetts	7/31/1973	302
New Hope, Georgia	4/4/1977	303
Los Angeles, California	4/12/1977	303
Kaysville, Utah	12/181977	303
Pensacola, Florida	5/8/1978	304
Portland, Oregon	12/28/1978	305
Luxembourg	11/11/1979	305
Atlanta, Georgia	10/7/1980	306
Washington, DC	1/13/1982	306
Sakhalin Island, USSR	9/3/1983	307
New York, New York	2/28/1984	307
Pacific Ocean	2/19/1985	308
San Francisco, California	3/31/1986	309
Los Angeles, California	6/30/1987	309
Atlantic Ocean	7/8/1987	309
Detroit, Michigan	8/16/1987	310
Mulhouse-Habsheim, France	6/26/1988	310
Dallas, Texas	8/31/1988	311
Dryden, Ontario	3/10/1989	311
London, England	11/21/1989	312
Cove Neck, New York	1/25/1990	313

Location	Date	Page
Atlantic Ocean	2/13/1990	313
Bangalore, India	2/14/1990	314
Detroit, Michigan	12/3/1990	314
Los Angeles, California	2/1/1991	315
Thailand	5/26/1991	315
Sydney, Australia	8/12/1991	315
Nakina, Ontario	12/12/1991	316
Strasbourg, France	1/??/1992	316
Colorado Springs, Colorado	12/8/1992	317
Warsaw, Poland	9/14/1993	317
Nagoya, Japan	4/26/1994	318
Hong Kong	6/6/1994	319
Manchester, England	6/21/1994	319
Toulouse, France	6/30/1994	320
Charlotte, North Carolina	7/2/1994	321
Pittsburgh, Pennsylvania	9/8/1994	322
Paris, France	9/24/1994	322
Roselawn, Indiana	10/31/1994	323

Grand Canyon, Arizona, 6/30/1956: Trans World Airlines L1049A and United Air Lines DC-7

At approximately 1031 hr PST, a TWA L-1049A and a United Air Lines DC-7 collided at about 21,000 ft over Grand Canyon, Arizona. Both aircraft fell into the Canyon; there were no survivors among the 128 persons aboard the two flights.

The Civil Aeronautics Board (CAB) determined that the flights were properly dispatched. In flight, the TWA crew requested 21,000 ft, or 1,000 ft on top (above cloud tops). The 21,000 ft request was denied by ATC because of UAL 718. TW then climbed to and flew at 21,000 ft above clouds. The last position report from each aircraft indicated that both were at 21,000 ft, estimating their next fix at 1031. The aircraft were in uncontrolled airspace and were not receiving traffic control services at the time of the collision.

The Board determined that the probable cause of the collision was that the pilots did not see each other in time to avoid the collision. The Board could not determine why the pilots did not see each other but suggested the following factors: intervening clouds, visual limitations due to cockpit visibility, preoccupation with matters unrelated to cockpit duties such as attempting to provide the passengers with a more scenic view of the Grand Canyon, physiological limits to human vision, or insufficiency of en route air traffic advisory information due to inadequacy of facilities and lack of personnel (CAB, 1957).

Atlantic Ocean, 2/3/1959: Pan American World Airways B-707

Pan American flight 115 was en route from London to New York when it entered an uncontrolled descent of approximately 29,000 ft. Following recovery from the upset, the airplane was flown to Gander, Newfoundland, where a safe landing was made. A few of the 129 persons on board suffered minor injuries; the aircraft incurred extensive structural damage during recovery from the dive.

The aircraft was at 35,000 ft in smooth air with the autopilot engaged when the captain left the cockpit and entered the main cabin During his absence the autopilot disengaged and the aircraft smoothly and slowly entered a steep descending spiral. The copilot was not properly monitoring the aircraft instruments and was unaware of the airplane's attitude until considerable speed had been gained and altitude lost. During the rapid descent the copilot was unable to effect recovery. When the captain became aware of the unusual attitude he returned to the cockpit with considerable difficulty. With the aid of the other crew members, he was finally able to regain control of the aircraft at an altitude of about 6,000 ft.

The Civil Aeronautics Board determined that the accident resulted from the inattention of the copilot to the flight instruments during the captain's absence from the cockpit, and the involuntary disengagement of the autopilot. Contributing factors were the autopilot disengage warning light in the "dim" position and the Mach trim switch in the "off" position. During analysis, which was hindered by the flight data recorder having exhausted its supply of metal recording foil, it was indicated that the airplane had reached Mach 0.95 in its abrupt descent. Very high G forces were indicated by the recorder and had been reported by the pilots during their attempts to recover from the spiral dive. After landing, the lower surface skin of the horizontal stabilizers was found to be buckled; both wing panels and both outboard ailerons were damaged; the wing-to-fuselage fairings were damaged and a 3 ft section of the right fairing had separated in flight. Both wing panels suffered a small amount of permanent set. All four wing-to-strut fairing sections of the engine nacelle struts were buckled and other damage was also evident (CAB, 1959).

Boston, Massachusetts, 10/4/60: Eastern Air Lines Lockheed Electra L-188, Logan International Airport

This airplane crashed into Winthrop Bay immediately following takeoff. Ten of 72 persons survived the accident. A few seconds after becoming airborne, the airplane struck a flock of starlings. A number of these birds were ingested in engines 1, 2, and 4. Engine 1 was shut down and its propeller was feathered; engines 2 and 4 experienced a substantial momentary loss of power. The airplane yawed to the left and decelerated to the stall speed; the left wing dropped, the nose pitched up, and the airplane rolled left into a spin. The Civil Aeronautics Board determined that "the unique and critical sequence of the loss and recovery of power" at an altitude of less than 150 ft resulted in loss of control and precluded recovery from the stall (CAB, 1962).

Ellicott City, Maryland, 11/23/1962: United Air Lines Vickers Viscount

This four-engine turboprop crashed after penetrating a flock of whistling swans while in cruising flight at 6,000 ft. At least two birds were struck by the aircraft. One swan collided with the right horizontal stabilizer; "the other bird punctured the left horizontal stabilizer, traveled through the structure, and dented the elevator as it egressed." The weakened structure failed in this area, rendered the aircraft uncontrollable, and resulted in the aircraft striking the ground in a nose-low inverted attitude (CAB, 1964).

London, England, 6/18/1972: British European Airways Trident, Heathrow Airport

This aircraft commenced its operation under the command of a very senior BEA captain. The first officer was relatively inexperienced and the second officer was a recent graduate of the airline's ab initio training school. The airline was undergoing a difficult labor–management conflict, and the captain had been involved in a heated altercation in the crew room before departure.

Shortly after takeoff, when the first reduction of flaps was ordered, it is thought that the first officer inadvertently actuated the wing leading edge slat handle as well, raising the slats at a speed too low to sustain flight. Based on postmortem evidence, it is believed that the captain had a severe cardiac event at about the same time. Many warning lights and aural signals were actuated by the premature retraction of the slats. The inexperienced first officer was unable to diagnose the problem or to regain control of the airplane, which crashed into a reservoir just west of the airport. There were no survivors (Department of Trade and Industry, 1973).

Miami, FL, 12/29/1972: Eastern Air Lines L-1011

The airplane crashed in the Everglades at night after an undetected autopilot disconnect. The airplane was flying at 2,000 ft after requesting and executing a missed approach at Miami because of a suspected landing gear malfunction. All three flight crewmembers and a jumpseat occupant became immersed in diagnosing the malfunction. The accident caused 99 fatalities among the 176 persons on board.

The NTSB believed that the airplane was being flown on manual throttle with the autopilot in control wheel steering mode, and that the altitude hold function was disengaged by light force on the yoke. The crew did not hear the altitude alert departing 2,000 ft and did not monitor the flight instruments until the final seconds before impact. The Board found the probable cause to be the crew's failure to monitor the flight instruments for the final 4 min of the flight and to detect an unexpected descent soon enough to prevent impact with the ground. The captain failed to assure that a pilot was monitoring the progress of the aircraft at all times. The Board discussed overreliance on automatic equipment in its report and pointed out the need for procedures to offset the effect of distractions such as the malfunction during this flight (NTSB, 1973).

Boston, Massachusetts, 7/31/1973: Delta Air Lines DC9-31

This airplane struck a seawall bounding Boston's Logan Airport during an approach for landing after a flight from Burlington, Vermont, killing all 89 persons on board. The point of impact was 165 ft right of the runway 4R centerline and 3000 ft short of the displaced runway threshold. The weather was sky obscured, 400 ft ceiling, visibility 1½ miles in fog.

The cockpit voice recorder (CVR) showed that 25 sec before impact, a crewmember had stated, "You better go to raw data; I don't trust that thing." The next airplane on the approach, 4 min later, made a missed approach due to visibility below minimums. The accident airplane had been converted from a Northeast Airlines to a Delta Air Lines configuration in April 1973, at which time the Collins flight director had been replaced with a Sperry device; there had been numerous writeups for mechanical deficiencies since that time. The flight director command bars were different (see Fig. 5.8 for the two presentations), as were the rotary switches controlling the flight director. The crew were former Northeast Airlines pilots. If the crew had been operating in the go-around mode, which required only a slight extra motion of the replacement rotary switch, they would have received steering and wing-leveling guidance only, instead of ILS guidance. Required altitude callouts were not made during the approach.

The NTSB found the probable cause to be the failure of the crew to monitor altitude and their passage through decision height during an unstabilized approach in rapidly changing meteorological conditions. The unstabilized approach was due to passage of the outer marker above the glide slope, fast, in part due to nonstandard

ATC procedures. This was compounded by the flight crew's preoccupation with questionable information presented by the flight director system.

The Board commented that, "An accumulation of discrepancies, none critical (in themselves), can rapidly deteriorate, without positive flight management, into a high-risk situation...the first officer, who was flying, was preoccupied with the information presented by his flight director system, to the detriment of his attention to altitude, heading and airspeed control" (NTSB, 1974).

New Hope, Georgia, 4/4/1977: Southern Airways DC-9

The airplane crashed while attempting to land on a highway after losing all power in both engines while in a severe thunderstorm. The "total and unique" power loss was thought by the NTSB to be due to "the ingestion of massive amounts of water and hail which in combination with thrust lever movement induced severe stalling in and major damage to the engine compressors." The Board found that contributing factors were the failure of the company's dispatching system to provide the flightcrew with up-to-date severe weather information, the captain's reliance on airborne weather radar, and limitations in the air traffic control system that precluded the timely distribution of real-time hazardous weather information to the flight crew (NTSB, 1978a).

Los Angeles, California, 4/12/1977: Delta Air Lines L-1011

This airplane landed safely at Los Angeles after its left elevator jammed in the full up position shortly after takeoff from San Diego. The flight crew found themselves unable to control the airplane by any normal or standard procedural means. They were able, after considerable difficulty, to restore a limited degree of pitch and roll control by using differential power on the three engines. Using power from the tail-mounted center engine to adjust pitch and wing engines differentially to maintain directional control, and verifying airplane performance at each successive configuration change during an emergency approach to Los Angeles, the crew succeeded in landing the airplane safely and without damage to the aircraft or injury to its occupants (McMahon, 1978).

Kaysville, Utah, 12/181977: United Airlines DC-8

A cargo aircraft encountered electrical problems during its approach to the Salt Lake City Airport. The flight requested and accepted a holding clearance from the approach controller. The flight then requested and received clearance to leave the approach control frequency in order to communicate with company maintenance (one of the two communications radios had failed due to the electrical problem). Flight 2860 was absent from the approach control frequency for more than 7 min, during which time the flight entered an area near hazardous terrain. The approach controller recognized the crew's predicament but was unable to contact the flight.

When the crew returned to his frequency, the controller told the flight that it was too close to terrain on its right and to make an immediate left turn. After the controller repeated the instructions, the flight began a left turn. About 15 sec later, the controller told the flight to climb immediately to 8,000 ft. Eleven seconds later, the flight reported that it was climbing from 6,000 to 8,000 ft. The airplane crashed into a 7,665 ft mountain near the 7,200 ft level.

The NTSB determined that the probable cause of the accident was the approach controller's issuance and the flight crew's acceptance of an incomplete and ambiguous holding clearance, in combination with the flight crew's failure to adhere to prescribed impairment-of-communications procedures and prescribed holding procedures. The controller's and flight crew's actions were attributed to probable habits of imprecise communication and of imprecise adherence to procedures, developed through years of exposure to operations in a radar environment. The Board noted that the GPWS would not have provided a warning until 7.7 to 10.2 sec before impact, which was too late because of the rapidly rising terrain (NTSB, 1978b).

Pensacola, Florida, 5/8/1978: National Airlines B727-235, Escambia Bay

Flight 193 crashed into Escambia Bay about 3 miles short of the runway while executing a surveillance radar approach to Pensacola Airport runway 25 at night in limited visibility. The aircraft came to rest in about 12 ft of water.

The NTSB determined that the probable cause of the accident was the flight crew's unprofessionally conducted nonprecision instrument approach, in that the captain and crew failed to monitor the descent rate and altitude and the first officer failed to provide the captain with required altitude and approach performance callouts. The crew failed to check and utilize all instruments available for altitude awareness, turned off the ground proximity warning system, and failed to configure the aircraft properly and in a timely manner for the approach. Contributing to the accident were the radar controller's failure to provide advance notice of the start-descent point, which accelerated the pace of the crew's cockpit activities after the passage of the final approach fix.

The Board noted that the approach was rushed, that final flaps were never extended, and that the captain was unable to establish a stable descent rate after descending below 1,300 ft. The captain either misread or did not read his altimeters during the latter stages of the approach; the first officer did not make any of the required altitude callouts. The flight engineer's inhibition of the GPWS coincided with the captain's raising the nose and decreasing the descent rate. The pilots were misled into believing the problem was solved (NTSB, 1978c).

Portland, Oregon, 12/28/1978: United Airlines DC-8-61

This airplane crashed into a wooded area during an approach to Portland International Airport. The airplane had delayed southeast of the airport for about an hour while the flight crew coped with a landing gear malfunction and prepared its passengers for a possible emergency landing. After failure of all four engines due to fuel exhaustion, the airplane crashed about 6 miles southeast of the airport, with a loss of 10 persons and injuries to 23.

The NTSB found the probable cause to be the failure of the captain to monitor the fuel state and to respond properly to a low fuel state and to crewmember advisories regarding the fuel state. His inattention resulted from preoccupation with the landing gear malfunction and preparations for the possible emergency landing. Contributing to the accident was the failure of the other two crew members to fully comprehend the criticality of the fuel state or to successfully communicate their concern to the captain. The Board discussed crew coordination, management, and teamwork in its report (NTSB, 1979a).

Luxembourg, 11/11/1979: Aeromexico DC-10-30

During an evening climb in good weather to 31,000 ft en route to Miami from Frankfurt, Germany, flight 945 entered prestall buffet and a sustained stall at 29,800 ft. Stall recovery was effected at 18,900 ft. The crew performed a functional check of the airplane, and after finding that it operated properly they continued to their intended destination. After arrival, it was discovered that parts of both outboard elevators and the lower fuselage tail maintenance access door were missing.

The flight data recorder showed that the airplane slowed to 226 kt during a climb on autopilot, quite possibly in vertical speed mode rather than indicated airspeed mode. Buffet speed was calculated to be 241 kt. After initial buffet, engine 3 was shut down and the airplane slowed to below stall speed.

The NTSB found the probable cause to be failure of the flight crew to follow standard climb procedures and to adequately monitor the airplane's flight instruments. This resulted in the aircraft entering into prolonged stall buffet that placed it outside its design envelope (NTSB, 1980).

Atlanta, Georgia, 10/7/1980: Aircraft Separation Incidents at Hartsfield Airport

This episode involved several conflicts among aircraft operating under the direction of air traffic control in the Atlanta terminal area. In at least two cases, evasive action was required to avoid collisions. The conflicts were caused by multiple failures of coordination and execution by several controllers during a very busy period.

The NTSB found that the near collisions were the result of inept traffic handling by control personnel. This ineptness was due in part to inadequacies in training, procedural deficiencies, and some difficulties imposed by the physical layout of the control room. The board also found that the design of the low altitude–conflict alert system contributed to the controller's not recognizing the conflicts. The report stated that

> The flashing visual conflict alert is not conspicuous when the data tag is also flashing in the handoff status. The low altitude warning and conflict alerts utilize the same audio signal which is audible to all control room personnel rather than being restricted to only those immediately concerned with the aircraft. This results in a "cry wolf" syndrome in which controllers are psychologically conditioned to disregard the alarms (NTSB, 1981).

Washington, DC, 1/13/1982: Air Florida B-737, Washington National Airport

This airplane crashed into the 14th Street bridge over the Potomac River shortly after takeoff from Washington National Airport in snow conditions, killing 74 of 79 persons on board. The airplane had been deiced 1 hr before departure, but a substantial period of time had elapsed since that operation before it reached takeoff position. The engines developed substantially less than takeoff power during the takeoff and thereafter due to incorrect setting of takeoff power by the pilots. It was believed that the differential pressure probes in both engines were iced over, providing incorrect (too high) EPR indications in the cockpit. This should have been detected by examination of the other engine instruments, but was not.

The NTSB found that the probable cause of the accident was the flight crew's failure to use engine anti-ice during ground operation and takeoff, their decision to take off with snow and ice on the airfoils, and the captain's failure to reject the takeoff at an early stage when his attention was called to anomalous engine instrument readings by the copilot. Contributing factors included the prolonged ground delay after deicing, the known inherent pitching characteristics of the B-737 when the wing leading edges are contaminated, and the limited experience of the flight crew in jet transport winter operations (NTSB, 1982).

Sakhalin Island, USSR, 9/3/1983: Korean Air Lines B-747

The airplane was destroyed in cruise flight by air-to-air missiles fired from a Soviet fighter after it strayed into a forbidden area en route from Anchorage, Alaska, to Seoul, Korea. The airplane had twice violated Soviet airspace during its flight. The flight data and cockpit voice recorders were not recovered from the sea. After extensive investigation by the International Civil Aviation Organization, it was believed that the airplane's aberrant flight path had been the result of one or more incorrect sets of waypoints loaded into the INS systems prior to departure from Anchorage.

Many years later, the Russian government made available further information on the flight, which supported a finding that the crew had inadvertently left the airplane's autopilot in heading mode rather than INS mode for an extended period of time. As a result, the flight path took the airplane over Soviet territory, where it was destroyed by a Soviet fighter (Stein, 1985; see also Atlantic Ocean, 2/13/90).

New York, New York, 2/28/1984: Scandinavian Airlines DC-10-30, J. F. Kennedy Airport

After crossing the runway threshold at proper height but 50 kt above reference speed, the airplane touched down 4,700 ft beyond the threshold of an 8,400 ft runway and could not be stopped on the runway. It was steered to the right and came to rest in water 600 ft from the runway end. A few passengers sustained minor injuries during evacuation. The weather was poor and the runway was wet.

The airplane's autothrottle system had been unreliable for approximately 1 month and had not reduced speed when commanded during the first (Stockholm–Oslo) leg of this flight. The captain had deliberately selected an approach speed of 168 kt to compensate for a threatened wind shear. The throttles did not retard passing 50 ft and did not respond to the autothrottle speed control system commands (the flight crew was not required to use the autothrottle speed control system for this approach).

The NTSB cited as the probable cause the flight crew's disregard for prescribed procedures for monitoring and controlling airspeed during the final stages of the approach, its decision to continue the landing rather than to execute a missed approach, and overreliance on the autothrottle speed control system, which had a history of recent malfunctions. It noted that "performance was either aberrant or represents a tendency for the crew to be complacent and over-rely on automated systems." It also noted that there were three speed indications available to the crew: its airspeed indications, the fast–slow indicators on the attitude directors, and an indicated vertical speed of 1,840 ft per minute on glide slope. In its report, the Board discussed the issue of overreliance on automated systems at length (report pp. 37–39) and cited several other examples of the phenomenon (NTSB, 1984).

Pacific Ocean, 2/19/1985: China Airlines B747-SP, 300 Miles Northwest of San Francisco

The airplane, flying at 41,000 ft en route to Los Angeles from Taipei, suffered an inflight upset[1] after an uneventful flight. The airplane was on autopilot when engine 4 lost power. During attempts to restart the engine, the airplane rolled to the right, nosed over, and began an uncontrollable descent. The captain was unable to restore the airplane to stable flight until it had descended to 9,500 ft.

The autopilot was operating in the performance management system (PMS) mode for pitch guidance and altitude hold. Roll commands were provided by the INS. In this mode, the autopilot uses only the ailerons and spoilers for lateral control; rudder and rudder trim are not used. In light turbulence, the airspeed began to fluctuate; the PMS followed the fluctuations and retarded the throttles when airspeed increased. As the airplane slowed, the PMS moved the throttles forward; engines 1, 2, and 3 accelerated but engine 4 did not. The INS caused the autopilot to hold the left wing down because it could not correct with rudder. The airplane decelerated due to the lack of power. After attempting to correct the situation with autopilot, the captain disengaged the autopilot at which time the airplane rolled to the right, yawed, then entered a steep descent in cloud, during which it exceeded maximum operating speed. It was extensively damaged during the descent and recovery; the landing gear deployed, 10 to 11 ft of the left horizontal stabilizer was torn off and the number 1 hydraulic system lines were severed. The right stabilizer and 75% of the right outboard elevator was missing when the airplane landed; the wings were also bent upward.

The NTSB determined that the probable cause was the captain's preoccupation with an inflight malfunction and his failure to monitor properly the airplane's flight

---

[1] A loss of control.

instruments, which resulted in his losing control of the airplane. Contributing to the accident was the captain's overreliance on the autopilot after a loss of thrust on engine 4. The Board noted that the autopilot effectively masked the approaching onset of loss of control of the airplane (NTSB, 1986).

San Francisco, California, 3/31/1986: United Airlines B-767

This airplane was passing through 3,100 ft on its climb from San Francisco when both engines lost power abruptly. The engines were restarted and the airplane returned to San Francisco, where it landed without incident. The crew reported that engine power was lost when the flight crew attempted to switch from manual operation to the engine electronic control (EEC) system, a procedure that prior to that time was normally carried out at 3,000 ft during the climb. The EEC switches were guarded but were located above the fuel valves. It is believed that the crew may have inadvertently shut off fuel to the engines when they intended to engage the EEC, as in the incident cited next (NTSB impounds, 1986).

Los Angeles, California, 6/30/1987: Delta Air Lines B-767

Over water, shortly after takeoff from Los Angeles, this twin-engine airplane suffered a double engine failure when the captain, attempting to deactivate an electronic engine controller in response to an EEC caution light, shut off the fuel valves instead. The crew was able to restart the engines within 1 min after an altitude loss of several hundred feet. The fuel valves were located immediately above the electronic engine control switches on the airplane center console, although the switches were dissimilar in shape.

The FAA thereafter issued an emergency airworthiness directive requiring installation of a guard device between the cockpit fuel control switches (Engine shutdown, 1987).

Atlantic Ocean, 7/8/1987: Delta Air Lines L-1011/Continental Airlines B-747

These two airplanes experienced a near midair collision over the north Atlantic ocean after the Delta airplane strayed 60 miles off its assigned oceanic route. The incident, which was observed by other aircraft in the area but not, apparently, by the Delta crew, was believed to have been caused by an incorrectly inserted waypoint in the Delta airplane's INS prior to departure (Preble, 1987).

Detroit, Michigan, 8/16/1987: Northwest Airlines DC9-82, Detroit Metro Airport

The airplane crashed almost immediately after takeoff from runway 3C[2] en route to Phoenix. The airplane began its rotation about 1,200-1,500 ft from the end of the 8,500-ft runway and lifted off near the end. After liftoff, the wings rolled to the left and right; it then collided with a light pole located ½ mi. beyond the end of the runway. One hundred and fifty-four persons were killed; one survived.

During the investigation, it was found that the trailing edge flaps and leading edge slats were fully retracted. Cockpit voice recorder readout indicated that the takeoff warning system did not function and thus did not warn the flight crew that the airplane was improperly configured for takeoff.

The NTSB attributed the accident to the flight crew's failure to use the taxi checklist to insure that the flaps and slats were extended. The failure of the takeoff warning system was a contributing factor. This airplane has a stall protection system that announces a stall and incorporates a stick pusher, but autoslat extension and poststall recovery are disabled if the slats are retracted. Its caution and warning system also provides tone and voice warning of a stall, but this is disabled in flight by nose gear extension (NTSB, 1988b).

Mulhouse-Habsheim, France, 6/26/1988: Air France Airbus A320

This airplane crashed into tall trees following a very slow, very low altitude flyover at a general aviation airfield during an air show. Three of 136 persons aboard the aircraft were killed; 36 were injured. The captain, an experienced A320 pilot, was demonstrating the slow-speed maneuverability of the then-new airplane.

The French Commission of Inquiry found that the flyover was conducted at an altitude considerably lower than the minimum of 170 ft specified by regulations and also lower than the intended 100-ft altitude level pass briefed to the crew by the captain prior to flight. It stated that "The training given to the pilots emphasized all the protections from which the A320 benefits with respect to its lift which could have given them the feeling, which indeed is justified, of increased safety.... *However, emphasis was perhaps not sufficiently placed on the fact that, if the [angle of attack] limit cannot be exceeded, it nevertheless exists and*

---

[2] Runways are numbered to indicate their magnetic heading to the nearest 10°; 3 = 30° (actually from 26–35°). Parallel runways also have letter designators: L = left, C = center, R = right.

still affects the performance" [italics added]. The commission noted that automatic go-around protection had been inhibited and that this decision was compatible with the captain's objective of maintaining 100 ft. In effect, below 100 ft, this protection was not active.

The commission attributed the cause of the accident to the very low flyover height, very slow and reducing speed, engine power at flight idle, and a late application of go-around power. It commented on insufficient flight preparation, inadequate task sharing in the cockpit, and possible overconfidence because of the envelope protection features of the A320 (Ministry of Planning, Housing, Transport and Maritime Affairs, 1989).

Dallas, Texas, 8/31/1988: Delta Airlines B727-232, Dallas-Fort Worth Airport

The airplane crashed shortly after takeoff from runway 18L en route to Salt Lake City. The takeoff roll was normal but as the main gear left the ground the crew heard two explosions and the airplane began to roll violently. It struck an ILS antenna 1,000 ft past the runway end after being airborne for about 22 sec. Fourteen persons were killed.

The investigation showed that the flaps and slats were fully retracted. Evidence suggested that there was an intermittent fault in the takeoff warning system that was not detected and corrected during the last maintenance action. This problem could have manifested itself during the takeoff.

The NTSB found the probable cause to be the captain's and first officer's inadequate cockpit discipline and failure of the takeoff configuration warning system to alert the crew that the airplane was not properly configured for takeoff. It found as contributing factors certain management and procedural deficiencies and lack of sufficiently aggressive action by FAA to correct known deficiencies in the air carrier. The board took note of extensive non-duty-related conversations and the lengthy presence in the cockpit of a flight attendant, which reduced the flight crew's vigilance in insuring that the airplane was properly prepared for flight (NTSB, 1988a).

Dryden, Ontario, 3/10/1989: Air Ontario Fokker F-28

This airplane was dispatched with an inoperative auxiliary power unit from Winnipeg, Manitoba, to Thunder Bay, Ontario, thence via Dryden, Ontario, back to

Winnipeg. While preparing for the return trip at Thunder Bay, the crew found more passengers than had been planned for or could be accommodated if enough fuel for the entire flight to Winnipeg was boarded, as it had been. The captain preferred to offload passengers rather than fuel; he was overruled by the airline. This action resulted in a delay for defueling at Thunder Bay and required a landing at Dryden to take on additional fuel. The company's system operations center did not inform the captain of freezing rain forecast for Dryden.

Upon arriving at Dryden, which had no ground power units with which to start the airplane's engines, the captain was required to take on fuel with one engine running. The airplane could not be deiced with engines running, however, and freezing rain was falling prior to his takeoff, which was also delayed by a lost aircraft trying to land. The airplane crashed immediately after takeoff; ice was noted on the wings by surviving passengers and cabin crew.

The captain in this accident was placed in a "triple bind." He could not uplift sufficient fuel to fly to Winnipeg with the full passenger load. The lack of a ground power unit at Dryden, and the inoperative APU, meant that he could not shut down both engines at that station. If he landed at Dryden, he could refuel with an engine running, but could not deice if that was required. The defueling at Thunder Bay had already made his flight over 1 hr late. He received inadequate information and no guidance from his company.

The subsequent Commission of Inquiry, in an exemplary investigation, found a large number of latent organizational factors at many levels within Air Ontario, its parent Air Canada, and Transport Canada, the regulatory authority (Moshansky, 1992).

London, England, 11/21/1989: British Airways B747, Heathrow Airport

The aircraft approached London in very bad weather after a flight from Bahrain. Fuel was low due to headwinds; the copilot had been incapacitated for part of the flight due to gastroenteritis and diarrhea. The copilot was not certified for category II or III landings. British Airways flight operations authorized the approach despite the copilot's lack of qualifications. The approach, to runway 27 instead of 9 as briefed, was hurried. When the aircraft captured the localizer and glide slope, the autopilots failed to stabilize the aircraft, possibly due to late capture of the radio beams. At 125 ft above ground, the runway was not in sight and the captain gently began a missed approach. The aircraft sank to 75 ft above ground before gaining altitude. After a second approach, the aircraft landed safely.

An investigation by British Airways disclosed that during the first approach, the aircraft had been seriously to the right of the localizer course and had overflown a hotel to the north of the airport only a few feet above the highest obstacle on its course. The pilot and crew were suspended; legal action was later taken against the captain for endangering the passengers and persons on the ground (Wilkinson, 1994).

Cove Neck, New York, 1/25/1990: Avianca B-707-321

Avianca flight 052 crashed in a wooded residential area during an approach to Kennedy International Airport after all engines failed due to fuel exhaustion. The flight from Medellin, Colombia, had been placed in holding patterns three times for a total of about 1.3 hr. During the third period of holding, the crew reported that the airplane could not hold longer than 5 min., that it was running out of fuel, and that it could not reach its alternate airport in Boston. Subsequently, the flight executed a missed approach at Kennedy. While trying to return to the airport, the airplane lost power in all four engines and crashed 16 miles from the runway.

The NTSB determined that the probable cause of the accident was the failure of the flightcrew to adequately manage the airplane's fuel load, and their failure to communicate an emergency fuel situation to air traffic control before fuel exhaustion occurred. Contributing to the accident was the flightcrew's failure to use an airline operational control dispatch system to assist them during the international flight into a high-density airport in poor weather. Also contributing was inadequate traffic flow management by the FAA and a lack of standardized understandable terminology for pilots and controllers for minimum and emergency fuel states. Wind shear, crew fatigue, and stress were other factors that led to the unsuccessful completion of the first approach and thus contributed to the accident (NTSB, 1991a).

Atlantic Ocean, 2/13/1990: El Al B747
and British Airways B747

An El Al B-747 en route from Tel Aviv to New York almost collided with a British Airways 747 in the Reykjavik Flight Information Region after its crew failed to switch back from heading mode to INS mode after being cleared by Shanwick control to a new oceanic track. The crew deviated 110 nm north of the new track before realizing their error. Upon recognizing the error, the flightcrew notified ATC but provided no information on the magnitude of their deviation. ATC cleared them to turn left to reintercept their cleared track, which they did.

The near collision occurred while the crew were navigating back to the correct track without descending 1,000 ft below the prevailing traffic flow, as prescribed by North Atlantic Special Procedures for In-flight Contingencies. The El Al 747, passed right-to-left ahead of a westbound British Airways 747 which took evasive action, missing El Al by approximately 600 ft (Pan American World Airways, 1990).

Bangalore, India, 2/14/1990: Indian Airlines Airbus A320

[Official report not available] This airplane crashed short of the runway during an approach to land in good weather, killing 94 of 146 persons aboard including the pilots. The best available data indicate that the airplane had descended at idle power in the "idle open descent" mode until shortly before the accident, when an attempt was made to recover by adding power but too late to permit engine spool-up prior to impact. The airplane was being flown by a captain in training undergoing a route check by a check airman.

The crew allowed the speed to decrease to 25 kt below the nominal approach speed late in the descent. The recovery from this condition was started at an altitude of only 140 ft, while flying at minimum speed and maximum angle of attack. The check captain noted that the flight director should be off, and the trainee responded that it was off. The check captain corrected him by stating, "But you did not put off mine." If either flight director is engaged, the selected autothrust mode will remain operative, in this case, the idle open descent mode. The alpha floor mode was automatically activated by the declining speed and increasing angle of attack; it caused the autothrust system to advance the power, but this occurred too late for recovery to be effected before the airplane impacted the ground (Lenorovitz, 1990).

Detroit, Michigan, 12/3/1990: Northwest Airlines B-727 and DC-9, Detroit Metro Airport

These two aircraft collided while the 727 was taking off and the DC-9 had just inadvertently taxiied onto the active runway. The DC-9 was lost on the airport in severely restricted visibility. Both aircraft were on the ground. The accident site was not visible from the tower due to severe fog; ASDE[3] was not available.

The NTSB determined that the probable cause of the accident was a lack of proper crew coordination, including a reversal of roles, on the part of the DC-9 pilots. This led to their failure to stop taxiing and alert the ground controller of their positional uncertainty in a timely manner before and after intruding onto the active runway. A number of contributing factors were also cited (NTSB, 1991b).

---

[3]Airport Surface Detection Equipment (radar).

Los Angeles, California, 2/1/1991: US Air B-737 and Skywest Fairchild Metro

This accident occurred after the US Air airplane was cleared to land on runway 24L at Los Angeles while the Skywest Metro was positioned on the runway at an intersection awaiting takeoff clearance. There were 34 fatalities and 67 survivors. The Metro may not have been easily visible from the control tower; airport surface detection radar equipment (ASDE) was available but was being used for surveillance of the south side of the airport. The controller was very busy just prior to the time of the accident.

The NTSB investigation indicated that the controller cleared the Metro into position at an intersection on runway 24L, 2,400 ft from the threshold, two minutes before the accident. One minute later, the 737 was given a clearance to land on runway 24L. The Board determined that the probable cause of the accident was the failure of Los Angeles Air Traffic Facility management to implement procedures that provided adequate redundancy and the failure of FAA's Air Traffic Management to provide adequate policy direction and oversight. These failures ultimately led to the failure of the local controller to maintain awareness of the traffic situation (Dornheim, 1991; NTSB, 1991c).

Thailand, 5/26/1991: Lauda Air (Austria) B767-300ER

This airplane was climbing to altitude on a flight between Bangkok and Vienna when its right engine thrust reverser actuated because of a mechanical failure. The flight crew was unable to control the airplane due to the high level of reverse thrust coming from the right engine. The airplane crashed after an uncontrolled descent. Simulation studies indicated that recovery from such an event was not possible for pilots without advance knowledge of the event (Ministry of Transport and Communications, Thailand, 1993).

Sydney, Australia, 8/12/1991: Ansett Australia A320 and Thai Airways DC-10

During simultaneous crossing runway operations at Kingsford Smith Airport, a Thai DC-10 was landing on runway 34 and an Ansett A320 was on short final approach for intersecting runway 25. Landing instructions for the DC-10 included a requirement for the aircraft to hold short of the runway 25 intersection. While observing the DC-10's landing roll during his landing, the A320 captain judged that the DC-10 might not stop before the runway intersection. He elected to initiate a missed approach from a low height above the runway. The go-around was successful; the

A320 passed the centerline of runway 34 at a radio altitude of 52 ft. Under heavy braking, the DC-10 slowed to about 2 kt ground speed when it reached the edge of runway 25.

During the A320 go-around, differing attitude command inputs were recorded from the left and right sidesticks for a period of 12 sec. Neither the captain, who had taken over control, or the copilot, was aware of control stick inputs from the copilot during this period. Activation of the "takeover button" on the control stick was not a part of Ansett's standard operating procedures. The incident analysis noted that "Although the A320 successfully avoided the DC-10, under different circumstances the cross controlling between the pilots could have jeopardized a safe go-around.... This simultaneous input situation would almost certainly have been immediately apparent, and corrected rapidly had there been a sense of movement between the two sidesticks" (Bureau of Air Safety Investigation, 1993).

Nakina, Ontario, 12/12/1991: Evergreen International Airways B-747

While in cruise flight at 31,000 ft, a cargo aircraft entered a steep right bank (greater than 90°) and descended more than 10,000 ft at speeds approaching Mach 1. During the recovery, with vertical accelerations greater than 3G, the right wing was damaged. About 20 ft of honeycomb structure from the underside of the wing was missing; a small honeycomb panel on the upper portion of the wing was damaged and some structure was protruding into the airstream. Upon recovery from the dive, the aircraft was experiencing control difficulties; the crew successfully diverted to Duluth, Minnesota. During the approach and landing, the left and right flaps, as well as the right horizontal stabilizer, were damaged by debris from the damaged right wing. There were no injuries.

The Transportation Safety Board of Canada determined that the flight upset was caused by an uncommanded, insidious roll input by the channel A autopilot roll computer; the roll went undetected by the crew until the aircraft had reached an excessive bank angle and consequential high rate of descent. The recovery action was delayed slightly because of the time required by the crew to determine the aircraft attitude (Aviation Occurrence Report, 1991; NTSB, 1992a).

Strasbourg, France, 1/20/1992: Air Inter Airbus A320

The airplane was being given radar vectors to a nonprecision (VOR-DME) approach to the airport at Strasbourg. It was given vectors that left little time for cockpit setup prior to intercepting the final approach course. It is believed that the pilots (both inexperienced in this aircraft) intended to make an automatic approach using a flight path angle of −3.3° from the final approach fix; this maneuver would have placed them at approximately the correct point for visual descent when they reached minimum descent altitude.

The pilots, however, appear to have executed the approach in heading/vertical speed mode instead of track/flight path angle mode. The Flight Control Unit setting of "−33" yields a vertical descent rate of −3,300 ft/min in this mode, and this is almost precisely the rate of descent the airplane realized until it crashed into mountainous terrain several miles short of the airport. A push button on the FCU panel cycles the automation between H/VSI and T/FPA mode.

Modifications to A320 vertical speed/flightpath angle displays (in vertical speed mode, four digits are shown; in flight path angle mode, only two digits are visible) were subsequently made available by the manufacturer to avoid this error. New production A320s have been modified in this manner since November 1993 (Ministère de l'Equipment, des Transports et du Tourisme, 1993; (Anonymous, 1994b).

Colorado Springs, Colorado, 12/8/1992: United Airlines B737-291

United Airlines flight 585 was on the final approach course following a flight from Denver, Colorado to Colorado Springs, Colorado under visual meteorological conditions when it was observed by numerous eyewitnesses to roll steadily to the right and pitch nose down, reaching a nearly vertical attitude when it impacted the ground.

Despite an exhaustive investigation that is continuing, the NTSB has thus far been unable to identify conclusive evidence to explain the loss of this aircraft. It is surmised by the Board that either a rudder control anomaly or a "rotor," a horizontal axis wind vortex, may have precipitated the loss of control, but this is not certain (NTSB, 1992b).

Warsaw, Poland, 9/14/1993: Lufthansa A320

The aircraft, carrying 70 persons, landed at Warsaw in a downpour with strong, gusty winds. The pilot carried extra airspeed because of the wind conditions; a probable wind shear late in the approach made its ground speed still faster at touchdown. The airplane landed gently despite the gusts. It continued for approximately 8 sec after touchdown before being able to activate ground spoilers and reverse thrust. The airplane overran the runway end, traversed an embankment beyond the departure end, and caught fire. Two persons, including the copilot, were killed; 55 were injured.

Preliminary findings of the Polish inquiry suggest that the crew, having been advised of wind shear and a wet runway, correctly added 20 kt to the approach speed. When the forecast crosswind unexpectedly became a tailwind, making ground speed about 170 kt, the wheel spinup and [landing gear] squat switches did not [activate]. For a critical 9 sec [during which the aircraft may have been aquaplaning] thrust reverse, wheelbraking and lift dumping [full spoiler deployment] remained disarmed.... Although the A320 was...still to have the softer landing double-oleo modification, which might have "made" the switches, the priority question raised by the accident is whether pilots should have manual override of safety locks (Anonymous, 1994c).

Nagoya, Japan, 4/26/1994: China Airlines A-300-600R

During a normal approach to landing at Nagoya runway 34 in visual meteorological conditions, the captain indicated he was going around but did not indicate why. Within the next 30 sec, witnesses saw the aircraft in a nose-up attitude, rolling to its right before crashing tail-first 300 ft to the right of the approach end of the runway.

During the approach, the copilot flying apparently triggered the autopilot TOGA (takeoff-go-around) switch, whereupon the automation added power and commanded a pitch-up. The captain warned the copilot of the mode change, but the copilot continued to attempt to guide the aircraft down the glide slope while the automation countered his inputs with nose-up elevator trim. Ultimately, with stabilizer trim in an extreme nose-up position, the copilot was unable to counteract the trim with nose-down elevator. The aircraft nosed up to an attitude in excess of 50°, stalled, and slid backward to the ground. Two hundred and sixty-four people were killed in the crash.

This accident is still under civil and criminal investigation. It is presently thought that the pilots failed "to realize that their decision [to continue the approach] contradicted the logic of the airplane's automated safety systems. In February, 1991, an Interflug A310 at Moscow experienced a sudden, steep pitch-up similar to the one observed in this accident" ("Police Enter Crash Probe," 1994, p. 29; "Autopilot Go-Around Key to CAL crash," 1994; see also "New CAL 140 Transcript," 1994; "Pilot's Go-Around Decision Puzzles China Air Investigators," 1994).

On 8/31/94, the NTSB issued Safety Recommendations A-94-164 through 166 to the FAA. Its recommendation stated, "the Safety Board is concerned that the possibility still exists for a pilot-induced 'runaway trim' situation at low altitude and that...such a situation could result in a stall or the airplane landing in a nose-down attitude" (p. 5). Referring to other transport category aircraft autopilot systems, the Board said,

It is noted that the (autopilot) disconnect and warning systems are fully functional, regardless of altitude, and with or without the autopilot in the land or go-around modes. The Safety Board believes that the autopilot disconnect systems in the Airbus A-300 and A-310 are significantly different...additionally, the lack of a stabilizer-in-motion warning appears to be unique to (these aircraft). The accident in Nagoya and the incident in Moscow indicate that pilots may not be aware that under some circumstances the autopilot will work against them if they try to manually control the airplane. (p. 5)

The Board recommended that these autopilot systems be modified to ensure that the autopilot would disconnect if the pilot applies a specified input to the flight controls or trim system, regardless of the altitude or operating mode of the autopilot, and also to provide a sufficient perceptual alert when the trimmable horizontal stabilizer is in motion, irrespective of the source of the trim command (NTSB, 1994).

Hong Kong, 6/6/1994: Dragonair A-320, Kai Tak Airport

The airplane was attempting a landing at Kai Tak Airport during a severe storm. As the aircraft banked at about 1,000 feet, it encountered a wind shear that registered −1.6G.[4] It lost 12 kt of airspeed in 1 sec. The buffeting triggered its automatic flap locking safety mechanism, which is set if there is more than a 40 mm difference between the positions of the flaps to prevent them from becoming asymmetrical. The flaps locked at a full setting of 40°, or "flaps 4" (the landing position). The airplane's (leading edge) slats were in the third position of 22°. Sensing an anomaly, the electronic centralized aircraft monitoring system (ECAM) flashed a warning message for the pilot to correct it by moving the flaps lever to Flaps 3.

Unable to do so, the pilot aborted the landing. On the fourth try, he landed on runway 31, which allowed an approach without a banking maneuver. Two passengers were slightly injured after the aircraft ran off the runway. The article notes that a similar incident apparently occurred to an Indian Airlines A320 in November 1993. Airbus Industrie has recommended since this incident that pilots disregard the ECAM warning message. The software is being rewritten to eliminate the message; changes are also to be made in the flight control computers to prevent discrepancies between the flap lever position and the position of the flaps (A320 flap advisory, 1994).

Manchester, England, 6/21/1994: Britannia Airways B757-200

The aircraft was at light weight and was conducting a full-power takeoff. An altitude of 5,000 ft had been selected. The autopilot switched autonomously to altitude

---

[4]Negative 1.6 times acceleration due to gravity (32 ft/sec^2).

acquisition mode passing 2,200 ft because of the rapid climb speed. Power was reduced by the autothrust system and the airplane's speed began to drop rapidly toward takeoff safety speed because of the high pitch angle. Flight director bars continued to command pitch up, then disappeared from view. The pilot reduced the pitch attitude to 10° nose-up and normal acceleration resumed (Civil Aviation Authority [UK] 1994). (This incident resembles in many respects the more serious occurrence of an A330 at Toulouse [6/30/94, discussed next], which also involved a rapid switch to altitude acquisition mode after takeoff.)

Toulouse, France, 6/30/1994: Airbus A330-322 test flight, Toulouse Blagnac Airport

This airplane was on a category III certification test flight to study various pitch transition control laws in the autopilot Speed Reference System mode during engine failure at low altitude, rearward center of gravity and light aircraft weight. The flight crew included an experienced test pilot flying as captain, a copilot from a customer company, a flight test engineer, and three passengers. The copilot was handling the aircraft. During the takeoff, the copilot rotated the airplane slightly rapidly; the landing gear was retracted. The autopilot was engaged 6 sec after takeoff at a speed of 150 kt and a pitch angle of almost 25° nose up. Immediately thereafter, the left engine was brought to idle power and one hydraulic system was shut down, as planned for the test.

When the airplane reached 25° pitch angle, autopilot and flight director mode information were automatically removed from the flight mode annunciator (FMA) panel on the PFD. A maximum pitch angle of 29° was reached 8 sec after takeoff; the airplane was decelerating. The angle of attack reached 14°, which activated the alpha protection mode of the flight controls. The captain disconnected the autopilot 19 sec after takeoff. Subsequent control actions by the captain, which included reducing power on the right engine to regain control, deactivated alpha floor protection on the left engine. The airplane slowed to 100 kt, appreciably below minimum single-engine control speed of 118 kt, and yawed to the left. The left wing then stalled; speed reached 77 kt with an increasing left bank. Pitch angle reached 43° nose down and the airplane crashed 36 sec after takeoff.

During investigation, it was found that the aircraft autopilot had gone into altitude acquisition (ALT*) mode shortly after takeoff. In this mode, there was no maximum pitch limitation in the autoflight system software. As a consequence, at low speed, if a major thrust change occurs, the autopilot can induce "irrelevant pitch attitudes" because it is still trying to follow an altitude acquisition path that it cannot achieve.

The investigating committee cited the planned and inadvertent conditions under which the flight test was undertaken (high thrust, very aft center of gravity, trim within limits but nose-up, a selected altitude of 2,000 feet, late and imprecise definition of respective tasks between the pilot and copilot regarding the test to be performed, firm and quick rotation by the copilot, captain busy with the test actions, taking him out of the piloting loop). They also noted that the lack of pitch protection in the ALT* mode of the autopilot played a key role. Contributing factors included the inability of the flight crew to identify the active autopilot mode (due to the FMA declutter action at 25° nose-up), crew confidence in the anticipated aircraft reactions, late reaction of the flight test engineer to the rapid evolution of flight parameters (particularly the airspeed), and a late captain reaction to an abnormal situation.

A subsequent published article noted that "Contradictory autopilot requirements appear as a key factor that contributed to the loss of control: the 2,000 ft altitude was selected while the autopilot also had to simultaneously manage the combination of very low speed, an extremely high angle of attack, and asymmetrical engine thrust" (Director General of Armaments [France], 1994).

Charlotte, North Carolina, 7/2/1994: US Air DC-9-31

The airplane was returning from Columbia, South Carolina to Charlotte, North Carolina, when it encountered a wind shear during a very heavy rainstorm late in its final approach to the Charlotte-Douglas Airport. A wind shear alert had been received and the crew had briefed a missed approach if necessary. The captain flying ordered a missed approach at 200 ft because of poor visibility and strong, gusty winds. The first officer initiated the missed approach; the landing gear was retracted and flaps reduced from 40° (landing position) to 15°. At 350 ft the crew applied full throttles, but full thrust occurred only about 3 sec before impact, too late to arrest the descent and impact about .2 nautical miles to the right of runway 18R. Thirty-seven occupants were killed.

The crewmembers were unable to recall whether they had heard an aural warning from the wind shear detection system; investigation later revealed that the system's sensitivity is sharply reduced while wing flaps are in transit, to minimize the likelihood of false or nuisance warnings when airflow over the wing is disturbed during the change of configuration. Data provided to the NTSB by the system's manufacturer indicated that an alert would have been furnished 12 sec after a wind shear was detected if flaps were in transit, whereas an alert would have been generated within 5 sec after encountering a severe shear under other circumstances. As a result, the time lag "rendered the system useless" because the warning "would have occurred too late" for the pilots to perform a successful escape maneuver, according to the NTSB. It is believed that the pilots were unaware of this automatic reduction in sensitivity during flap transit.

The NTSB recommended that the FAA issue a flight standards bulletin informing pilots that wind shear warnings will be unavailable when flaps are in transit, and require modifications in the standard wind shear alert system to delete the delay feature, thereby ensuring "prompt warning activation" when flaps are transitioning between settings. The Board did not speak to the fact that this delay was incorporated in the system's software specifically to avoid nuisance warnings caused by temporary airflow disturbances. Honeywell had stated that such false alarms could cause pilots to "overreact or lose confidence" in the system's detection capabilities (Hughes, 1994; Phillips, 1994c, 1995b; NTSB, 1995).

Pittsburgh, Pennsylvania, 9/8/1994: US Air B737-300

During a routine approach to Pittsburgh International Airport, US Air flight 427 was cleared to turn left to a heading of 100°, reduce speed to 190 kt, and descend to 6,000 ft in preparation for a right downwind on a visual approach to runway 28R. The pilots extended their slats and flaps to the "Flaps 1" position. As the airplane began its turn, it rolled left, then decreased its bank angle, then increased it again to at least 100° as the nose pitched downward. The airplane struck the ground 23 sec later at an angle of about 80° and an airspeed in excess of 260 kt. The accident was not survivable.

The NTSB has undertaken extensive investigations of this accident, which thus far remains unexplained. Data collection is continuing. The similarity between certain aspects of this accident and a B737-291 accident at Colorado Springs, Colorado, on 12/8/92, also unexplained, has prompted intensive studies of rudder control and other aircraft mechanisms by the Board, the Boeing Company, and component manufacturers. In both cases, the Board has been hampered by the unavailability of control surface position indications in the flight data recorders, which were older models with limited parameter recording capability (Phillips, 1994b).

Paris, France, 9/24/1994: Tarom (Romanian) Airlines A310-300, Orly Airport

The airplane, carrying 182 persons on a flight from Bucharest to Paris, was on final approach to Orly Airport under visual meteorological conditions when it suddenly assumed a steep, nose-high attitude, then rolled into a dive before the pilots regained control at 800 ft above ground. No one was seriously injured and the airplane landed safely. A videotape taken by a witness showed the airplane in a steep nose-up attitude, then rolling off on one wing and descending in a nose-down attitude for several seconds before recovery. The digital flight data recorder was apparently inoperative during the incident, but data were obtained from the cockpit voice recorder and a direct access recorder used for maintenance purposes.

It is believed that the autopilot "suddenly went into the 'level change' mode" because flap limit speed was exceeded by 2 kt during the approach; this resulted in the pitch-up. "According to one report, the electric trim countered the pilot's action" during the attempt to recover from the pitch-up (Comments on Romanian Airlines, 1994; Anonymous, 1994a; see also Automated cockpits, 1995b).

Roselawn, Indiana, 10/31/1994: American Eagle Airlines ATR72

The airplane went out of control and crashed after flying at 10,000 ft at relatively low airspeed in a holding pattern for an extended period under icing conditions. The airplane carried a highly capable digital flight data recorder, whose data indicated that severe lateral control instability occurred, due, it is thought, to an accretion of ice ahead of the ailerons but aft of the wing leading edge deicer boots. The airplane was being flown on autopilot when control was first lost.

The accident is still under investigation, but the NTSB has issued urgent safety recommendations. The FAA has warned ATR42/72 pilots to avoid prolonged flight under icing conditions and to avoid high angles of attack if lateral instability occurs. Autopilot use under such conditions is proscribed, because autopilot corrective actions can mask the onset of the controllability problem. NTSB was aware of "similar, uncommanded autopilot disengagements and uncommanded lateral excursions" that have occurred on ATR42 aircraft in the past 6 years (Phillips, 1994a).

APPENDIX B

Wiener and Curry Guidelines for Aircraft Automation

In a landmark paper in 1980, Earl Wiener and Renwick Curry discussed "Flight-Deck Automation: Promises and Problems." Their contribution has been the stimulus for a great deal of research during the 15 years since it was published. This appendix summarizes these authors' thoughts on this subject.

Wiener and Curry pointed out that even in 1980, the question was "no longer whether one or another function can be automated, but, rather, whether it should be" (p. 2). They questioned the assumption that automation can eliminate human error. They pointed out failures in the interaction of humans with automation and in automation itself. They discussed control and monitoring automation and emphasized the independence of these two forms of automation: "It is possible to have various levels of automation in one dimension independent of the other" (Fig. B.1).

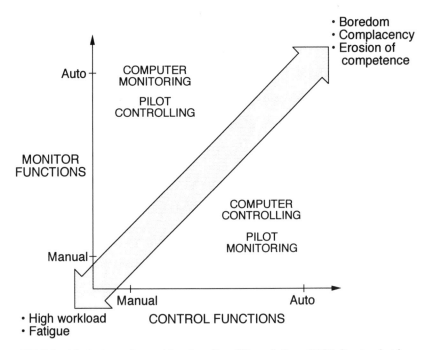

FIG. B.1.   Monitoring and control functions. From Wiener & Curry (1980). Reprinted with permission.

324

The authors discussed system goals and design philosophies for control and monitoring automation. They offered some generalizations about advantages and disadvantages of automating human–machine systems and went on to propose some guidelines for the design and use of automated systems in aircraft.

It is worth reviewing Wiener and Curry's guidelines because they foresaw many of the advantages and disadvantages of automation as it is used today. The following are abstracted from their guideline statements.

Control Tasks

1. System operation should be easily interpretable by the operator to facilitate the detection of improper operation and to facilitate the diagnosis of malfunctions.
2. Design the automatic system to perform the task the way the user wants it done ... this may require user control of certain parameters, such as system gains (see guideline 7). Many users of automated systems find that the systems do not perform the function in the manner desired by the operator. For example, autopilots, especially older designs, have too much "wing waggle" for passenger comfort when tracking ground-based navigation stations.... Thus, many airline pilots do not use this feature.
3. Design the automation to prevent peak levels of task demand from becoming excessive ... keeping task demand at reasonable levels will insure available time for monitoring.
4. The operator must be trained and motivated to use automation as an additional resource (i.e., as a helper).
5. Operators should be trained, motivated, and evaluated to monitor effectively.
6. If automation reduces task demands to low levels, provide meaningful duties to maintain operator involvement and resistance to distraction...it is extremely important that any additional duties be meaningful (not "make-work").
7. Allow for different operator "styles" (choice of automation) when feasible.
8. Insure that overall system performance will be insensitive to different options, or styles of operation.
9. Provide a means for checking the setup and information input to automatic systems. Many automatic system failures have been and will continue to be due to setup error, rather than hardware failures. The automatic system itself can check some of the setup, but independent error-checking equipment and procedures should be provided when appropriate.

10. Extensive training is required for operators working with automated equipment, not only to insure proper operation and setup, but to impart a
knowledge of correct operation (for anomaly detection) and malfunction
procedures (for diagnosis and treatment).

Monitoring Tasks

11. Keep false alarm rates within acceptable limits (recognize the behavioral
effect of excessive false alarms).
12. Alarms with more than one mode, or more than one condition that can
trigger the alarm for a mode, must clearly indicate which condition is
responsible for the alarm display.
13. When response time is not critical, most operators will attempt to check the
validity of the alarm. Provide information in the proper format so that this
validity check can be made quickly and accurately.... Also, provide the
operator with information and controls to diagnose the automatic system and
warning system operation.
14. The format of the alarm should indicate the degree of emergency. Multiple
levels of urgency of the same condition may be beneficial.
15. Devise training techniques and possibly training hardware...to insure
that flightcrews are exposed to all forms of alerts and to many of the
possible combinations of alerts, and that they understand how to deal with
them.

The authors concluded in 1980 that

the rapid pace of automation is outstripping one's ability to comprehend all the
implications for crew performance. It is unrealistic to call for a halt to cockpit
automation until the manifestations are completely understood. We do, however,
call for those designing, analyzing, and installing automatic systems in the cockpit
to do so carefully; *to recognize the behavioral effects of automation* [italics added]:
to avail themselves of present and future guidelines; and to be watchful for
symptoms that might appear in training and operational settings.

GLOSSARY OF ACRONYMS
AND ABBREVIATIONS

Where appropriate, acronyms and abbreviations used here conform to FAA-approved acronyms as used in the *Airman's Information Manual* and other regulatory and advisory material. Acronyms and abbreviations used for cockpit devices by specific manufacturers or in specific aircraft are indicated.

AAS
: Advanced automation system (FAA): the constellation of hardware, software, and procedures to be implemented during the 1990s for air traffic control and management in United States national airspace.

AATT
: Advanced Air Transportation Technology program: a NASA research and development initiative.

A/C
: (Also, AC) Abbreviation for "aircraft."

AC
: Alternating current.

ACARS
: ARINC Communications and Address Reporting System.

ADI
: Attitude director indicator: a gyroscopic aircraft attitude display, also known as an artificial horizon. See also EADI.

ADS
: Automatic dependent surveillance: means whereby an airplane's position, altitude, and velocity vector data are automatically reported to ground control stations at frequent intervals.

AERA
: Automated en route air traffic control, the FAA's advanced ATC system concept. There is no longer as clear-cut a separation between en route and terminal automation and use of this term is declining; see also AAS, FAS.

AI
: Artificial intelligence.

ALPA
: Air Line Pilots Association, a labor organization for air carrier pilots.

ALT*
: (ALT-STAR): Altitude acquisition mode of flight management system, in which the airplane is commanded to climb to and level off at a preselected altitude.

AOC
: Airline operations center; also called SOC, system operations center.

AOPA
: Aircraft Owners and Pilots Association, a representative organization for general aviation pilots.

327

APU   Auxiliary power unit, a small turbine that provides electrical power, compressed air, and a source of power for airplane hydraulic systems.

ARINC   Aeronautical Radio, Incorporated, provides international and domestic data transmission, receiving and forwarding services for air carriers and other subscribers.

ARTCC   Air Route Traffic Control Center (USA): provides en route tactical control of air traffic.

ASC   Aircraft systems controller: a computer that controls the operation of an aircraft subsystem (McDonnell-Douglas MD-11).

ASD   Aircraft situation display, an information element of the U.S. traffic management system.

ASDE   Airport surface detection equipment (radar).

ASRS   NASA Aviation Safety Reporting System, a voluntary, confidential incident reporting system operated by NASA for FAA.

ATA   Air Transport Association of America, the U.S. air carrier industry organization.

ATC   Air traffic control system: tactical control of air movements (in United States) by towers and air route traffic control centers.

ATCRBS   Air traffic control radar beacon system: a surface transponder–interrogator system that obtains information from aircraft.

ATCSCC   Air Traffic Control System Command Center (USA): FAA organization whose mission is to balance air traffic demand with system capacity (strategic traffic management). Also referred to here as SCC.

ATCU   Air traffic control unit: the forerunner of today's air route traffic control centers.

ATM   Air traffic management: strategic direction of air movements (in United States, by ATC System Command Center).

Avianca   The national flag airline of Colombia.

AWST   *Aviation Week and Space Technology*: an aerospace industry technical periodical.

BA   British Airways.

BAC   British Aerospace Corporation, formerly British Aircraft Corporation.

CAA	United Kingdom Civil Aviation Authority; the UK equivalent of the FAA.
CAB	Civil Aeronautics Board (U.S.), the organization that formerly controlled air transport in the United States and also investigated aircraft accidents. Now defunct.
CD-ROM	A means of storage of documents on laser computer disks with read-only memory.
CDU	Control and display unit: the flight management system human–system interface (in general usage).
CFMU	Central Flow Management Unit: the European equivalent of the U.S. System Command Center.
CRM	Crew or cockpit resource management: a concept to improve the resource management skills of pilots, cabin crews, and others in the aviation system.
CRT	Cathode ray tube.
CTAS	Center-Tracon Automation System: a set of software modules designed to assist terminal area controllers in the management of air traffic.
CVR	Cockpit voice recorder, a device that preserves 30 min of voice comments and transmissions to, from, and within the cockpit.
CWS	Control wheel steering: an autopilot mode that permits pilot input to the autoflight system using the control yoke.
DA	Descent advisor, a component of CTAS that assists controllers to order descending traffic.
Dead reckoning:	A means of navigating using time, estimated distance traveled, and headings, all corrected for estimated winds.
DME	Distance measuring equipment, an element in the common navigation system.
Doppler	Aircraft-based navigation system making use of Doppler radar to sense rate of change of position. Also a ground radar system that detects wind shears or microbursts associated with thunderstorms.
DOD, DoD	U.S. Department of Defense.
DOT	Department of Transportation, the Cabinet agency that supervises the FAA.
DRAPhyS	Diagnostic reasoning about physical systems: a model-based AI diagnostic system for aircraft faults.
E-MACS	Engine monitoring and control system.
EAD	Engine and alert display (McDonnell-Douglas MD-11).

EADI	Electronic attitude director indicator: provides aircraft attitude information on an electronic display (CRT or other EDU).
EATCHIP	European Air Traffic Control Harmonization and Integration Program.
ECAM	Electronic centralized aircraft monitoring system (Airbus Industrie term).
EDU	Electronic display unit (generic): a screen that displays data or graphics by any means, including CRTs, light-emitting diodes, liquid crystal or plasma displays, or other display technology.
EEC	Electronic engine controller (Boeing 757/767).
EGT	Exhaust gas temperature.
EICAS	Engine indication and crew alerting system (Boeing 757/767, 747-400, 777).
ELS	Electronic library system: an automated system for the storage and retrieval of documents in an airplane.
EM	Electromagnetic.
ENG OUT	A flight management system mode that takes account of changed circumstances when an engine fails.
EPR	Exhaust pressure ratio: the ratio of heated exhaust gas pressure to intake air pressure; a measure of power being developed by a jet engine.
ES	Expert system: a type of artificial intelligence reasoning and inference system.
ETMS	Enhanced Traffic Management System: the advanced software system to be utilized by the FAA's System Command Center.
F-PLN	Abbreviation for "flight plan."
FAA	Federal Aviation Administration.
FADEC	Full-authority digital engine controller.
FANS	Future air navigation system: a specification developed by ICAO.
FAR	Federal aviation regulation(s).
FAS	Full automation system (FAA): the advanced air traffic control system to be implemented in U.S. airspace, including conflict detection and resolution.
FAST	Final approach spacing tool: a component of CTAS for control of aircraft during final approach to landing.
FCC	Flight control computer (Airbus A320/330/340 aircraft).

FCU	Flight control unit (Airbus Industrie): the tactical mode and input data control panel for the autoflight system; located centrally at the top of the aircraft instrument panel. See also MCP.
FDP	Flight data processor: a computer component used in air traffic control facilities.
FMA	Flight mode annunciation panel or function: in older aircraft, a dedicated panel, usually above or near the attitude indicator; in glass cockpit aircraft, a display of flight modes located at the top of the primary flight display.
FMC	Flight management computer.
FMS	Flight management system.
FSF	Flight Safety Foundation: an international voluntary, user-supported air safety research and educational organization.
GA	General aviation: all civil aviation other than air transport.
Glonass	Global positioning system, a satellite-based navigation system (Russia).
GNSS	Global navigation system by satellite systems, a generic term.
GPS	Global positioning system, a satellite-based navigation system (United States).
GPWS	Ground proximity warning system.
HDG/VS	Heading/vertical speed, a flight management system mode in which the airplane's flight path is determined by these two parameters.
HF	High frequency, a portion of the electromagnetic spectrum used for aeronautical voice and data communications. Unlike VHF and UHF bands, HF is not limited to line-of-sight; it is, however, much more susceptible to weather and solar event disruption. Until the advent of satellite communications systems, HF communications were virtually the only form of real-time voice communications in transoceanic flying.
HSCT	High-speed civil transport: a future supersonic transport airplane (generic).
HSI	Horizontal situation indicator, either electromechanical or glass cockpit display. EHSI: an electronic horizontal situation representation.

HUD	Head-up display of flight path information.
IATA	International Air Transport Association, the representative organization of international air carriers, headquartered in Montreal, Quebec, Canada.
ICARUS	A Flight Safety Foundation technical committee set up to explore ways to reduce human factors accidents in aviation.
ICAO	International Civil Aviation Organization, an arm of the United Nations; headquarters in Montreal, Quebec, Canada.
IFR	Instrument flight rules: a system of rules for the conduct of air traffic under conditions of limited visibility. Essentially all transport flying is done under these rules.
ILS	Instrument landing system, consisting of localizer and glide slope transmitters on the ground. Also used to describe an approach conducted using ILS guidance. (Obsolete: ILAS).
IMC	Instrument meteorological conditions: visibility below specified minima, requiring that aviation operations be conducted under instrument flight rules (IFR).
INIT	Initialize: a flight management system mode and function.
Inmarsat	International Maritime Satellite: a system of communications satellites in geosynchronous orbits; also the organization that controls this system.
INS	Inertial navigation system, an airborne system of gyroscopes and accelerometers that keeps track of aircraft movement in three spatial axes.
IR	Infrared portion of the electromagnetic spectrum.
IRS	Inertial reference system: provides inertial data for navigation, as does INS, but also provides other data to pilot and aircraft systems.
IVSI	Instantaneous vertical speed indicator: an electromechanical instrument using air data quickened by acceleration data; also the display of such information on a primary flight display in a glass cockpit aircraft.
JAA	Joint Airworthiness Authority, an agency of the European Union charged with the task of preparing new supra-national aviation regulations for the Union.
KBE	Knowledge-based engineering system: an expert system for design of aircraft components.

KLM	The Royal Dutch flag airline.
LCD	Liquid crystal display.
LNAV	Lateral navigation; also a navigation mode in flight management systems.
Locus of control (pl., loci of control)	The site from which control of a system is exercised, or the agent responsible for system control.
LORAN	Long-range navigation system: uses ground-based low-frequency radio aids to provide triangulation-based position derivation for aircraft, marine and surface vehicles. The LORAN system in the United States is operated by the U.S. Coast Guard.
Mach, M.	A scale put forward by Ernst Mach that states speed relative to the speed of sound in air. M. 1 = the speed of sound. The speed of sound varies with absolute temperature.
MCP	Mode control panel: the tactical control panel for the autoflight system; almost always located centrally at the top of the aircraft instrument panel. Most airframe and avionics manufacturers except Airbus use this acronym. (See also FCU.)
MeV	Millions of electron volts: a measure of the energy of particulate or photon radiation.
MIT	Massachusetts Institute of Technology.
MITRE	MITRE Corporation, an engineering firm that conducts systems analyses and provides engineering technical support and guidance to the FAA, Department of Defense, and others.
MLS	Microwave landing system: a high-precision landing aid that provides the capability for curved as well as straight-in approaches to a runway, and conveys certain other advantages. The system, in advanced development by FAA until 1995, is the future standard precision landing system presently endorsed by ICAO.
MMW	Millimeter-wave portion of the electromagnetic spectrum.
Mode A	An ATCRBS transponder mode that provides position and range information, but no altitude data.
Mode C	Transponder mode that responds to ATC interrogations with position and altitude data.
Mode S	An enhanced ATCRBS transponder system that permits data link information to be communicated between ATC

	and aircraft, in addition to providing position and altitude information to the ground from aircraft in flight. Mode S transponders can also communicate collision avoidance information between aircraft in conflict.
MONITAUR	The monitoring "front end" of the DRAPhyS and related fault diagnosis systems.
MSAW	Minimum safe altitude warning: a software module in air traffic control computers that warns of aircraft operating below a safe altitude above the ground.
NAS	National Airspace System.
NASA	National Aeronautics and Space Administration.
NATS	National Air Traffic System (United Kingdom): equivalent of the U.S. Air Traffic Control system.
NC	Numerical control: automated machine control system using input data for production processes.
NM	Nautical Miles.
NTSB	National Transportation Safety Board (United States), investigates all aircraft accidents.
PERF	Abbreviation for "performance."
PFD	Primary flight display, usually electronic. See also ADI, EADI.
PIC	Pilot in command.
PMS	Performance management system: a forerunner of the flight management system.
QRH	Quick reference handbook: a booklet containing aircraft operating procedures, especially abnormal and emergency procedures.
RA	Resolution advisory: an avoidance maneuver provided by TCAS systems when another aircraft poses a serious threat.
RAF	Royal Air Force (United Kingdom).
RBES	Rule-based expert system; see ES.
RDP	Radar data processor: ATC computer modules that synthesize, from a number of radar sources, plan view displays for air traffic control.
RMI	Radio magnetic indicator, an electromechanical instrument showing magnetic heading and bearing to VOR or low-frequency nondirectional radio beacons. Also, this information presented on an electronic display.

RNAV

Area navigation system, a generic acronym for any device capable of aircraft guidance between pilot-defined waypoints, such as LORAN, Doppler, INS, and so on.

RVR

Runway visual range: a measure of visibility in a runway's landing zone.

SAS

Scandinavian Airlines System.

SBO

Specific behavioral objectives: a method of constructing training programs oriented toward specific tasks and activities rather than general system knowledge.

SCC

System Command Center (FAA): the strategic air traffic flow management organization and facility. Also ATCSCC.

SID

Standard instrument departure procedure.

SOC

Systems Operations Center (air carriers): flight operations management organization. Also called AOC.

SSR

Secondary surveillance radar: radar that makes use of ATCRBS to obtain data from aircraft.

STAR

Standard arrival route: an FAA-approved arrival route and procedure (see also SID).

Synoptic

A diagrammatic representation of an aircraft system, presented on an electronic display unit in the cockpit.

T/FPA

Track/flight path angle: a flight management system mode in which the airplane's flight path is guided by these two parameters. See also HDG/VS.

TA

Traffic advisory: an indication that another aircraft poses a potential threat, provided by TCAS systems.

TATCA

Terminal air traffic control automation research and development program (FAA).

TCA

Terminal control area: the former designation for Class B airspace.

TCAS

Traffic alert and collision avoidance system: TCAS-II, now installed in most U.S. and many foreign air carrier aircraft, provides vertical maneuver guidance for the resolution of serious potential conflicts; this system is mandated for U.S. transport aircraft. TCAS-III, in development, will provide both vertical and horizontal avoidance maneuvers. TCAS-I, a less expensive system, provides information concerning potential conflicts but does not provide resolution advisories.

TMA

Traffic management advisor, a component of CTAS.

TOGA	Take off go around: an aircraft automation mode that controls and displays information about the takeoff and go-around maneuvers.
TRACON	Terminal radar approach control facility (FAA).
TV	Television.
UHF	Ultra-high frequency: a portion of the electromagnetic spectrum used for aeronautical communications and navigation. It is limited to line-of-sight.
UK	United Kingdom.
VDU	Video display unit, a display device.
VFR	Visual flight rules: the rules that govern aircraft operations under conditions of good visibility (see also IFR).
VHF	Very high frequency: a portion of the electromagnetic spectrum used for line-of-sight aeronautical communications and navigation.
VMC	Visual meteorological conditions: visibility conditions that permit VFR flight.
VNAV	Vertical navigation; ordinarily refers to a navigation mode used for climbs and descents in flight management systems.
VOR	Very high frequency omnidirectional range: a surface radio navigation beacon transmitter that forms the core of the common overland navigation system for aircraft.
VSI	Vertical speed indicator (generic).
Waypoint	A geographic point that defines a route of flight. It may be a natural or cultural feature on the ground, or a point defined by longitude and latitude coordinates entered into a flight management system.
WSAS	Wind shear advisory system: a system that provides warnings of wind shear to pilots. The system may be passive (reactive), using airborne inertial sensors that react to accelerational forces on an airplane, or active, searching the environment for evidence of shears. If the latter, it may be located either on the ground (e.g., Doppler radar) or in an airplane (Lidar and radar are both under study).

REFERENCES

A320 flap advisory. (1994, June 27). *Aviation Week & Space Technology, 140*(26), 32.

Abbott, K. H. (1990). *Robust fault diagnosis of physical systems in operation.* Unpublished doctoral dissertation, Rutgers University, New Brunswick, NJ.

Abbott, K. H., & Rogers, W. H. (1992). *Presenting information for fault management.* Hampton, VA: NASA Langley Research Center.

Abbott, T. S. (1989). *Task-oriented display design: Concept and example* (NASA Tech. Mem. 101685). Hampton, VA: NASA Langley Research Center.

Abbott, T. S. (1990). *A simulation evaluation of the engine monitoring and control system display* (NASA Tech. Paper 2960). Washington, DC: NASA.

Air Transport Association of America. (1989, April). *National plan to enhance aviation safety through human factors improvements.* Washington, DC: Author.

Airbus Industrie. (1989, September). *A320 Flight Deck and Systems Briefing for Pilots,* issue 4 (Rep. No. AIEV-O 473 774/89. Toulouse, France: Author.

Aircraft Owners and Pilots Association. (1994). *AOPA's Aviation USA* (24th ed.). Frederick, MD: Author.

Anonymous (1994a, November). [News item concerning excursion of Romanian Airlines A310 while on final approach to landing at Orly Airport, Paris] *Royal Aeronautical Society Aerospace,* 1994, p. 5.

Anonymous (1994b, February). [Unattributed current news item concerning Air Inter A320 accident at Strasbourg, France] *Royal Aeronautical Society Aerospace,* p. 5.

Anonymous (1994c, February). [Unattributed news item concerning Lufthansa A320 accident on landing at Warsaw, Poland] *Royal Aeronautical Society Aerospace,* p. 6.

Ashby, W. R. (1956). *An introduction to cybernetics.* London: Methuen & Co.

Automated cockpits special report, part 1. (1995a, January 30). *Aviation Week & Space Technology, 142*(5), 52–65.

Automated cockpits special report, part 2. (1995b, February 6). *Aviation Week and Space Technology 142*(6), 48–57.

Aviation Occurrence Report. (1991). *Flight Control System Malfunction, Evergreen International Airlines, Inc. Boeing 747-121 N475EV, Nakina, Ontario, 12 December 1991* (Rep. No. A91H0014). Ottawa, ON: Transportation Safety Board of Canada. Also see National Transportation Safety Board Safety Recommendations A-92-31 through -35, May 14, 1992. Washington, DC: NTSB.

Barnhart, W., Billings C. E., Cooper G. E., Gilstrap R., Lauber J. K., Orlady H. W., Puskas B. & Stephens, W. (1975). *A method for the study of human factors in aircraft operations.* Moffett Field, CA: Ames Research Center. NASA TM X-62, 472, September.

Begault, D. R. (1993). Head-up auditory displays for traffic collision avoidance system advisories: A preliminary investigation. *Human Factors, 35*(4), 707–717.

Begault, D. R., & Wenzel, E. M. (1992). Techniques and applications for binaural sound manipulation in human-machine interfaces. *International Journal of Aviation Psychology, 2*(1), 1–22.

Billings, C. E. (1991). *Human-centered aircraft automation: A concept and guidelines* (NASA Techn. Mem. 103885). Moffett Field, CA: NASA-Ames Research Center.

337

Billings, C. E. (1996). Human-centered aviation automation: Principles and guidelines. (NASA Tech. Mem., 110381) Moffett Field, CA: NASA-Ames Research Center.

Billings, C. E., & O'Hara, D. B. (1978). Human factors associated with runway incursions (ASRS Eighth quarterly report, NASA Tech. Mem. 78540). Moffett Field, CA: NASA-Ames Research Center.

Billings, C. E., & Woods, D. D. (1994, April). Concerns about adaptive automation in aviation systems. Paper presented at the First Automation Technology and Human Performance Conference, Washington, DC.

Boeing Commercial Airplane Group. (1993). Accident prevention strategies: Removing links in the accident chain: Commercial Jet Aircraft Accidents world wide operations 1982–1991 (Boeing Airplane Safety Engineering B-210B). Seattle, WA: Author.

Boeing Commercial Airplane Group. (1994). Statistical summary of commercial jet aircraft accidents: Worldwide operations 1959–1993 (Boeing Airplane Safety Engineering B-210B). Seattle, WA: Author.

Braune, R., & Fadden, D. M. (1987, October). Flight deck automation today: Where do we go from here? Presented at SAE Aerotech '87. Long Beach, CA.

Broadbent, D. E. (1971). Decision and stress. London: Academic Press.

Bureau of Air Safety Investigation. (1993). Near Collision at Sydney (Kingsford Smith) Airport, 12 August 1991 (BASI report B/916/3032). Civic Square, ACT, Australia: Transport and Communications.

Butterworth-Hayes, P. (1995). Europe launches GPS counteroffensive. Aerospace America, 33(8), 4.

Bylander, T. (1988). Diagnosis by integrating model-based reasoning with knowledge-based reasoning. OSU LAIR Rep. 88-TB-DIAG. Columbus, OH: Ohio State University.

Byrnes, R. E., & Black, R. (1993). Developing and implementing CRM programs: The Delta experience. In E. L. Wiener, B. G. Kanki, & R. L. Helmreich (Eds.), Cockpit resource management (pp. 421–443). San Diego, CA: Academic Press.

Caracena, R. L., Holle, R. L., & Doswell, C., III. (1989). Microbursts: A handbook for visual identification. Washington, DC: National Oceanic and Atmospheric Administration.

Celio, J. C. (1990). Controller perspective of AERA 2 (MITRE Corp. Rep. MP-88W00015, Rev. 1). McLean, VA: MITRE Corporation.

Chapanis, A. (1965). On the allocation of functions between men and machines. Occupational Psychology, 39, 1–11.

Chappell, S. L., Billings, C. E., Scott, B. C., Tuttell, R. J., Olson, M. C., & Kozon, T. E. (1988). Pilot's use of a traffic-alert and collision avoidance system (T/CAS II) in simulated air carrier operations (NASA TM 100094, 2 Vols.). Moffett Field, CA: NASA-Ames Research Center.

Cheaney, E. S., & Billings, C. E. (1981, October). Application of the epidemiological model in studying human error in aviation. 1980 Aircraft Safety and Operating Problems Conference, NASA CP 2170, Hampton, VA.

Chidester, T. R. (1990). Trends and individual differences in response to short-haul flight operations. Aviation, Space and Environmental Medicine, 61, 132–138.

Chidester, T. R., Kanki, B. G., Foushee, H. C., Dickinson, C. L., & Bowles, S. V. (1990). Personality factors in flight operations: Volume I. Leader characteristics and crew performance in a full-mission air transport simulation (NASA TM 102259). Moffett Field, CA: NASA-Ames Research Center.

Civil Aeronautics Board. (1957). Trans World Airlines Lockheed 1049A and United Air Lines Douglas DC-7, Grand Canyon, AZ, June 30, 1956 (CAB Rep. No. SA-320). Washington, DC: Author.

Civil Aeronautics Board. (1959). Pan American World Airways Boeing 707 over the Atlantic between London, England, and Gander, Newfoundland, February 3, 1959. Washington, DC: Author.

Civil Aeronautics Board. (1962). Eastern Air Lines Lockheed Electra L-188, Logan International Airport, Boston, MA, October 4, 1960. (CAA Report SA-358, July 31). Washington, DC: Author.

Civil Aeronautics Board. (1964). *United Air Lines Vickers-Armstrong Viscount near Ellicott City, MD, November 23, 1963* (CAA Report, March 22). Washington, DC: Author.

Civil Aviation Authority (UK). (1994). *Narrative report on B 757-200 incident at Manchester, England* (Occurrence No. 9402551G). London: Author.

Clancey, W. J. (1983). The epistemology of a rule-based expert system—a framework for explanation. *Artificial Intelligence, 20*, 215–251.

Comments on Romanian Airlines A310 excursion on approach to Orly Airport, Paris. (1994, October 3). *Aviation Week & Space Technology*, p. 37.

Cooley, M. (1987). Human centred systems: An urgent problem for systems designers. *AI & Society, 1*, 37–46.

Cooper, G. (1994a). Euro Flow Control. *Royal Aeronautical Society Aerospace, 21*(3), 8–11.

Cooper, G. (1994b): Seeing through the fog. *Royal Aeronautical Society Aerospace 21*(9), 16–18.

Corker, K. M., & Reinhardt, T. (1990). *Procedural representation techniques applied to flight management computer systems: A software representation system* (BBN Contract No. W288-386; BBN Rep. Ref. 5653). Cambridge, MA: BBN Systems and Technologies.

Costley, J., Johnson, D., & Lawson, D. (1989, April). A comparison of cockpit communication B737-B757. In *Proceedings of the Fifth International Symposium on Aviation Psychology* (pp. 413–418). Columbus, OH: Ohio State University.

Craik, K. J. W. (1947a). Theory of the human operator in control systems I: The operator as an engineering system. *British Journal of Psychology, 38*, 56–61.

Craik, K. J. W. (1947b). Theory of the human operator in control systems II: Man as an element in a control system. *British Journal of Psychology, 38*, 142–148.

Curran, J. A. (1992). *Trends in advanced avionics*. Ames: Iowa State University Press.

Curry, R. E. (1985). *The introduction of new cockpit technology: A human factors study* (NASA Tech. Mem. 86659). Moffett Field, CA: NASA-Ames Research Center.

Davis, R. (1993, October). *Human factors in the global marketplace*. Keynote address to the Human Factors and Ergonomics Society, Seattle, WA.

Davis, R., & Hamscher, W. (1988). Model-based reasoning: Troubleshooting. In H. E. Shrobe (Ed.), *Exploring artificial intelligence* (pp. 297–346). San Mateo, CA: Morgan Kaufman.

Degani, A., & Wiener, E. L. (1991, April). *Philosophy, policies and procedures: The three P's of flight-deck operations*. Paper presented at the Sixth International Symposium on Aviation Psychology, Columbus, OH.

Del Balzo, J. M. (1992). Worldwide aviation after the year 2000: A new era begins. In G. M. Crook (Ed.), *Airports and automation* (pp. 107–108). London: Thomas Telford.

Department of Trade and Industry. (1973). *Trident G-ARPI. Report of the Public Inquiry into the Causes and Circumstances of the Accident near Staines on 18 June, 1972*. London: Her Majesty's Stationery Office, Civil Aviation Accident Report, April.

de Sitter, U. (1989). Moderne Sociotechniek. *Gedrag en Organisatie, 4/5*, 222–252.

Director General of Armaments (France). (1994). *Investigation Committee report on A330 Accident in Toulouse on 30 June 1994*. Unpublished report. See also Sparaco, P. (1994, July 11). *Aviation Week & Space Technology, 141*(2), 26–27.

Dornheim, M. A. (1991, February 11). 737-Metro crash raises controller workload issues. *Aviation Week & Space Technology, 134*, 27.

Douglas Aircraft Co. (1990). *Functional decomposition of the commercial flight domain for function allocation*. Final report to NASA-Langley Research Center, Contract No. NAS1-18028. Briefing Paper, December 10.

Duke, T. (1996, March). Conquering CFIT. *Air Line Pilot 65*(3), 10–13.

Edwards, E., & Lees, F. P. (1972). *Man and computer in process control*. London: Institute of Chemical Engineers.

Emery, F. (1977). *Futures we are in*. Leiden, Netherlands: Martinus Nijhoff.

Endsley, M. R. (1994). Situation awareness in dynamic human decision-making: Theory. In R. D. Gilson, D. J. Garland, & J. M. Koonce (Eds.), *Situational awareness in complex systems* (pp. 27–58). Daytona Beach, FL: Embry-Riddle Aeronautical University Press.

Engine shutdown prompts FAA directive. (1987, July 6). *Aviation Week & Space Technology*, p. 36.

Erzberger, H., & Nedell, W. (1988). *Design of automation tools for management of descent traffic* (NASA Tech. Mem. 101078). Moffett Field, CA: NASA-Ames Research Center.

Erzberger, H., & Nedell, W. (1989). *Design of automated system for management of arrival traffic* (NASA Tech. Mem. 102201). Moffett Field, CA: NASA-Ames Research Center.

European Coal and Steel Community. (1976). *Human factors evaluation at Hoogovens No. 2 Hot Strip Mill.* Proceedings of the meeting on 26–27 October. Secretariat of Community Ergonomics Action, European Coal and Steel Community, Luxembourg.

FAA/NASA/MIT Symposium. (1988, September). *Aviation systems concepts for the 21st century.* Cambridge, MA: Transportation Systems Center.

Fadden, D. M. (1990). Aircraft automation challenges. In *Abstracts of AIAA-NASA-FAA-HFS Symposium, Challenges in Aviation Human Factors: The National Plan.* Washington, DC: American Institute of Aeronautics and Astronautics.

Federal Aviation Administration. (n.d.). *14 CFR Chapter 1, Subchapter F: Air Traffic and General Operating Rules, Part 91: General Operating and Flight Rules, Subpart A, Section 91.3 (a, b, c).* Washington, DC: Author.

Federal Aviation Administration. (1987). *Introduction to MLS.* Washington, DC: Author.

Federal Aviation Administration. (1994). *1994 Aviation Capacity Enhancement Plan.* Federal Aviation Administration report no. DOT/FAA/ASC-94-1, November 1. Washington, DC.

Fitts, P. M. (1951). *Human engineering for an effective air navigation and traffic control system.* Washington, DC: National Research Council.

Fitzsimons, B. (1993). ATC by numbers. *Royal Aeronautical Society Aerospace, 20*(5), 22.

Flach, J. M., & Dominguez, C. O. (1995, July). Use-centered design: Integrating the user, instrument and goal. *Ergonomics in Design*, pp. 19–24.

Flight Safety Foundation. (1982, January). *A safety appraisal of the Air Traffic Control system* (Rep. No. FSF-ATC 1142-1-82U). Arlington, VA: Author.

Folkerts, H. H. & Jorna, P. G. A. M. (1994). *Pilot performance in automated cockpits: A comparison of moving and non-moving thrust levers* (Rep. No. NLR-TP-94005U). Amsterdam: National Aerospace Laboratory NLR.

Foushee, H. C. (1984). Dyads and triads at 35.000 feet: Factors affecting group process and aircrew performance. *American Psychologist, 39,* 886–893.

Gilbreth, F. B. (1985). *Primer of Scientific Management.* (2nd ed.). Easton, MD: Hive Publishing. (Original work published 1914)

Gilson, R. D., Garland, D. J., & Koonce, J. M. (Eds.). (1994). *Situational awareness in complex systems.* Daytona Beach, FL: Embry-Riddle Aeronautical University Press.

Golaszewski, R. (1983). *The influence of total flight time, recent flight time and age on pilot accident rates* (FAA Safety Analysis Division Final Rep. DTRS57-83-P-80750). Washington, DC: Department of Transportation/FAA.

Golaszewski, R. (1991). *General aviation safety studies: Preliminary analysis of pilot proficiency.* Chevy Chase, MD: Abacus Technology Corp.

Gorham, J. A. (1973, July). *Automatic flight control and navigation systems on the L-1011: Capabilities and experiences.* Paper presented at the USSR/US Aeronautical Technology Symposium, Moscow.

Graeber, R. C., Lauber, J. K., Connell, L. J., & Gander, P. A. (1986). International aircrew sleep and wakefulness after multiple time zone flights: A cooperative study. *Aerospace Medicine, 57* (12(II)), B3–B9.

Growth projections for air transport. (1993, March 15). *Aviation Week and Space Technology, 140;* 65.

Grunwald, A. J., Robertson J. B., & Hatfield J. J. (1980). *Evaluation of a computer-generated perspective tunnel display for flight path following* (NASA Tech. Rep. 1736). Hampton, VA: NASA Langley Research Center.

Habakkuk, H. J. (1962). *American and British technology in the nineteenth century.* Cambridge, England: Cambridge University Press.

Hackman, J. R. (1993). Teams, leaders, and organizations: New directions for crew-oriented flight training. In E. L. Wiener, B. G. Kanki, & R. L. Helmreich (Eds.), *Cockpit resource management* (pp. 47–69). San Diego, CA: Academic Press.

Hard limits, soft options [Editorial], *Flight International* (1990, October 31–November 6). *138,* p. 3.

Harwood, K., & Sanford, B. D. (1993). *Denver TMA assessment* (NASA Contractor Rep. 4554). Moffett Field, CA: NASA-Ames Research Center.

Helmreich, R. L., & Foushee, H. C. (1993). Why crew resource management? Empirical and theoretical bases of human factors training in aviation. In E. L. Wiener, B. G. Kanki, & R. L. Helmreich (Eds.), *Cockpit resource management* (pp. 3–45). San Diego, CA: Academic Press.

Helmreich, R. L., & Wilhelm, J. A. (1989). *When training boomerangs: Negative outcomes associated with cockpit resource management programs.* Proceedings of the Fifth Symposium on Aviation Psychology. Columbus, OH: Ohio State University.

Helmreich, R. L., & Wilhelm, J. A. (1991). Outcomes of crew resource management training. *International Journal of Aviation Psychology, 1*(4), 287–300.

Hollnagel, E. (1993). *Reliability of cognition: Foundations of human reliability analysis.* London: Academic Press.

Honeywell Commercial Flight Systems Group. (1990, September). *MD-11 Flight Management System Pilot's Guide* (PUB No. 28-3643-01-00). Phoenix, AZ: Author.

Hopkin, V. D. (1994a, April). Human factors in air traffic system automation. In R. Parasuraman & M. Mouloua (Eds.), *Human performance in automated systems: Current research and trends* (pp. 319–336). Mahwah, NJ: Lawrence Erlbaum Associates.

Hopkin, V. D. (1994b). Situational awareness in air traffic control. In, R. D. Gilson, D. J. Garland, & J. M. Koonce (Eds.), *Situational awareness in complex systems* (pp. 171–178) Daytona Beach, FL: Embry-Riddle Aeronautical University Press.

Hopkins, H. (1990, October 24–30). Masterfully Digital (MD-) 11. *Flight International,* p. 138.

Hopkins, R. (1993, March 31–April 6). Backing up of approaches [Letter to the editor]. *Flight International,* p. 40.

Hughes, D. (1994). More wind shear detail might have aided DC-9. *Aviation Week & Space Technology, 141*(2), 24–25.

Hughes, J. A., Randall, D., & Shapiro, D. (1992, November). *Faltering from ethnography toward design.* CSCW 92 Proceedings.

Hutchins, E. (1993, August). *An integrated Mode Management Display.* Paper presented at the NASA-Ames Mode meeting. Moffett Field, CA: NASA-Ames Research Center.

Inspector of Marine Accidents. (1993, November 17): *Report into the grounding of the German container ship* Berlin Express *in Port Phillip Bay, on 2 May 1992.* Report No. 53. Canberra, Australia: Department of Transport and Communications.

International Air Transport Association. (1994, December). *IATA Draft Requirements for Air Traffic Management in the Future Air Navigation System* Unpublished report. Montreal: Author.

Jordan, N. (1963). Allocation of functions between man and machines in automated systems. *Journal of Applied Psychology, 47*(3), 161–165.

Kelly, B. D., Graeber, R. C., & Fadden, D. M. (1992, November). *Applying crew-centered concepts to flight deck technology: The Boeing 777.* Paper presented at the Flight Safety Foundation 45th International Air Safety Seminar, Long Beach, CA.

Kerns, K. (1994, September). *Human factors in ATC/flight deck integration: Implications of data link simulation research.* Report MP 94W0000098. McLean, VA: MITRE Corp.

Killbridge, M. N., & Wester, L. (1963, June). The assembly line model-mix sequencing problem. *Proceedings of the Third International Conference on O.R.,* Oslo.

Kinney, G. C., Spahn, J., & Amato, R. A. (1977). *The human element in air traffic control: Observations and analyses of the performance of controllers and supervisors in providing ATC separation services* (METREK Division of the MITRE Corporation, Rep. MTR-7655). McLean, VA: MITRE Corp.

Kraft, C. L., & Elworth, C. L. (1969, March–April). Night visual approaches. *Boeing Airliner,* pp. 2–4.

Kuipers, H. (1989). Zelforganisatie als ontwerpprincipe: Sociotechnisch organisatie-ontwerp in vijftien stellingen. *Gedrag en Organisatie, 4/5,* 199–221.

Laming, J. (1993, June 9–15). Who controls the aircraft? [Letter to the editor]. *Flight International,* p. 140.

Langer, H. A. (1992, January/February). Meeting the challenges together. *The Cockpit.* Chicago: United Airlines.

Last, S., & Alder, M. (1991, September). *British Airways Airbus A320 Pilots' Autothrust Survey.* Paper presented at the SAE Aerospace Technology Conference, Long Beach, CA.

L.A. Tower tapes show controller unaware of aircraft holding on runway. (1991, April 8). *Aviation Week & Space Technology, 134,* 61.

Lauber, J. K. (1989, March). *Remarks of John K. Lauber, member NTSB, before the Aero Club of Washington March Luncheon* [NTSB safety information pamphlet]. Washington, DC: National Transportation Safety Board.

Lauber, J. K. (1993): A *Safety culture perspective.* Flight Safety Foundation 38th Corporate Aviation Safety Seminar, Irving, TX, April 14–16.

Lauber, J. K., Bray, R. S., Harrison, R. L., Hemingway, J. C., & Scott, B. C. (1982). *An operational evaluation of head-up displays for civil transport operations: NASA/FAA Phase III final report* (NASA Tech. Paper 1815). Moffett Field, CA: NASA-Ames Research Center.

Lautmann, L. G., & Gallimore, P. L. (1987, April–June). Control of the crew-caused accident: Results of a 12-operator survey. *Boeing Airliner.* Seattle: Boeing Commercial Airplane Company. Also in *Proceedings of the 40th Flight Safety Foundation International Air Safety Seminar* (October), Tokyo.

Layton, C., Smith, P. J., & McCoy, E. (1994). Design of a cooperative problem solving system for enroute flight planning: An empirical evaluation. *Human Factors, 36*(1), 94–119.

Lee, L. (1992). *The day the phones stopped.* New York: Donald I. Fine.

Lenorovitz, J. M. (1990, June 25). Indian A320 crash probe data show crew improperly configured aircraft. *Aviation Week & Space Technology, 132,* 84–85.

Leveson, N. G.. (1995). SAFEWARE: *System safety and computers.* Reading, MA: Addison-Wesley.

Mackworth, N. H. (1950). *Researches on the measurement of human performance* (Medical Research Council Special Rep. Ser., No. 268). London: Her Majesty's Stationery Office.

Malin, J. T., Schreckenghost, D. L., Woods, D. D., Potter, S. S., Johannesen, L., & Holloway, M. (1991). *Making intelligent systems team players: Case studies and design issues, Volume 2: Fault management system cases* (NASA Tech. Mem. 104738). Moffett Field, CA: NASA-Ames Research Center.

Mårtensson, L. (1995). The aircraft crash at Gottröra: Experiences of the cockpit crew. *International Journal of Aviation Psychology, 5*(3), 305–326.

Marthinson, H. F., & Hagy, H. K. (1993a, October–November). Boeing 737 overruns: A case history, part I. *Air Line Pilot, 62*(9), 27–31.

Marthinson, H. F., Hagy, H. K. (1993b, December). Boeing 737 overruns: A case history, part II. *Air Line Pilot, 62*(10), 25–27.

Maxim, H. S. (1908). *Artificial and natural flight.* New York: Macmillan.

McClumpha, A. J., James, M., Green, R. G., Belyavin, A. J. (1991). Pilots' attitudes to cockpit automation. *Proceedings of Human Factors Society 35th Annual Meeting* (pp. 107–111). Santa Monica, CA: Human Factors and Ergonomics Society.

McFarland, R. A. (1946). *Human factors in air transport design.* New York: McGraw-Hill.

McMahon, J. (1978, July). Flight 1080. *Air Line Pilot*. Reprinted in National Aeronautics and Space Administration. (1983). *Restructurable Controls Conference* (NASA Conference Paper 2277). Washington, DC.

Mecham, M. (1994, May 9). Autopilot go-around key to CAL crash. *Aviation Week and Space Technology*, pp. 31–32.

Mellone, V. J. (1993). TCAS Incident reports analysis. Paper presented at the Second International TCAS Conference, FAA, Washington, DC.

Meredith, J. (1992). The cost of airport congestion. In G. M. Crook (Ed.), *Airports and automation* (pp. 109–112). London: Thomas Telford.

Mertes, F., & Jenney, L. (1974). *Automation applications in an advanced air traffic management system: Vol. III, Methodology for man-machine task allocation* (Rep. No. DOT-TSC-OST-74-14-III). McLean, VA: TRW, Inc.

Ministère de l'equipment, des transports et du tourisme. (1993, November). *Rapport de la commission d'enquete sur l'accident survenu le 20 Janvier 1992 près du Mont Sainte Odile (Bas Rhin) a l'airbus A320 immatricule F-GGED exploite par la compagnie Air Inter*. Paris, France: Author.

Ministry of Planning, Housing, Transport and Maritime Affairs. (1989). *Investigation Commission Final Report concerning the accident which occurred on June 26th 1988 at Mulhouse-Habsheim (68) to the Airbus A320, registered F-GFKC* (Rep. 11/29/1989). Bangkok, Thailand: Author. Also extracted in *Aviation Week & Space Technology, 132*, 6/4/90, p. 107; 6/18/90, p. 99; 6/25/90, p. 98; *133:* 7/9/90, p. 60; 7/23/90, p. 90; 7/30/90, p. 90.

Ministry of Transport and Communications, Thailand. (1993). *Lauda Air Boeing 767-300ER (accident at) Dan Chang Province, Thailand, 26 May 1991*.

Mohler, S. R., & Johnson, B. H. (1971). *Wiley Post, his Winnie Mae, and the world's first pressure suit*. Washington, DC: Smithsonian Institution Press.

Mokyr, J. (1990). *The lever of riches: Technological creativity and economic progress*. Oxford: Oxford University Press.

Monan, W. P. (1986, March). *Readback related problems in ATC communications: The hearback problem* (NASA CR 177398). Mountain View, CA: Aviation Safety Reporting System Office.

Moray, N., Lee, J., & Hiskes, D. (1994, April). *Why do people intervene in the control of automated systems?* Paper presented at the First Automation Technology and Human Performance Conference, Washington, DC.

Moshansky, V. P. (1992). *Commission of Inquiry into the Air Ontario accident at Dryden, Ontario* (Final Report, Vol. 1–4). Ottawa, ON: Minister of Supply and Services, Canada.

Nagel, D. C. (1988). Human error in aviation operations. In E. L. Wiener & D. C. Nagel (Eds.), *Human factors in aviation* (pp. 263–303). San Diego: Academic Press.

NASA Aviation Safety Reporting System. (1976). *Incident report ACN 00362*. Mountain View, CA: Aviation Safety Reporting System Office.

NASA Aviation Safety Reporting System. (1986). *Incident report ACN 59282*, Mountain View, CA: Aviation Safety Reporting System Office.

NASA Aviation Safety Reporting System. (1994, May). *Callback, 180*, 1.

National Aeronautics and Space Administration. (1990). *Aviation Safety/Automation Program Plan* (NASA Information Sciences and Human Factors Division, unpublished document). Washington, DC: Author.

National Aeronautics and Space Administration. (1991). *Advanced subsonic aircraft (capacity) initiative* (Information Sciences and Human Factors Division, unpublished briefing document). Washington, DC: Author.

National Transportation Safety Board. (1973). *Eastern Air Lines L-1011, Miami, FL, December 29, 1972* (NTSB Rep. No. AAR-73-14). Washington, DC: Author.

National Transportation Safety Board. (1974). *Delta Air Lines Douglas DC-9-31, Boston, MA, 7/31/73* (NTSB Rep. No. AAR-74/03). Washington, DC: Author.

National Transportation Safety Board. (1978a). *Southern Airways DC-9-31, New Hope, GA, April 4, 1977* (NTSB Rep. No. AAR-78-3). Washington, DC: Author.

National Transportation Safety Board. (1978b). *United Airlines Douglas DC-8 near Kaysville, UT, December 18, 1977* (NTSB Report no. AAR-78-8). Washington, DC: Author.

National Transportation Safety Board. (1978c). *National Airlines Boeing 727, Escambia Bay, Pensacola, FL, May 8, 1978* (NTSB Rep. No. AAR-78-13). Washington, DC: Author.

National Transportation Safety Board. (1979): *United Airlines DC-8-61, Portland, OR, 12/28/78* (NTSB Rep. No. AAR-79/07). Washington, DC: Author.

National Transportation Safety Board. (1980). *Aeromexico DC-10-30 over Luxembourg, 11/11/79* (NTSB Rep. No. AAR-80-10). Washington, DC: Author.

National Transportation Safety Board. (1981). *Aircraft Separation Incidents at Hartsfield Atlanta International Airport, Atlanta, GA, 10/7/80* (NTSB Rep. No. SIR-81-6). Washington, DC: Author._

National Transportation Safety Board. (1982). *Air Florida B-737-222, Collision with 14th Street Bridge, near Washington National Airport, DC, 1/13/82* (NTSB Rep. No. AAR-82-8). Washington, DC: Author.

National Transportation Safety Board. (1984). *Scandinavian Airlines DC-10-30, J. F. Kennedy Airport, New York, 2/28/84* (NTSB Rep. No. AAR-84-15). Washington, DC: Author.

National Transportation Safety Board. (1986). *China Airlines B-747-SP, 300 NM northwest of San Francisco, CA, 2/19/85* (NTSB Rep. No. AAR-86/03).Washington, DC: Author.

National Transportation Safety Board. (1988a). *Delta Airlines B727-232, Dallas Fort Worth Airport, Texas, 8/31/88* (NTSB Rep. No. AAR-89/04). Washington, DC: Author.

National Transportation Safety Board. (1988b). *Northwest Airlines DC-9-82, Detroit Metro Wayne County Airport, Romulus, Michigan, 8/16/87* (NTSB Rep. No. AAR-88/05). Washington, DC: Author.

National Transportation Safety Board. (1991a). *Avianca, the Airline of Colombia, Boeing 707-321B, fuel exhaustion, Cove Neck, New York, Jan. 25, 1990.* (NTSB Rep. No. AAR/91/04). Washington, DC: Author.

National Transportation Safety Board. (1991b). *Northwest Airlines B-727 and DC-9, Detroit Metro Airport, Romulus, MI, 12/3/90* (NTSB Report no. AAR/91/05). Washington, DC: Author.

National Transportation Safety Board. (1991c). *Runway Collision of US Air Boeing 737 and Skywest Fairchild Metroliner, Los Angeles, 2/1/91.* (NTSB Rep. No. AAR-91/08). Washington, DC: Author.

National Transportation Safety Board. (1992a, May 14). Safety Recommendations A-92-31 through -35 (concerning *Evergreen International Airlines B747-100 upset at Nakina, Ontario, December 12, 1991.*) Washington, DC, National Transportation Safety Board. See also Aviation Occurrence Rep. No. A91H0014, *Flight Control System Malfunction.* Ottawa, ON: Transportation Safety Board of Canada.

National Transportation Safety Board. (1992b). *United Airlines Boeing 737 collision with terrain for undetermined reasons, Colorado Springs, CO* (NTSB Rep. No. AAR-92/06). Washington, DC: Author.

National Transportation Safety Board. (1994, August 31). *Safety Recommendations A-94-164 through -166 (concerning China Airlines Airbus A-300-600R accident at Nagoya, Japan, April 26, 1994).* Washington, DC: Author.

National Transportation Safety Board. (1995). Flight into terrain during missed approach, USAir Flight 1016, DC-9-31, Charlotte/Douglas International Airport, Charlotte, NC, 7/2/94 (NTSB Rep. No. AAR/95/03). Washington, DC: Author.

National Transportation Safety Board impounds United 767 after engine power loss. (1986, April 7). *Aviation Week & Space Technology*, p. 36.

Noble, D. F. (1983). *Forces of production: A social history of industrial automation.* New York: Knopf.

Nolan, M. S. (1994). *Fundamentals of air traffic control.* Belmont, CA: Wadsworth.

Nordwall, B. D. (1995). New FMS to offer 737 fuel savings. *Aviation Week & Space Technology*, 142(13), p. 48.

Norman, D. A. (1981). Categorization of action slips. *Psychological Review, 88*, 1–15.

Norman, D. A. (1988). *Psychology of everyday things*. New York: Basic Books.

Norman, D. A. (1989, June). *The problem of automation: Inappropriate feedback and interaction, not "over-automation."* Paper prepared for the discussion meeting, Human Factors in High-Risk situations, the Royal Society (Great Britain), London.

Norman, D. A. (1993). *Things that make us smart: Defending human attributes in the age of the machine.* Reading, MA: Addison-Wesley.

Ovenden, C. R. (1991). *Model-Based Reasoning applied to Cockpit Warning Systems* (Smiths Industries Aerospace and Defense Systems report). Cheltenham, UK.

Owen, R. (1927). *A new view of society*. London: Everyman. (Original work published 1830)

Palmer, E. A., Mitchell C. M., & Govindaraj, T. (1990). *Human-centered automation in the cockpit: Some design tenets and related research projects.* Paper presented at the ACM SIGCHI Workshop on Computer–Human Interaction in Aerospace Systems. Washington, DC.

Pan American World Airways. (1990, September). *Flight Check, 68*, 32.

Paulson, G. (1994). Global navigational satellite system for aviation, RAeS lecture (edited by B. Fitzsimons). *Aerospace, 21*(6), 12–15.

Perrow, C. (1984). *Normal accidents*. New York: Basic Books.

Phillips, E. H. (1994a). ATR42/72 review focuses on icing. *Aviation Week & Space Technology, 141*(20), 28.

Phillips, E. H. (1994b). NTSB slates hearing on Pittsburgh crash. *Aviation Week & Space Technology, 141*(23), 28.

Phillips, E. H. (1994c), NTSB Links Detection-Delay Feature with DC-9 Crash. *Aviation Week & Space Technology, 141*(24), 30. See also *Aviation Week & Space Technology*, July 11, 1994, pp. 25–27.

Phillips, E. H. (1995a). ACE plan offers fixes for capacity shortfall. *Aviation Week & Space Technology, 142*(11), 40.

Phillips, E. H. (1995b). Pilots, ATC blamed for USAir accident. *Aviation Week & Space Technology, 142*(15), 32.

Pilot's go-around decision puzzles China Air Investigators. (1994, May 2). *Aviation Week & Space Technology, 140*(18), 26.

Pinet, J., & Enders, J. H. (1994, December). *Human factors in aviation: A consolidated approach.* Arlington, VA: Flight Safety Digest, pp. 7–12.

Police enter crash probe. (1994, December 5). *Aviation Week & Space Technology, 141*(23), 29.

Porter, R. F., & Loomis, J. P. (1981). *An investigation of reports of controlled flight toward terrain* (NASA Contractor Rep. 166230). Mountain View, CA: Aviation Safety Reporting System Office.

Preble, C. (1987, July). Delta Air Lines officials baffled by series of unrelated mishaps. *Aviation Week & Space Technology, 127*, 31–32.

President's Task Force on Crew Complement. (1981, July). *Report of the President's Task Force on Crew Complement.* Washington, DC.

Price, H. E. (1985). Allocation of functions. *Human Factors, 27*(1), 33–45.

Proctor, P. (1994a). Boeing predicts recovery in transport orders. *Aviation Week & Space Technology, 140*(22), 32.

Proctor, P. (1994b). HGS offers airlines safety, economic edge. *Aviation Week & Space Technology, 141*(24), 50.

Proctor, P. (1995). Boeing adopts "expert" design system. *Aviation Week & Space Technology, 142*(17), 27.

Radio Technical Commission on Aeronautics. (1995, January 18). *Report of the RTCA Board of Directors' Select Committee on Free Flight.* Copyright 1995. Washington, DC: Author.

Randle, R. J., Larsen, W. E., & Williams, D. H. (1980). *Some human factors issues in the development and evaluation of cockpit alerting and warning systems* (NASA Ref. Pub. 1055). Washington, DC: NASA.

Rasmussen, J. (1988). *Information processing and human-machine interaction: An approach to cognitive engineering.* New York: North Holland.

Rauner, F., Rasmussen, L., & Corbett, J. M. (1988). The social shaping of technology and work: Human centred CIM systems. *AI & Society*, 2, 47–61.

Reason, J. T. (1990). *Human error*. Cambridge: Cambridge University Press.

Reeve, R. M. (1971). *The Industrial Revolution 1750–1850*. London: University of London Press.

Reintjes, J. F. (1991). *Numerical control: Making a new technology*. Oxford: Oxford University Press.

Rodgers, M. D., & Nye, L. G. (1993). Factors associated with the severity of operational errors at air route traffic control centers. In M. D. Rodgers (Ed.), *An examination of the operational error database for air route traffic control centers* (Office of Aviation Medicine, Final Rep. AM-93/22) (pp. 11–25). Washington, DC: Department of Transportation/FAA.

Rogers, W. H. (1990). *Flight crew information requirements for fault management on a commercial flight deck*. NASA Draft Information Requests Task Plan, Langley Research Center, Hampton, VA.

Roscoe, S. N. (1979). *Ground-referenced visual orientation with imaging displays: Final report* (Tech. Rep. Eng. Psy-79-4/AFOSR-79-4). Urbana-Champaign: University of Illinois.

Rose, R. M., Jenkins, C. D., & Hurst, M. W. (1978). *Air traffic controller health change study* (Rep. FAA-AM-78-39). Washington, DC: FAA Office of Aviation Medicine.

Roth, E. M., Bennett, K. B., & Woods, D. D. (1987). Human interaction with an "intelligent" machine. *International Journal of Man-Machine Studies, 27(5–6)* 479–525.

Rouse, W. B. (1980). *Systems engineering models of human-machine interaction*. New York: North Holland.

Rouse, W. B. (1988). Adaptive aiding for human/computer control. *Human Factors, 30*(4), 431–443.

Rouse, W. B. (1991). *Design for success*. New York: Wiley.

Rouse, W. B., & Rouse, S. H. (1983). *A framework for research on adaptive decision aids* (Tech. Rep. AFAMRL-TR-83-082). Wright-Patterson Air Force Base, OH: Air Force Aerospace Medical Research Laboratory.

Rouse, W. B., Geddes, N. D., & Curry R. E. (1987). An architecture for intelligent interfaces: Outline for an approach supporting operators of complex systems. *Human Computer Interaction, 3*(2), 87–122.

Rudisill, M. (1994, April). *Flight crew experience with automation technologies on commercial transport flight decks*. Paper presented at the First Automation Technology and Human Performance Conference, Washington, DC.

Rudisill, M. (1995, April). Line pilots' attitudes about and experience with flight deck automation: Results of an international survey and proposed guidelines. *Proc. Eighth International Symposium on Aviation Psychology*. Columbus, OH: The Ohio State University Press.

Ruffell Smith, H. P. (1974, May). Pilots' activities immediately preceding a fatal accident. *Proceedings of the Annual Scientific Meeting, Aerospace Medical Association*, Washington, DC.

Ruffell Smith, H. P. (1979). *A simulator study of the interaction of pilot workload with errors, vigilance and decisions* (NASA Tech. Mem. 78482). Moffett Field, CA: NASA-Ames Research Center.

Russell, P. (1992, May/June). The Boeing safety agenda. *ARMS Newsletter, 3*, 1. Renton, WA: Boeing Commercial Airplane Group.

Sanford, B. D., Harwood, K., Nowlin, S., Bergeron, H., Heinrichs, H., Wells, G., & Hart, M. (1993, October). Center/TRACON automation system: Development and evaluation in the field. *Proceedings of the Air Traffic Control Association Conference*, Nashville, TN.

Sarter, N. B. (1994): *"Strong, silent, and out of the loop": Properties of advanced cockpit automation and their impact on human-automation interaction*. Unpublished doctoral dissertation, The Ohio State University, Columbus.

Sarter, N. B., & Woods, D. D. (1991). Situation awareness: A critical but ill-defined phenomenon. *International Journal of Aviation Psychology, 1* (1), 45–57.

Sarter, N. B., & Woods, D. D. (1992a). Mode Error in Supervisory Control of Automated Systems. In *Proceedings of the Human Factors Society 36th Annual meeting* (pp. 26–29)

Sarter, N. B., & Woods, D. D. (1992b). Pilot interaction with cockpit automation: Operational experiences with the flight management system. *International Journal of Aviation Psychology*, *2*, 303–321.

Sarter, N. B., & Woods, D. D. (1994, April). *Decomposing automation: Autonomy, authority, observability and perceived animacy*. Paper presented at the First Automation Technology and Human Performance Conference, Washington, DC.

Sarter, N. B., & Woods, D. D. (1995). How in the world did we ever get into that mode? Mode error and awareness in supervisory control. *Human Factors*, *37*(1), 5–19.

Schroeder, D. J. (1982). The loss of prescribed separation between aircraft: How does it occur? In *Proceedings of the behavioral objectives in aviation automated systems symposium*, P-114 (pp. 257–269). Warrendale, PA: Society of Automotive Engineers.

Schroeder, D. J., & Nye, L. G. (1993). An examination of the workload conditions associated with operational errors/deviations at air route traffic control centers. In M. D. Rodgers (Ed.), *An examination of the operational error database for air route traffic control centers* (pp. 1–10). Office of Aviation Medicine, Final Rep. No. AM-93/22. Washington, DC: Department of Transportation/FAA.

Scott, B., Goka, T., & Gates, D (1987, October). *Design, development and operational evaluation of an MLS/RNAV control display unit*. Paper presented at the Tenth IEEE/AIAA Digital Avionics Systems Conference. Los Angeles. See also Federal Aviation Administration (1987).

Sheridan, T. B. (1984). Supervisory control of remote manipulators, vehicles and dynamic processes. In W. B. Rouse (Ed.), *Advances in man–machine systems research* (Vol. 1, pp. 49–137). Greenwich, CT: JAI Press.

Sheridan, T. B. (1987). Supervisory control. In G. Salvendy (Ed.), *Handbook of human factors* (pp. 1243–1268). New York: Wiley.

Sheridan, T. B. (1988). Task allocation and supervisory control. In M. Helander, (Ed), *Handbook of human-computer interaction* (pp. 159–174). North-Holland: Elsevier.

Sheridan, T. B. (1995, August). *Human centered automation: Oxymoron or common sense?* Lecture delivered at industrial summer school on Human-Centered Automation, Saint-Lary, France.

Shontz, W. D., Records, R. M., & Antonelli, D. R. (1992). *Flight deck engine advisor: Final report*. NASA contract NAS1-18027. Boeing Commercial Airplane Group, Flight Deck Research, Seattle WA.

Simpson, T. R. (1995). A second revolution for air navigation and landing. *Aerospace America*, *33* (8), 18–20.

Smith, P. J., McCoy, E., Layton, C., & Bihari, T. (1993). *Design concepts for the development of cooperative problem-solving systems* (CSEL paper). Columbus, OH: Ohio State University.

Sperry Gyroscope Company. (n.d.). *The Sperry automatic pilot*. Brooklyn, NY: Author.

Stein, K. J. (1985, October 3). Human factors analyzed in 007 navigation error. *Aviation Week & Space Technology*, *119*, 165. (Also see relevant navigation routes in *Aviation Week & Space Technology*, *119*, 9/12/83 18–23.)

Stein, K. J. (1986, December 1). Complementary displays could provide panoramic, detailed battle area views. *Aviation Week & Space Technology*, *125*, 40–44.

Stout, C. L., & Stephens, W. A. (1975, November). *Results of simulator experimentation for approach and landing safety*. Paper presented at the International Air Transport Association 20th Technical Conference on Safety in Flight Operations, Istanbul, Turkey.

Swade, D. D. (1993, February). Redeeming Charles Babbage's Mechanical computer. *Scientific American*, pp. 86–91.

Swain, A. D., & Guttman, H. F. (1980). *Handbook of human reliability analysis with the emphasis on nuclear power plant applications* (NRC NUREG-CR-1278). Albuquerque, NM: Sandia Laboratories.

Taplin, H. J. (1969). "George": An experiment with a mechanical autopilot. *Journal of the American Aviation Historical Society*, *4*(4). 234–235.

Tarrell, R. J. (1985). *Non-airborne conflicts: The causes and effects of runway transgressions* (NASA Contractor Rep. No. 177372). Mountain View, CA: Aviation Safety Reporting System Office.

Taylor, F. W. (1906). *On the art of cutting metals.* New York: ASME.

Taylor, F. W. (1911). *The principles of scientific management.* New York: Harper.

Tenney, Y. J., Rogers, W. H., & Pew, R. W. (1995). *Pilot opinion on high level flight deck automation issues: Toward the development of a design philosophy.* (NASA Contractor Rep. No. 4669) Hampton, VA: NASA Langley Research Center.

Tobias, L., & Scoggins, J. L. (1986, June). Time-based traffic management using expert systems. *Proceedings of the American control conference* (pp. 693–700). Seattle, WA.

Uchtdorf, D., & Heldt, P. (1989, April). *Flight Crew Info, Special issue: Survey on cockpit systems B737-200 and A310-200, 1986* (Lufthansa Flight operations division publication). Frankfurt, Germany: Lufthansa German Airlines.

van Beinum, H. J. J. (1989). Sociotechniek: Voorwoord [Socioengineering]. *Gedrag en Organisatie, 4/5,* 195–197.

Vortac, O. U., & Manning, C. A. (1994). *Automation and cognition in air traffic control: An empirical investigation.* Washington, DC: FAA Report DOT/FAA/AM-94/3.

Wiener, E. L. (1985a, October). *Cockpit automation: In need of a philosophy.* Paper presented at the Aerospace Technology Conference, Long Beach, CA.

Wiener, E. L. (1985b). *Human factors of cockpit automation: A field study of flight crew transition* (NASA Contractor Rep. No. 177333). Moffett Field, CA: NASA-Ames Research Center.

Wiener, E. L. (1987). Fallible humans and vulnerable systems: Lessons learned from aviation. In J. A. Wise & A. Debons (Eds.), *Information systems: Failure analysis* (NATO ASI Series, Vol. F-32, pp. 164-181). Berlin: Springer-Verlag.

Wiener, E. L. (1989). *Human factors of advanced technology (glass cockpit) transport aircraft* (NASA Contractor Rep. No. 177528). Moffett Field, CA: NASA-Ames Research Center.

Wiener, E. L. (1993). *Intervention strategies for the management of human error* (NASA Contractor Rep. No. 4547. Moffett Field, CA: NASA-Ames Research Center.

Wiener, E. L., & Curry, R. E. (1980). *Flight-deck automation: Promises and problems* (NASA Tech. Mem. 81206). Moffett Field, CA: NASA-Ames Research Center.

Wiener, E. L., Kanki, B. G., & Helmreich, R. L. (1993). *Cockpit resource management.* San Diego: Academic Press.

Wilkinson, S. (1994, February/March). The November Oscar incident. *Air & Space,* pp. 81–87.

Woods, D. D. (in press). *Visualizing function: The theory and practice of representation design in the computer medium.* Columbus: Ohio State University.

Woods, D. D. (1984). Visual momentum: A concept to improve the cognitive coupling of person and computer. *International Journal of Man-Machine Studies, 21,* 229–244.

Woods, D. D. (1993a). *Cognitive activities and aiding strategies in dynamic fault Management* (Cognitive Systems Engineering Laboratory Paper). Columbus, OH: Ohio State University.

Woods, D. D. (1993b). *Cognitive systems in context* (Cognitive Systems Engineering Laboratory Paper). Columbus, OH: Ohio State University.

Woods, D. D. (1995). Towards a theoretical base for representation design in the computer medium: Ecological perception and aiding in human cognition. In J. Flach, P. Hancock, J. Caird, & K. Vicente (Eds.), *Global Perspectives on the Ecology of Human–Machine Systems* (Vol. 1, pp. 157–188). Hillsdale, NJ: Lawrence Erlbaum Associates.

Woods, D. D. (1996). *Decomposing automation: Apparent simplicity, real complexity.* In R. Parasuraman & M. Mouloua (Eds.), *Automation and human performance: Theory and application* (pp. 3–18). Mahwah, NJ: Lawrence Erlbaum Associates.

Woods, D. D., Johannesen, L. J., Cook, R. I., & Sarter, N. B. (1994). *Behind human error: Cognitive systems, computers and hindsight.* State-of-the-art report for CSERIAC, Dayton, OH.

Author Index

Subject Index